Numerical Methods

Scientific and Engineering Applications

Numerical Methods

Scientific and Engineering Applications

H.S.G. Rao
Professor & Head
Department of Mathematics
Dr. Ambedkar Institute of Technology
Bangalore

TECHSAR™
NEW DELHI

Published by

I.K. International Pvt. Ltd.
4435-36/7, Ansari Road, Daryaganj,
New Delhi–110 002 (India)
E-mail: tech_info@techsarworld.com
Website: www.techsarworld.com

ISBN: 978-93-80578-54-5

Reprint 2024

Published/Printed by Krishan Makhijani for I.K. International Pvt. Ltd., 4435-36/7, Ansari Road, Daryaganj, New Delhi - 110 002.

Dedicated to

My dearest Mother
Smt. Padmavathamma

Preface

This book is designed to meet the requirements of students of Engineering and Science. The study of methods of obtaining approximations for solutions of mathematical problems is known as numerical methods. When we study the different topics in applied mathematics, then we encounter the computation of numerical results for extracting useful information from available solutions. Although the numerical data used in solving the practical problems are approximate, moreover the methods and processes used to find the derived results are approximate, but the information derived from such approximates is very useful. The subject matter has been so arranged that even a layman can understand how to apply the several methods to numerical computation of various quantities required to estimate and to approximate to different branches of mathematics, physics, engineering and computer science.

H.S.G. Rao

Acknowledgements

The satisfaction and euphoria that accompanies the successful completion of any task would be incomplete without the mention of the people who made it possible, whose constant encouragement crowned my efforts with success.

I consider my privilege to express my gratitude and thanks to the following persons for their help, encouragement and intellectual influence during the course of this dissertation work.

I am indebted to Dr. P. Martin Jebaraj, Principal Dr. Ambedkar Institute of Technology, Bangalore for his invaluable support without which I would not have been able to do any justice to my work. The discussions with him have always been fruitful and stimulating.

I am very thankful to Mr. Krishan Makhijani M/s I.K. International Publishing House Pvt. Ltd. for their patience and useful formatting suggestions to bring out this book.

H.S.G. Rao

Contents

Chapter 1

Errors and Approximations

INTRODUCTION

In this chapter, we shall discuss various kinds of errors and their determination in the case of various interpolation and quadrature formulae.

The numbers like, 2, 5, 8, 11, $\frac{2}{5}$, $\frac{3}{2}$,..., ... π, e,..., etc., ... are 'exact' numbers and those numbers that represent the numbers to a certain degree of accuracy are 'approximate' numbers.

Consider the numbers

$$\sqrt{2} \equiv 1.41421,...,$$
$$e \equiv 2.7183...,$$
$$\pi \equiv 3.1416, \text{ etc. are not exact}$$

Since they contain infinitely many non-recurring digits, such numbers which can be obtained by retaining a few digits, are called 'APPROXIMATE' numbers.

For example, an approximate value of π is 3.1416, or if we want a better approximation, it is 3.14159265. But we are unable to write the exact value of 'π'.

ERROR IN ROUNDING OFF A NUMBER

Error: The difference between an exact number x and its approximation x' is called an error.

Consider the number $\frac{3}{7}$, by actual division of this is given by $\frac{3}{7} \equiv 0.428571...$ a non-terminating decimal fraction.

In practical computation, such numbers will be reduced to a convenient size such as 43, 0.428, etc.

Thus, the process of cutting-off superfluous digits and retaining as many of them as desired (required) is known as "rounding off a number".

General Rule to Round off a Number

(a) If the digit in the decimal places to be discarded is more than 5, we add 1 to the last retaining digit (previous digit to the discarding digit).
For

$$0.9267 \approx 0.927; \quad 6 > 5$$
$$0.9256 \approx 0.926; \quad 5 \equiv 5$$

(b) If the discarding digit is less than 5, then the previous digit is retained.
For

$$0.9247 \approx 0.924; \quad 4 < 5$$
$$0.9236 \approx 0.923; \quad 3 < 5$$

(c) If the discarding digit is exactly 5, then we add 1 to the previous digit if it is odd and do not alter if it is even or zero.
For

$$0.9325 \approx 0.932;$$
$$0.9335 \approx 0.934;$$

Example 1. Round off the following numbers correct to four significant figures.

58.3643, 979.267, 7.7265, 56.395, 0.065738, 6657692000

Solution: (i) $58.3643 \approx 58.36$

(ii) $979.267 \approx 979.3$

(iii) $7.7265 \approx 7.726$

(iv) $56.395 \approx 56.40$

(v) $0.065738 \approx 0.06574$

(Here '0' on the left is not significant)

(vi) $6657692000 \approx 6658 \times 10^6$

Different Types of Errors

Errors are classified into three categories:

(a) Absolute error
(b) Relative error
(c) Percentage error

Definition:

(a) The absolute or numerical difference between the true value of a quantity and its approximate value is known as ABSOLUTE ERROR.

If x' is the approximate value of a quantity and "x" be its true or (EXACT) value, then the absolute error E_a is

$$E_a \equiv \delta x \equiv |x - x'|$$

Thus,
$$\boxed{E_a \equiv |x - x'|}$$

(b) The relative error is defined as the ratio of the absolute error and the exact value of the quantity and is denoted by E_r.

Thus,
$$E_r \equiv \left|\frac{E_a}{x}\right|$$
$$\equiv \left|\frac{x - x'}{x}\right|$$

(c) If we multiply the relative error (E_r) by 100, we get the percentage error, it is defined as "100 times the relative error E_r"

or
$$E_p \equiv \text{Percentage error}$$
$$\equiv 100 \times E_r$$
$$\equiv 100 \times \left|\frac{x - x'}{x}\right|$$

Thus,
$$\boxed{E_p \equiv 100 \times \left|\frac{x - x'}{x}\right|}$$

Example 2. If 0.333 is the approximate value of $\frac{1}{3}$, find absolute, relative and percentage errors.

Solution: (i) $E_a \equiv$ Absolute error

$$\equiv |x - x'| \equiv \left|\frac{1}{3} - 0.333\right|$$

$$\equiv \left|\frac{1}{3} - \frac{333}{1000}\right| \equiv \frac{1}{3000} = 0.00033$$

(ii) $E_r \equiv$ Relative error

$$\equiv \left|\frac{x - x'}{x}\right| \equiv \frac{0.00033}{\frac{1}{3}} \equiv 0.00099 \text{ and}$$

(iii) $E_p \equiv$ Percentage error

$$\equiv 100 \times \left|\frac{x - x'}{x}\right| \equiv 100 \times 0.00099$$

$$\equiv 0.099.$$

Error Formula

Let $y \equiv f(x_1, x_2, \ldots x_n)$ be the function of n-variables x_i $(i = 1, 2, \ldots, n)$ and let Δx_i be a error in x_i for each i and let Δy be the corresponding error in y, then we have

$$y + \Delta y \equiv f(x_1 + \Delta x_1, x_2 + \Delta x_2, \ldots, x_n + \Delta x_n)$$

$$\equiv f(x_1, x_2, \ldots, x_n) + \sum_{i=1}^{n} \frac{\partial f}{\partial x_i} \Delta x_i + \text{terms containing } (\Delta x_i)^2 \text{ etc., } \ldots \text{ on expand-}$$

ing by Taylor's series.

If we assume that the error in x_i is small then we have $\frac{\Delta x_i}{x_i} \leq 1$.

Neglecting the squares and higher powers of Δx_i, the above relation reduces to

$$\Delta y \approx \sum_{i=1}^{n} \frac{\partial f}{\partial x_i} \Delta x_i$$

$$\equiv \frac{\partial f}{\partial x_1} \Delta x_1 + \frac{\partial f}{\partial x_2} \Delta x_2 + \ldots + \frac{\partial f}{\partial x_n} \Delta x_n \qquad \text{(A)}$$

Dividing by y, the relative error is

$$E_r \equiv \frac{\Delta y}{y} \equiv \frac{\partial f}{\partial x_1} \cdot \frac{\Delta x_1}{y} + \frac{\partial f}{\partial x_2} \cdot \frac{\Delta x_2}{y} + \ldots + \frac{\partial f}{\partial x_n} \cdot \frac{\Delta x_n}{y} \qquad \text{(B)}$$

Taking Mod of (A) and (B), the maximum absolute error and relative error are given as $|\Delta y| \leq \left| \Delta x_1 \frac{\partial f}{\partial x_1} \right| + \left| \Delta x_2 \frac{\partial f}{\partial x_2} \right| + \ldots + \left| \Delta x_n \frac{\partial f}{\partial x_n} \right|$

and $\left| \frac{\Delta y}{y} \right| \leq \left| \frac{\Delta x_1}{y} \cdot \frac{\partial f}{\partial x_1} \right| + \left| \frac{\Delta x_2}{y} \cdot \frac{\partial f}{\partial x_2} \right| + \ldots + \left| \frac{\Delta x_n}{y} \cdot \frac{\partial f}{\partial x_n} \right|$

As an illustration, let us consider the function

$$u \equiv \frac{5xy^2}{z^3}$$

Then

$$\frac{\partial u}{\partial x} \equiv \frac{5y^2}{z^3}, \quad \frac{\partial x}{\partial y} = \frac{10xy}{z^3}$$

$$\frac{\partial u}{\partial z} \equiv -\frac{15xy^2}{z^4}$$

$$\therefore \qquad \Delta u \equiv \frac{5y^2}{z^3} \Delta x + \frac{10xy}{z^3} \Delta y - \frac{15xy^2}{z^4} \Delta z$$

The errors, Δx, Δy and Δz may be positive or negative.

$$\therefore \qquad |\Delta u| \le \left|\frac{5y^2}{z^3}\Delta x\right| + \left|\frac{10xy}{z^3}\Delta y\right| + \left|\frac{15xy^2}{z^4}\Delta z\right|$$

or
$$(\Delta u)_{\max} \approx \left|\frac{5y^2}{z^3}\Delta x\right| + \left|\frac{10xy}{z^3}\Delta y\right| + \left|\frac{15xy^2}{z^4}\Delta z\right|$$

If we choose $\Delta x \equiv \Delta y \equiv \Delta z \equiv 0.001$

and $\qquad x = y = z = 1,$

then the relative maximum error is given by

$$(E_r)_{\max} \equiv \frac{(\Delta u)_{\max}}{u}$$

$$\equiv \frac{0.03}{5} \equiv 0.006$$

We have

Absolute error $\equiv$ | true value – approximate value |

Also, If $E_a \equiv$ Absolute error, $Q \equiv$ true value,

then $E_r \equiv$ Relative error is given by

$$E_r \equiv \frac{E_a}{Q} \qquad \text{(A)}$$

and
$$E_p \equiv \text{Percentage error}$$
$$\equiv \text{Relative error} \times 100$$
$$\equiv E_r \times 100$$
$$\equiv \frac{E_a}{Q} \times 100 \quad \text{[By (A)]}$$

If a number is correct to 'n' significant figures, then its E_a (absolute error) cannot be greater than half a unit in the nth place.

For example, if the number in 32.68 is correct to four figures, its

absolute error $\not> 0.01 \times \frac{1}{2} \equiv 0.005$

(A) *Error in addition of numbers*

If $y \equiv x_1 + x_2 + x_3 + \ldots + x_n$

$$\therefore \qquad y + \Delta y \equiv (x_1 + \Delta x_1) + (x_2 + \Delta x_2) + \ldots + (x_n + \Delta x_n)$$
$$\equiv (x_1 + x_2 + \ldots + x_n) + (\Delta x_1 + \Delta x_2 + \ldots + \Delta x_n)$$

$$\therefore \qquad \Delta y \equiv \Delta x_1 + \Delta x_2 + \ldots + \Delta x_n$$

($\div y$) (Divide by y)

$$\therefore \qquad \frac{\Delta y}{y} \equiv \frac{\Delta x_1}{y} + \frac{\Delta x_2}{y} + \ldots + \frac{\Delta x_n}{y}$$

or

$$\left|\frac{\Delta y}{y}\right| \le \left|\frac{\Delta x_1}{y}\right| + \left|\frac{\Delta x_2}{y}\right| + \ldots + \left|\frac{\Delta x_n}{y}\right|$$

Absolute error is obtained by

$$\Delta y \equiv \sum_{i=1}^{n} \left|\frac{\Delta x_i}{y}\right| \cdot y$$

$\equiv$ sum of products of relative error and the number 'y'.

Example 3. Approximate values of $\frac{1}{7}$ and $\frac{1}{11}$ (correct to 4 decimal places are 0.1429 and 0.0909 respectively). Find the possible relative error and absolute error in the sum of 0.1429 and 0.0909.

Solution: We have

$$0.1429 \equiv \text{correct to four places of decimal}$$

and

$$0.0909 \equiv \text{correct to four places of decimal}$$

$\therefore$ The maximum error is $\frac{1}{2} \times 0.0001 = 0.00005$

(i) $E_r \equiv$ Relative error $= \left|\frac{\Delta y}{y}\right| < \left|\frac{0.0005}{0.2338}\right| + \left|\frac{0.0005}{0.2338}\right|$

$$\therefore \qquad y \equiv 0.1429 - 0.0909$$
$$\equiv 0.2338$$

$$\therefore \left|\frac{\Delta y}{y}\right| < \frac{0.0001}{0.2338} = 0.00043$$

(ii) $E_a \equiv$ Absolute error $\equiv \left|\frac{\Delta y}{y}\right| \cdot y$

$$\equiv \frac{0.0001}{0.2338} \times 0.2338$$
$$\equiv 0.0001$$

(B) ***To find the relative error and absolute error in subtraction of numbers.***

Solution: Let $y = x_1 - x_2$

$$\therefore \qquad y + \Delta y \equiv (x_1 + \Delta x_1) - (x_2 + \Delta x_2)$$
$$\equiv (x_1 - x_2) + (\Delta x_1 - \Delta x_2)$$
$$\therefore \qquad \Delta y \equiv \Delta x_1 - \Delta x_2$$

and $$\frac{\Delta y}{y} = \frac{\Delta x_1}{y} - \frac{\Delta x_2}{y}$$

but $$|\Delta y| \le |\Delta x_1| + |\Delta x_2| \text{ and}$$

$\therefore$ $$\left|\frac{\Delta y}{y}\right| \le \left|\frac{\Delta x_1}{y}\right| + \left|\frac{\Delta x_2}{y}\right|$$

$$\therefore \text{Relative error} \equiv \left|\frac{\Delta y}{y}\right| \le \left|\frac{\Delta x_1}{y}\right| + \left|\frac{\Delta x_2}{y}\right|$$
$$\text{Absolute error } |\Delta y| \le |\Delta x_1| + |\Delta x_2|$$

Error in Product of Numbers

(C) ***To find the relative error and absolute error in product of numbers.***

Suppose $y = x_1 x_2 \dots x_n$.

Then we have $$\frac{\Delta y}{y} = \frac{\Delta x_1}{y} \cdot \frac{\partial y}{\partial x_1} + \frac{\Delta x_2}{\partial y} \cdot \frac{\partial y}{\partial x_2} + \dots + \dots + \frac{\Delta x_n}{y} \cdot \frac{\partial y}{\partial x_n} \quad \text{[By (A)]}$$

Now $$\frac{1}{y} \cdot \frac{\partial y}{\partial x_1} \equiv \frac{x_2 x_3 \dots x_n}{x_1 x_2 \dots x_n} \equiv \frac{1}{x_1}$$

$$\frac{1}{y} \cdot \frac{\partial y}{\partial x_2} = \frac{x_2 x_3 \dots x_n}{x_1 x_2 \dots x_n} \equiv \frac{1}{x_2}$$

$$\dots \dots \dots \dots$$
$$\dots \dots \dots \dots$$

$$\frac{1}{y} \cdot \frac{\partial y}{\partial x_n} \equiv \frac{x_1 x_2 \dots x_{n-1}}{x_1 x_2 \dots x_{n-1} x_n} \equiv \frac{1}{x_n}$$

$\therefore$ $$\frac{\Delta y}{y} = \frac{\Delta x_1}{x_1} + \frac{\Delta x_2}{x_2} + \dots + \frac{\Delta x_n}{x_n}$$

$\therefore$ Relative error and absolute error (maximum) are given by

$$E_r \equiv \text{Relative error}$$
$$\equiv \left|\frac{\Delta y}{y}\right| \le \left|\frac{\Delta x_1}{x_1}\right| + \left|\frac{\Delta x_2}{x_2}\right| + \ldots + \left|\frac{\Delta x_n}{x_n}\right|$$
$$E_r = \text{Absolute error}$$
$$= \left|\frac{\Delta y}{y}\right| y \equiv \left|\frac{\Delta y}{y}\right| \times (x_1 x_2 \ldots x_n)$$

(D) ***To find the error in division of two numbers.***

Suppose $$y = \frac{x_1}{x_2}$$

We have (by error formula)

$$\frac{\Delta y}{y} = \frac{\Delta x_1}{y} \cdot \frac{\partial y}{\partial x_1} + \frac{\Delta x_2}{y} \cdot \frac{\partial y}{\partial x_2}$$

$$\equiv \frac{\Delta x_1}{\frac{x_1}{x_2}} \times \frac{1}{x_2} + \frac{\Delta x_2}{\frac{x_1}{x_2}}\left(-\frac{x_1}{x_2^2}\right)$$

$$\equiv \frac{\Delta x_1}{x_1} - \frac{\Delta x_2}{x_2}$$

$\therefore$ $$\left|\frac{\Delta y}{y}\right| \le \left|\frac{\Delta x_1}{x_1}\right| + \left|\frac{\Delta x_2}{x_2}\right|$$

Thus,

Relative error (E_r) is given by
$$E_r \le \left|\frac{\Delta x_1}{x_1}\right| + \left|\frac{\Delta x_2}{x_2}\right|$$

(E) ***To find the error in evaluating x^k***

Let $y \equiv x^k$, (k, integer or fraction)

Then $E_r \equiv$ Relative error

$$\left|\frac{\Delta y}{y}\right| \le \frac{\Delta x}{y} \cdot \frac{\partial y}{\partial x} \quad \text{(error formula)}$$

$$\le \frac{\Delta x}{x^k} \cdot k \cdot x^{k-1}$$

$$\equiv k \cdot \frac{\Delta x}{x}$$

$\therefore$ Relative error in evaluating x^k $\equiv k \cdot \left|\dfrac{\Delta x}{x}\right|$

Example 4. $\sqrt{5.5} \equiv 2.345$ and $\sqrt{6.1} = 2.470$, correct to four significant figures. Find the relative error in taking the difference of these numbers.

Solution: Relative error $\equiv E_r$

$$\left|\frac{\Delta y}{y}\right| \equiv E_r < \left|\frac{\Delta x_1}{y}\right| + \left|\frac{\Delta x_2}{y}\right|$$

Here $\Delta x_1 \equiv \Delta x_2 \equiv \dfrac{1}{2}(0.001) = 0.0005$

$\therefore$ Relative error $\equiv 2\left|\dfrac{\Delta x_1}{y}\right|$

$$\left|\frac{\Delta y}{y}\right| \equiv 2\left|\frac{0.0005}{2.470-2.345}\right|$$

$$\equiv 2\left|\frac{0.0005}{0.125}\right|$$

$$\equiv \frac{0.001}{0.125} = 0.008$$

Example 5. $\sqrt{10} = 3.162$ and $e \approx 2.718$, correct to decimal places. Find the percentage error in their difference.

Solution:

$$\text{Relative error} \equiv E_r \equiv 2 \times \frac{0.0005}{3.162-2.718}$$

$$\left[\because\ E_r < \left|\frac{\Delta x_1}{y}\right| + \left|\frac{\Delta x_2}{y}\right|\right]$$

Here $\Delta x_1 = \Delta x_2 \equiv \dfrac{1}{2}(0.001) = 0.0005$

$\therefore$ Relative error $\equiv E_r$

$$\equiv 2 \times \frac{0.0005}{3.162-2.718}$$

$$\equiv \frac{0.001}{0.444}$$

∴ Percentage error

$$\equiv E_p \equiv \frac{0.001 \times 100}{0.444} \equiv 0.2252252$$

$$\approx 0.23$$

Example 6. Find the product of 346.1 and 865.2 and state how many figures of the result are trustworthy, given that the numbers are correct to four significant figures.

Solution:

$$\Delta x_1 \equiv 0.05 \text{ and } \Delta x_2 \text{ is also} \equiv 0.05$$

$$y \equiv 346.1 \times 865.2$$

$$\equiv 299446 \text{ (correct to 6 significant figures)}$$

$$E_r \leq \left|\frac{\Delta x_1}{x_1}\right| + \left|\frac{\Delta x_2}{x_2}\right|$$

$$= \left|\frac{0.05}{346.1}\right| + \left|\frac{0.05}{865.2}\right|$$

$$\equiv 0.000144 + 0.000058$$

$$\equiv 0.000202$$

∴ $$E_a \equiv E_r\, y \leq 0.000202 \times 299446 \approx 60$$

True value of the product of the numbers given lies between

$$299446 - 60 = 299386$$

and

$$299446 + 60 = 299506$$

The mean of these values is

$$\frac{299386 + 299506}{2} = 299446$$

which is 299.4×10^3 correct to 4 significant figures.

There is uncertainty about the last figure 4.

Example 7. Find the relative error in the calculation of 3.724×4.312 and determine the interval in which the true result lies. Given that the numbers 3.724 and 4.312 are correct to last digit.

Solution:

$$E_r = \text{Relative error}$$

$$\equiv \left|\frac{0.0005}{3.724}\right| + \left|\frac{0.0005}{4.312}\right|$$

$$\equiv 0.0002501$$

$$\begin{aligned}
\text{Absolute error} &\equiv E_r \cdot y \\
&\equiv 0.0002501 \times 3.724 \times 4.312 \\
&\equiv 0.0040157 \\
\text{Product } x_1 x_2 &\equiv 3.724 \times 4.312 \\
&\equiv 16.057888 \\
\text{Lower limit} &\equiv 16.057888 - 0.004016 \\
&\equiv 16.053872 \\
\text{Upper limit} &\equiv 16.057888 + 0.004016 = 16.061904
\end{aligned}$$

Thus, true value of the product lies between 16.0539 and 16.0619.

Example 8. Find the relative error in calculation of $\frac{7.342}{0.241}$. Numbers 7.342 and 0.241 are correct to three decimal places. Determine the smallest interval in which the true result lies.

Solution: Relative error

$$\equiv E_r \le \left|\frac{\Delta x_1}{x_1}\right| + \left|\frac{\Delta x_2}{x_2}\right|$$

Here $\Delta x_1 \equiv \Delta x_2 \equiv 0.0005$

$x_1 = 7.342,\ x_2 = 0.241$

$\therefore$

$$E_r \equiv \text{Relative error} \le \left|\frac{0.0005}{7.342}\right| + \left|\frac{0.0005}{0.241}\right|$$

$$\le 0.0005\left(\frac{1}{7.342} + \frac{1}{0.241}\right)$$

$$\frac{0.0005 \times 7.583}{7.342 \times 0.241} \equiv 0.0021$$

$$\text{Absolute error} \equiv E_r \le 0.0021 \times \left(\frac{x_1}{x_2}\right)$$

$$\equiv \frac{0.0021 \times 7.342}{0.241} \equiv 0.0639$$

Now

$$\frac{x_1}{x_2} \equiv \frac{7.342}{0.241} \equiv 30.4647$$

$\therefore$ True value of $\frac{7.342}{0.241}$ lies between 30.4647 and 0.0639

$\equiv 30.4008$ and $30.4647 + 0.0639$

$= 30.5286.$

Example 9. Find the number of correct figures in the quotient $\frac{65.3}{\sqrt{7}}$, assuming that the numerator is correct to the last figure.

Solution: Relative error

$$\equiv E_r \leq \left|\frac{\Delta x_1}{x_1}\right| + \left|\frac{\Delta x_2}{x_2}\right|$$

Here $$\Delta x_1 \equiv \Delta x_2 \equiv 0.05$$

$$x_1 \equiv 65.7,\ x_3 \equiv 2.6$$

$$E_r \leq \left|\frac{0.05}{65.7}\right| + \left|\frac{0.05}{2.6}\right|$$

$$\leq \frac{0.05 \times 68.3}{65.7 \times 2.6} = 0.01999$$

$$\text{Absolute error } (E_a) \leq 0.01999 \times \frac{65.3}{2.6}$$

$$\equiv 0.502$$

Relations Connecting the Errors with Illustrations

***Property* 1.** If the first significant figure of a number is 'r' and the number is correct to 'n' significant figures, then $(E_r) \equiv$ relative error is less than

$$\frac{1}{r \times 10^{n-1}}$$

i.e. $$E_r < \frac{1}{r \times 10^{n-1}}$$

***Property* 2.** If the relative error in an approximate number is less than

$$\frac{1}{\left[(r+1) \times 10^{n-1}\right]}$$

then the number is correct to n significant figures.

Note: (i) If $E_r \ngtr$ (not greater than) $\frac{1}{2 \times 10^n}$, then the number is correct to n-significant figures.

Note: (ii) Then E_a ($\equiv$ absolute error) is related to the number of decimal places whereas the relative error is related to the number of significant figures.

Example 10. Find the number of trustworthy figures in $(367)^{1/5}$ where 367 is correct to 3 significant figures.

Solution: Relative error $\equiv E_r < \frac{1}{5} \cdot \frac{\Delta x}{x}$

$$\therefore \quad \frac{1}{5} \cdot \frac{\Delta x}{x} \equiv \frac{1}{5} \times \frac{0.5}{367} \equiv 0.0003$$

$$\therefore \quad \text{Absolute error} \equiv E_a < (367)^{1/5} \times 0.0003$$

$$(367)^{1/5} \times 0.0003 \equiv 3.258 \times 0.0003$$

$$\equiv 0.001$$

$$\therefore \quad \text{Absolute error} < 0.001$$

The error effects fourth significant figures and hence $(367)^{1/5} \approx 3.26$ is correct to three figures.

EXERCISES

1. If 0.667 is the approximate value of $\frac{2}{3}$, find the absolute, relative and percentage errors.

[Ans: $E_a \equiv 0.00033$
$E_r \equiv 0.000495$
$E_p \equiv 0.0495$]

2. The round-off value of an exact number is 32.10. How many valid digits are there in the number 32.10? Find the absolute error in this case.

[Ans: $E_a \equiv 0.005$]

3. The number $x = 17.031753$ contains five valid digits. Find its absolute error.

[Ans: 0.0005]

4. Find the absolute error, relative error and percentage error in the following approximate numbers, x whose exact number X being given.

(i) $X = \frac{1}{3}, x = 0.333$

(ii) $X = \frac{1}{7}, x = 0.143$

(iii) $X = \frac{2}{11}, x = 0.182$

(iv) $X = \frac{3}{7}, x = 0.428$

(v) $X = \frac{1}{6}$, $x = 0.167$

(vi) $X = \frac{2}{9}$, $x = 0.222$

5. The round-off value of an exact number is x. How many valid digits are there in the number x^2? Find the absolute error in each case.
(i) 17.30 (ii) 16.814 (iii) 132.50 (iv) 0.445

Ans: (4)

(i) $E_a \equiv 0.00033$; $E_r \equiv 0.00099$; $E_p \equiv 0.099$
(ii) $E_a \equiv 0.00014$; $E_r \equiv 0.00098$; $E_p \equiv 0.098$
(iii) $E_a \equiv 0.00018$; $E_r \equiv 0.00099$; $E_p \equiv 0.099$
(iv) $E_a \equiv 0.00057$; $E_r \equiv 0.00133$; $E_p \equiv 0.133$
(v) $E_a \equiv 0.00033$; $E_r \equiv 0.00198$; $E_p \equiv 0.198$
(vi) $E_a \equiv 0.00022$; $E_r \equiv 0.00099$; $E_p \equiv 0.099$

Ans: (5)

(i) 0.005
(ii) 0.0005
(iii) 0.0005
(iv) 0.005

Example 11. The value of the function $f = \cos x \sin y$ is required to four places of decimal. Estimate the premature error in x and y at $x = 2.9$, $y = 3.1$ (measured in scales).

Solution: Given:

$$f = \cos x \sin y$$

$$\frac{\partial f}{\partial x} = -\sin x \cos y$$

$$\frac{\partial f}{\partial y} = \cos x \cos y \qquad (1)$$

The errors δx and δy are given by

$$\delta x = \frac{\delta f}{2\left(\frac{\partial f}{\partial x}\right)}$$

$$\delta y = \frac{\partial f}{\partial\left(\frac{\partial f}{\partial y}\right)} \qquad (2)$$

at $x = 2.9$, $y = 3.1$, we get

$$\frac{\partial f}{\partial x} = -0.009948$$

$$\frac{\partial f}{\partial y} = 0.970118$$

Then $$\delta x = \frac{-\delta f}{0.019896},\ \delta y = \frac{\delta f}{1.940236} \quad (3)$$

Since it is required to 4 decimal places, we substitute $|\delta f| < 0.5 \times 10^{-4}$

From (3), we get $|\delta x| < 2.513 \times 10^{-3}$

$$|\delta y| < 2.577 \times 10^{-5}$$

Example 12. Prove that the error in the common logarithm of a number is less than one half the relative error in the number.

Solution: Let $L = \log_{10}^{N}$ and let N_A be an approximate to N and L_A be an approximate to L.

Then the error in L is given by

$$\begin{aligned}\in_L &= L - L_A \\ &= \log_{10}^{N} - \log_{10}^{N_A} \\ &= \log_{10}\left(\frac{N}{N_A}\right) \\ &= \log_{10}\left(1+\frac{\in_N}{N_A}\right) \\ &= k \log_e \left(1+\frac{\in_N}{N_A}\right) \quad (1)\end{aligned}$$

Thus, $$k = \frac{1}{\log_e^{10}} \approx 0.43429$$

$$\boxed{\log_{10}^{N} = \frac{\log_e^{N}}{\log_e^{10}}}$$

Expanding $\log_e (1 + \in_N/N)$ on a logarithmic series, and neglecting the 2nd and higher process $\in_N/N$.

We get from (i)

$$\in_L = k\,\frac{\in_N}{N} < \frac{1}{2}\,\frac{\in_N}{N} \quad (2)$$

Thus, the error in $L = \log_{10}^{N}$ is less than one-half the relative error in N.

$$\log(1+1) = x - \frac{x^2}{2} + \frac{x^3}{3} - \frac{x^4}{4} + \frac{x^5}{5}$$

$$\therefore \qquad \in_L \approx k \log_e\left(1+\frac{\in_n}{N}\right)$$

$$k \approx \frac{1}{\log_e^{10}} \approx 0.43429$$

Error in a Function

Let us consider $f = f(x)$ and it is required to find the error in $y = f(x)$ corresponding to an error δx in x and vice versa.

Let δx be the error in x and δf be the error in f.

We find: $$\delta f = f'(x)\ \delta x \qquad (1)$$

(1) is used to compute δf when δx is given or vice versa.

Now consider, $f(x_1, x_2, \ldots x_n)$

Let $\delta x_1, \delta x_2, \ldots$ denote the errors in $x_1, x_2, \ldots x_n$ respectively, we use the formula to compute δf

i.e. $$\delta f = \frac{\partial f}{\partial x_1}\delta x_1 + \frac{\partial f}{\partial x_2}\delta x_2 + \ldots + \frac{\partial f}{\partial x_n}\delta x_n$$

$$= \sum_{k=1}^{n} \frac{\partial f}{\partial x_k}\delta x_k \qquad (2)$$

It is assumed that all the partial derivatives in (2) are equal.

Then, we get

$$\delta f = n\frac{\partial f}{\partial x_i}\delta x_i \qquad (3)$$

When $\frac{\partial f}{\partial x_i}$ is any of the derivatives present in (2)

From (3), we get

$$\boxed{\delta x_i = \frac{\delta f}{n(\partial f/\partial x_i)}} \qquad (4)$$

This formula is used to find errors $\delta x_1, \delta x_2, \ldots \delta x_n$, in $x_1, x_2, \ldots x_n$ corresponding to an error δf in f.

By the rounding off rule, it follows that an approximation is correct to n places of decimals iff

$$|\text{round-off error}| < 0.5 \times 10^{-n}$$

Example 13. If $f = \frac{1}{2}\left(\frac{x^2}{y} - y\right)$ and the error in f is at most 0.4%, find the percentage errors allowable in x and y at $x = 5.1$ and $y = 5.8$.

Solution: From the given f, we get

$$\frac{\partial f}{\partial x} = \frac{x}{y}, \frac{\partial f}{\partial y} = -\frac{1}{2}\left(1 + \frac{x^2}{y^2}\right)$$

∴ The allowable error in x and y are given by

$$\delta x = \frac{\delta f}{2\,(\partial f/\partial x)} = \frac{\delta f}{2\,(x/y)}$$

$$\delta y = \frac{\delta f}{2\,(\partial f/\partial y)} = -\frac{\delta f}{(1 + x^2/y^2)}$$

For $\delta f = 0.4$

$x = 5.1$

and $y = 5.8$

These yield

$$\delta x = \frac{0.4}{1.7586}$$
$$= 0.22745$$

$$\delta y = -\frac{0.4}{1.773186}$$
$$= -0.22558$$

Example 14. Evaluate e to 4 significant figures and determine the absolute and relative errors.

Solution: We have $e = 2.718281828$

$$= 0.2718281828 \times 10^1$$

Rounding off error to four significant figures,

we get $e = 0.2718 \times 10^1 = 2.718$

The E_a in this approve

$$|\epsilon| = (0.2718281828 - 0.2718) \times 10^1$$
$$= (0.0000281828) \times 10^1$$
$$= 0.281828 \times 10^{-3}$$

and $\frac{|\epsilon|}{e} = 0.103678 \times 10^{-3}$

Example 15. Obtain a second degree polynomial approximation to $f(x) = (1+x)^{½}\ x + [0,\ 0.1]$.

Using Taylor's sins above $x = 0$

Use this expansion to approximate $f(0.05)$.

Solution: We have the expansion for Taylor for about $x = 0$.

$$f(x) = f(0) + xf^{1}(0) + \frac{x^2}{2!} f''(0) + \epsilon_3$$

where $\boxed{\epsilon_3 = \frac{x^3}{3!} f'''(\xi)}$, $0 < \xi < x$

$$f(x) = (1+x)^{1/2},\ f(0) = 1$$

$$f'(1) = \frac{1}{2}(1+x)^{-1/2},\ f'(0) = \frac{1}{2}$$

$$f''(0) = -\frac{1}{4}(1+x)^{-3/2},\ f''(0) = -\frac{1}{4}$$

$$f'''(\xi) = \frac{3}{8}(1+x)^{-5/2}$$

But

$$\epsilon_3 = \frac{x^3}{3!} f'''(\xi)\ 0 < \xi < x$$

$$(1+x)^{1/2} \approx 1 + \frac{x}{2} - \frac{x^2}{8} \tag{1}$$

For $x = 0.05$

$$\sqrt{1.05} = 1 + \frac{1}{2}(0.05) - \frac{(0.05)^2}{8}$$

$$= 1.0246875$$

The truncation error $= \epsilon_3 = \frac{x^3}{3!}\left[\frac{3}{5}(1+\xi)^{-5/2}\right]$

$$= \frac{x^3}{10}(1+\xi)^{-3/2}$$

This is maximum when $\xi \to 0 \quad 0 < \xi < 0.05$

Max $\epsilon_3 = (0.05)^{3/16} = 0.7812 \times 10^{-5}$

The value of log 1.1 is computed by using the Taylor's expansion of log $(1+x)$ upto the terms containing x^4.

To which place of the decimal is the approximation correct?

The Taylor's series expansion of log (1 + x) about $x = 0$ is $\log(1+x) = x - \frac{x^2}{2} + \frac{x^3}{3} - \frac{x^4}{4} + \frac{x^5}{8} - + -$

Clearly, this is a logarithm, say if we compute by taking the terms upto x^4,

we get, $$\log(1+x) \simeq x - \frac{x^2}{2} + \frac{x^3}{3} - \frac{x^4}{4} \qquad (2)$$

This approximation is correct to n places of decimal, if the next term in the expansion of log (1 + x) is less than 0.5×10^{-n},

i.e. if $\frac{x^5}{5} < 0.5 \times 10^{-n}$

or $$5 \log_{10}^{x} - \log_{10}^{5} < \log_{10}^{5}(n+1)$$

$$n + 1 < 2 \log_{10}^{5} - 5 \log_{10}^{x}$$

$$[\log_{10}^{5} = 0.19897]$$

or $$n < 0.39794 - 5 \log_{10}^{x}$$

For $x = 0.1$, this becomes

$$n < 0.39794 + 5$$
$$= 5.39794$$

which is correct to 5 places of decimals.

Approximation by Taylor's Series

Consider $f(x)$ which is differentiable upto a desired order at any specified point $x = a$, then

Expanding it by Taylor's theorem, we get

$$f(x) = f(a + x - a)$$

$$= f(a) + (x-a)\frac{f'(a)}{\lfloor 1} + \frac{(x-a)^2}{\lfloor 2} f''(a)$$

$$+ \ldots + \frac{(n-a)^{n-1}}{\lfloor (n-1)} f_{(a)}^{(n-1)} + \in_n \qquad (1)$$

where $$\in_n = \frac{(x-a)^n}{\lfloor n} f^{(n)}(\xi)$$

$$a < \xi < x \qquad (2)$$

($\in_n$ is called the remainder term)

If this function $f(x)$ is approximated by the sum of 'n' terms, we get

$$f(x) \simeq f(a) + (x - a) f'(a) + \ldots + \frac{(x - a)^{n-1}}{(x-1)} f^{(n)}(a) \quad (3)$$

then $\in_n$ serves as the **truncation error** in $f(x)$ at $x = a$.

We note that:

$$|\in_n| \leq \frac{|x - a|^n}{\lfloor n} \cdot M$$

where $$M = \text{Max } |f^{(n)}(\xi)|$$

$$a < \xi < x$$

This is the maximum truncation error at $x = a$.

Chapter **2**

Fundamental Concepts from the Theory of Equations

INTRODUCTION

Numerical methods deal with the process by which mathematical problems can be solved by arithmetical operations. Sometimes this will involve the development of algorithms to solve a problem. While solving problems we usually start with some initial data and then after some intermediate steps we compute the final desired result. We start with the concepts from the theory of equations.

The function $f(x)$ of the form

$$f(x) = a_0\, x^n + a_1\, x^{n-1} + a_2\, x^{n-2} + \dots + a_{n-1} + a_n \tag{1}$$

where n is a positive integer and $a_0, a_1, a_2, \dots a_n$ are independent of x (i.e. are constants), with $a_0 \neq 0$ although some of $a_1, a_2, \dots, a_n$ may be zero, is known as a polynomial of degree n.

(a) If $f(x)$ is exactly divisible by x-α, then α is a root of $f(x) = 0$.

Proof: Divide $f(x)$ by $x-\alpha$ and let $q(x)$ be the quotient.

Then $\quad f(x) = (x-\alpha)\, q(x)$

putting $x = \alpha$, $\therefore\ f(\alpha) = (\alpha-\alpha)\, q(\alpha) = 0$

which shows that α is a root of $f(x) = 0$.

(b) Every equation of the nth degree has only n roots (real or imaginary).

Conversely if $\alpha_1, \alpha_2, \dots, \alpha_n$ are the roots of the nth degree equation $f(x) = 0$, then

$$f(x) = A(x-\alpha_1)\,(x-\alpha_2)\,\dots\,(x-\alpha_n)$$

where A is a constant.

Observations: If a polynomial of degree n vanishes for more than n values of x, it must be identically zero.

(c) **Intermediate-value property**

If $f(x)$ is continuous in the interval (a, b) and if $f(a)$ and $f(b)$ are of opposite signs, then the equation $f(x) = 0$ will have atleast one real root between a and b.

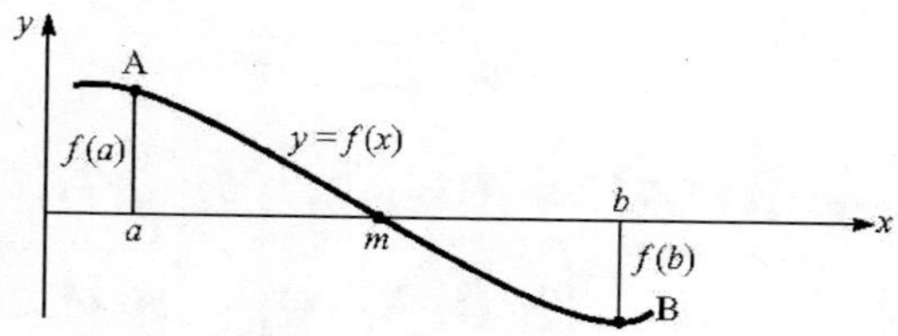

Fig. 1

Suppose we have an equation of the form $f(x) = 0$ whose solution in the range (a, b) is to be searched. Since $f(x)$ is continuous between a and b, $f(x)$ must pass through all the values from $f(a)$ and $f(b)$.

Geometrically, this means that if $f(a)$ and $f(b)$ are of opposite signs, then the curve $y = f(x)$ cuts the x-axis alteast at one point m that lies between a and b. This point m is a real root of the equation $f(x) = 0$. The intermediate value property of a continuous function $f(x)$ may be employed to find an initial approximation for a real root of $f(x) = 0$.

(d) In an algebraic equation with real coefficients complex roots occur in conjugate pairs. Meaning thereby that if $\alpha + i\beta$ is a root of an equation then $\alpha - i\beta$ will also be its root.

Similarly if $\alpha + \sqrt{\beta}$ is an irrational root of an equation, then $\alpha - \sqrt{\beta}$ must also be its root.

Observation: Every equation of the odd degree has atleast one real root. This follows from the fact that imaginary roots occur in conjugate pairs.

Example 1. Solve the equation $x^3 + 6x + 20 = 0$, one root being $1 + 3i$.

Since one root of the equation is given to be $1 + 3i$, another root must be $1 - 3i$, because imaginary roots occur in conjugate pairs.

Solution: Let
$$x = 1 + 3i$$
$\Rightarrow$
$$x - 1 = 3i$$
squaring,
$$x^2 - 2x + 1 = -9$$
or
$$x^2 - 2x + 10 = 0$$

$\therefore$ $x^2 - 2x + 10$ is a factor of $x^3 + 6x + 20$

or $(x - 1 + 3i)(x - 1 - 3i)$
$$= (x - 1)^2 + 9 = x^2 - 2x + 10$$

is a divisor of $x^3 + 6x + 20$

By actual division, the quotient $= x + 2$

$$x^2 - 2x + 10\overline{)x^3 + 6x + 20}(x + 2$$
$$\frac{-(x^3 - 2x^2 + 10x)}{2x^2 - 4x + 20}$$
$$\frac{-(2x^2 - 4x + 20)}{0}$$

$\therefore$ Depressed equation is

$$x + 2 = 0$$

or

$$x = -2$$

Hence, the roots of the given equation are $1 \pm 3i, -2$.

(e) **Descartes' rule of signs**

The equation $f(x) = 0$ cannot have more positive roots than the changes of signs in $f(x)$; and more negative roots than the changes of sign in $f(-x)$.

Alternatively, we can state the rule of signs as follows:

(1) The number of real positive roots of an equation is not more than the number of change of sign in the coefficients of $f(x)$.

(2) The number of real negative roots of $f(x) = 0$, is not more than the number of changes of sign in the coefficients of $f(-x)$.

For instance, consider the equation

$$f(x) = 2x^5 + 8x^4 - 7x^3 - 7x^2 + 5x + 4 = 0$$

The sign of the coefficients $f(x)$ are $(+ + - - + +)$.

There are only two changes, at the second term to the third term, and fourth term to the fifth term. So, the equation has at the most two real positive roots.

$$f(-x) = -2x^5 + 8x^4 + 7x^3 - 7x^2 - 5x + 4$$
$$= - + + - - +$$

There are 3 changes of sign.

So, there cannot be more than three real negative roots of the equation.

Observation: Existence of imaginary roots: If an equation of nth degree has at the most 'p' positive roots and at the most 'q' negative roots, then it follows that the equation has at least $n - (p + q)$ imaginary roots.

Example 2. Apply Descartes' rule of sign to locate the positive root of

$$x^3 - 2x^2 - 1 = 0$$

Solution: Let $f(x) = x^3 - 2x^2 - 1 = 0$ be the given equations. $f(x)$ has only one change of sign.

$\Rightarrow$ $f(x) = 0$ has only one positive root

$f(0) = -1 < 0;\ f(1) = -2 < 0$

$f(2) = -1 < 0;\ f(3) = 8 > 0$

Since $f(2)$ and $f(3)$ have opposite signs, the only positive root lies between 2 and 3.

Example 3. Solve $x^4 - 10x^2 + 35\,x^2 - 50x + 24 = 0$.

Solution:

1	– 10	35	– 50	24
1	100 – 70	1225 – 1000 48	2500 – 1680	576
1	30	273	820	576
1	900 – 546	74529 – 49200 1152	672400 – 314496	331776
1	354	26481	357904	331776

$$|\alpha_1| = \left(\frac{354}{1}\right)^{1/4} = 4.3375$$

$$|\alpha_2| = \left(\frac{26481}{354}\right)^{1/4} = 2.9409$$

$$|\alpha_3| = \left(\frac{357904}{26481}\right)^{1/4} = 1.9174$$

$$|\alpha_4| = \left(\frac{331776}{357904}\right)^{1/4} = 0.9812$$

Determining the signs, we find all are positive.

The exact roots are 4, 3, 2, 1.

Example 4. Solve the biquadratic $x^4 + 2x^3 - 16x^2 - 22x + 7 = 0$ which has a root $2 + \sqrt{3}$.

Solution: The given equation is

$$x^4 + 2x^3 - 16x^2 - 22x + 7 = 0 \qquad (1)$$

The given root is $2 + \sqrt{3}$.

Since irrational roots occur in pairs.

$\therefore$ $2 - \sqrt{3}$ must also be a root.

Let $x = 2 + \sqrt{3}$

$\Rightarrow$ $x - 2 = \sqrt{3}$ only irrational quantity on r.h.s.

Squaring, $x^2 - 4x + 4 = 3$

$\Rightarrow$ $x^2 - 4x + 1 = 0$

$\therefore$ $x^2 - 4x + 1$ is a factor of $x^4 + 2x^3 - 16x^2 - 22x + 7$

By actual division, the quotient $= x^2 + 6x + 7 = 0$.

$$x = \frac{-6 \pm \sqrt{36-28}}{2} = -3 \pm \sqrt{2}$$

Hence, the roots of the given equation are $2 \pm \sqrt{3}$, $-3 \pm \sqrt{2}$.

(f) **Relation between roots and coefficients of an equation.**

If $\alpha, \beta, \gamma, \delta$ ... be the roots of an equation

$a_0x^n + a_1x^{n-1} + a_2x^{n-2} + \ldots + a_n = 0$, then

If $\sqrt{a} + \sqrt{b}$ is a root of an equation, with rational coefficients, then,

$\sqrt{a} - \sqrt{b}$, $-\sqrt{a} + \sqrt{b}$ and $-\sqrt{a} - \sqrt{b}$ are also roots.

$$\Sigma\alpha = (\alpha + \beta + \gamma + \delta + \ldots) = -a_1/a_0$$
$$\Sigma\alpha = +a_2/a_0$$
$$\Sigma\alpha = -a_3/a_0, \ldots \ldots \ldots$$
$$\alpha.\beta.\gamma\ldots = \text{product of roots} = (-1)^n (a_n/a_0)$$

i.e. negative if n is odd, positive if n is even.

Two Important Results

(1) If the roots of $f(x) = 0$ are all positive, then the coefficients (including that of the highest power of x) will be alternatively positive and negative.

Let the leading coefficient be positive since all the roots $\alpha, \beta, \gamma, \ldots$ are positive.

$\therefore$ $\Sigma\alpha$ is positive, i.e, $\frac{-a_1}{a_0}$ is positive

Since a_0 is assumed to be positive,

$\therefore$ a_1 must be negative.

$\Sigma\alpha\beta$ is positive i.e, $+\frac{a_2}{a_0}$ is positive

Since a_0 is positive

$\therefore$ a_2 is also positive and so on.

(2) If the roots of an equation $f(x) = 0$ are all negative, the coefficients will all be positive.
Since the roots $\alpha, \beta, \gamma, \delta$ ... are all negative.

$\therefore$ $\Sigma\alpha$ is negative i.e, $\frac{-a_1}{a_0}$ is negative or $\frac{a_1}{a_0}$ is positive.

Since a_0 is positive $\therefore$ a_1 is positive.

Also $\Sigma\alpha\beta$ is positive i.e, $\frac{a_2}{a_0}$ is positive

Since a_0 is positive $\therefore$ a_2 is positive and so on.

Solve the Equation

$x^4 - 2x^3 + 4x^2 + 6x - 21 = 0$, given that the sum of two of its roots is zero.

Solution: The equation has rational coefficients
Let $\alpha, \beta, \gamma, \delta$ be the roots of $(x^4 - 2x^3 + 4x^2 + 6x - 21) = 0$

Also $$\alpha + \beta + \gamma + \delta = 2$$

$$\therefore \quad \gamma + \delta = 2$$

Thus the quadratic factor corresponding to α, β is of the form $x^2 - ox + p$ and that corresponds to γ, δ is of the form $x^2 - 2x + q$.

$$\therefore \quad x^4 - 2x^3 + 4x^2 + 6x - 21 = (x^2 + p)(x^2 - 2x + q)$$

By equating the coefficients, we get (1)

$$4 = p + q, \ 6 = 2p$$

$$\therefore \quad p = -3, \ q = 7$$

$\therefore$ The given equation is equivalent to $(x^2 - 3)(x^2 - 2x + 7) = 0$

$\therefore$ The roots are $\pm\sqrt{3}$, $1 \pm i\sqrt{6}$.

(g) **Transformation of equations**

(1) To transform an equation into another whose roots are m times the root of the given equation.

Let $f(x) = a_0x^n + a_1x^{n-1} + a_2x^{n-2} + \ldots + a_{n-1}x + a_n = 0$ (1)

be the given equation.

If x is a root of (1) and y, a root of the required equation, then $y = mx$

or $x = \frac{y}{m}$ putting $x = \frac{y}{m}$ in (1), we get

$$a_0\left(\frac{y}{m}\right)^n + a_1\left(\frac{y}{m}\right)^{n-1} + \ldots + a_n = 0$$

Multiplying throughout by m^n, we have

$$a_0 y^n + m a_1 y^{n-1} + m^2 a_2 y^{n-2} + \dots + m^n a_n = 0$$

Expressing it in x.

$$a_0 x^n + m a_1 x^{n-1} + m^2 a_2 x^{n-2} + \dots + m^n a_n = 0$$

Rule: To multiply the roots of a given equation by a given quantity m, multiply the successive coefficients beginning with the second by m, third term by m^2, and so on (all missing terms supplied with zero coefficients).

(2) A few elementary cases are shown below.
It is given that (α, β, γ, ...) are the roots of $f(x) = 0$

To form the equation whose roots are:	We put, in $f(x) = 0$
(i) $-\alpha, -\beta, -\gamma, \dots$	$y = -x$; or $x = -y$;
(ii) $k\alpha, k\beta, \dots$	$y = kx$; or $x = (y/k)$
(iii) $\alpha^2, \beta^2, \gamma^2, \dots$	$y = x^2$; or $x = \sqrt{y}$
(iv) $\alpha-h, \beta-h, \gamma-h, \dots$	$y = x-h$; or $x = y + h$
(v) $\frac{1}{\alpha}, \frac{1}{\beta}, \frac{1}{\gamma}, \dots$	$y = (1/x)$; or $x = (1/y)$.

(3) Sometimes the new roots are not functions of a single root. Thus, let the new root be $\beta + \gamma - \alpha$, $\gamma + \alpha - p$, $\alpha + \beta - \gamma$, where α, β, γ are the roots of the cubic,

$$3x^3 - 6x^2 + 7x + 8 = 0$$

$$\alpha + \beta + \gamma = -6/3 = 2$$

$$\therefore \quad \alpha + \beta - \gamma = \alpha + \beta + \gamma - 2\gamma$$

$$= 2 - 2\gamma = 2(1 - \gamma)$$

therefore, the new roots are $2(1 - \alpha)$, $2(1 - \beta)$, $2(1 - \gamma)$.
This is reduced to the form of a case where the new roots are the functions of a single root. So, we put

$$y = 2(1 - x); \qquad \text{or } x = 1 - y/2$$

In general, the principle underlying the transformation is as follows: We try to reduce a new root to a function of a single root say α; say $g(\alpha)$. Then we put $y = g(\alpha)$.
This may be one of the two types:

(a) It may be possible to solve for α, say $\alpha = \phi(y)$. In this case we put $x = \phi(y)$.

in $f(x) = 0$, to get the new equation, as in (2) and (3) herein above.

(b) It may not be possible to solve for α in terms of y. In this case we retain it as $\alpha = g(y)$ or say $\phi(\alpha, y) = 0$.
Then replace α by x.
$\therefore$ We have $f(x) = 0$ and $\phi(x, y) = 0$.
Eliminate x between the two. We get $f(y/2)$ or $f(y) = 0$.

Example 5. Transform the equation $x^3 + 3x^2 + x - 4 = 0$ into the equation whose roots are multiplied by 10.

Solution: The required equation is obtained by multiplying the successive coefficients 1, 3, 1, – 4 by 1, 10, 10^2, 10^3 respectively. Hence, the required equation is $x^3 + 30x^2 + 100x - 4000 = 0$.

Example 6. Form the equation whose roots are negative of the roots of $x^3 - x^2 + x - 4 = 0$.

Solution: The required equation is $f(-x) = 0$.

$\therefore$ It is given by $(-x)^3 - (-x)^2 + (-x) - 4 = 0$

$\Rightarrow$ $x^3 + x^2 + x + 4 = 0$

(h) **Reciprocal equation:** For an equation $f(x) = 0$, if for every α which is a root of the equation, $1/\alpha$ is also a root, the equation is said to be a reciprocal equation. That is, if $f(\alpha) = 0$, then $f(1/\alpha) = 0$ also. Hence, if the two equations $f(x) = 0$, and $f(1/x) = 0$ are "equivalent", then the equation is said to be a reciprocal equation.

1.
$$f(x) = 0;\ a_0x^n + a_1x^{n-1} \dots a_{n-1}\,x + a_n = 0 \quad (1)$$
$$f(1/x) = 0;\ a_nx^n + a_{n-1}x^{n-1} \dots a_1x + a_0 = 0 \quad (2)$$

For a reciprocal equation, (1) and (2) are equivalent hence, the coefficients are proportional.

$\therefore$ $$a_0/a_n = a_1/a_{n-1} \dots = a_{n-1}/a_1 = a_n/a_0$$

$\therefore$ $$a_0^2 = a_n^2;\ a_1^2 = a_{n-1}^2;\ \text{etc.}$$

$\therefore$ $$a_0 = a_n;\ a_1 = a_{n-1};\ a_2 = a_{n-2} \quad (3)$$

or $$a_0 = -a_n;\ a_1 = -a_{n-1};\ a_2 = -\,a_{n-2} \quad (4)$$

If (3) is the case, that is, the coefficients of the equation (1), equidistant from either extremity are equal in 'magnitude' and 'sign'. The equation is said to be reciprocal and of the "first class". If they are equal in magnitude, but "opposite" in sign, then the equation is "reciprocal" and of the "second class".

Example 7.

1. (a) $3x^4 - 7x^3 + 6x^2 - 7x + 3 = 0$; first class, even degree
   (b) $3x^5 - 7x^4 + 6x^3 + 6x^2 - 7x + 3 = 0$; first class odd degree
   (c) $3x^5 - 7x^4 + 6x^2 + 7x - 3 = 0$; second class, odd degree
   (d) $3x^4 - 7x^3 + 7x - 3 = 0$, "Not reciprocal".
   (e) $3x^4 - 7x^3 + 6x^2 + 7x - 3 = 0$, "Not reciprocal".

   It can be verified that $f(x) = 0$ and $f(1/x) = 0$ are not equivalent equations (since the middle term is present).
2. If the equation $f(x) = 0$ is reciprocal and of the
   (i) First class and odd degree, as in (*b*), $f(-1) = 0$.
   Hence, (–1) is an obvious root; and so $(x + 1)$ is an obvious factor of $f(x)$.
   (ii) Second class and odd degree (as in (c)), then $f(+1) = 0$. Hence, (+1) is an obvious root and so $(x - 1)$ is an obvious factor of $f(x)$.
   (iii) Second class and even degree (as in (d)), $f(+1) = 0$ and $f(-1) = 0$. Hence, (–1) and (+1) are obvious roots; and so $(x + 1)$ and $(x - 1)$ are obvious factors of $f(x)$.
   (iv) First class and even degree (as is (c)), there are no obvious roots. The equation is said to be irreducible.
3. **Working rule to solve a reciprocal equation**
   (a) Note the class and degree
   (b) Note the obvious roots if any
   (c) Divide by the obvious factors like we get an equation of the first class and even degree.
   (d) Divide by x^m if the equation is of degree $2m$; i.e., divide by x^2 if the irreducible equation is of degree four.
   (e) Put $\left(x+\dfrac{1}{x}\right) = y$; $x^2 + \dfrac{1}{x^2} = y^2 - 2$; etc.
   (f) Solve the resulting equation for y. Then taking $x + \dfrac{1}{x} = y$ for each of the values, solve for x. The roots are those obtained in (ii) and in (vi).

   We state the following theorems without proof.

   Theorem: If $\alpha_1, \alpha_2, \ldots \alpha_n$ are the roots of $f(x) = 0$, then $\dfrac{1}{\alpha_1}, \dfrac{1}{\alpha_2}, \ldots, \dfrac{1}{\alpha_n}$ are the roots of $x^n f\left(\dfrac{1}{x}\right) = 0$.

Note: If $f(x) = 0$ is a reciprocal equation then $f(x) = 0$ and $x^n f\left(\frac{1}{x}\right) = 0$ both have the same roots.

Theorem: $f(x) = a_0x^n + a_1x^{n-1} + \ldots + a_n = 0$ is a reciprocal equation if and only if either $a_{n-r} = -a_r$ $(0 \le r \le n)$ or $a_{n-r} = a_r$ $(0 \le r \le n)$.

Example 8. If 2 and 3 are the roots of the equation $6x^6 - 35x^5 + 56x^2 + 35x - 6 = 0$ find the remaining roots.

Solution: The given equation is a reciprocal equation of second class and even degree.

$\therefore$ $x^2 - 1$ is a factor and hence, 1, – 1 are two roots.

Since 2 and 3 are the roots of the given reciprocal equation $\frac{1}{2}, \frac{1}{3}$ are also the roots.

Hence, the six roots are $1, -1, 2, \frac{1}{2}, 3, \frac{1}{3}$.

Example 9. It α, β, γ are the roots of the cubic $x^3 - px^2 + qx - v = 0$, from the equation whose roots are $\beta\gamma + \frac{1}{\alpha}, \gamma\alpha + \frac{1}{\beta}, \alpha\beta + \frac{1}{\gamma}$.

Solution: If x is a root of the given equation and y, a root of the required equation, then

$$y = \beta\gamma + \frac{1}{\alpha} = \frac{\alpha\beta\gamma + 1}{\alpha} = \frac{\gamma + 1}{\alpha} \quad (\because \alpha\beta\gamma = \gamma)$$

$$= \frac{\gamma + 1}{x}$$

$$\Rightarrow \qquad x = \frac{(\gamma + 1)}{y}$$

Substituting in the given equation we get

$$\left(\frac{\gamma+1}{y}\right)^2 - p\left(\frac{\gamma+1}{y}\right)^2 + q\left(\frac{\gamma+1}{y}\right) - \gamma = 0$$

$$\Rightarrow \qquad ry^3 - q(r+1)y^2 + p(r+1)y^2 - (r+1)^2 = 0$$

is the required equation.

Example 10. If α, β, γ are the roots of $x^3 + 2x^2 + 3x - 1 = 0$ what are the roots of $x^3 + 20x^2 + 300x - 1000 = 0$?

Solution: Given $x^3 + 2x^2 + 3x - 1 = 0$

The successive coefficients are multiplied by 1, 10, 10^2, 10^3 respectively, we get the equation: $x^3 + 20x^2 + 300x - 1000 = 0$, whose roots are 10 times the roots of the equation.

$\therefore$ $10\alpha, 10\beta, 10\gamma$ are the roots of $x^3 + 20x^2 + 300x - 1000 = 0$

(i) Synthetic division of a polynomial by a linear expression.

Let us assume a polynomial of degree 'n' in x in the form $f(x) = a_0 x^n + a_1x^{n-1} + \ldots + a_{n-1} x + a_n$ (1)

and if it is divided by $x - \alpha$, we get a polynomial of degree $n - 1$, say,

$$g(x) = b_0 x^{n-1} + b_1 x^{n-2} + \ldots + b_{n-2} x + b_{n-1} \quad (2)$$

as quotient and a remainder b_n. Hence

$$f(x) = (x - d)\, g(x) + b_n \quad (3)$$

Setting $x = \alpha$, we find $f(\alpha) = b_n$, i.e., if a polynomial $f(x)$ is divided by $x - d$, then the remainder is $f(d)$. This is the remainder theorem. Substituting (1) and (2) in (3), we get on carrying out the multiplication and equating the like powers of x,

$$\begin{aligned} a_0 &= b_0 & \qquad b_0 &= a_0 \\ a_1 &= b_1 - \alpha b_0 & \qquad b_1 &= a_1 + \alpha b_0 \\ a_2 &= b_2 - \alpha b_1 \text{ or} & \qquad b_2 &= a_1 + \alpha b_1 \\ &\ldots\ldots & &\ldots\ldots \\ &\ldots\ldots & &\ldots\ldots \\ a_n &= b_n - \alpha b_{n-1} & \qquad b_n &= a_n + \alpha b_{n-1} \end{aligned}$$

The above equations can be exhibited in a tabular form =

$$\begin{array}{c|cccccc} & a_0 & a_1 & a_2 & \ldots & a_{n-1} & a_n \\ \alpha & & \alpha b_0 & \alpha b_1 & \ldots & \alpha b_{n-2} & \alpha b_{n-1} \\ \hline & b_0 & b_1 & b_2 & \ldots & b_{n-1} & \multicolumn{1}{|l}{b_n = f(\alpha)} \\ & & & & & & \multicolumn{1}{|l}{\text{The remainder}} \end{array}$$

This scheme of getting the quotient and remainder is called synthetic division or Horner's method.

Computation of f(x)

To compute $f(x)$ by this method only n multiplications are made whereas in the direct substituting of α for x in $f(x)$ we have to make $n + (n - 1) + (n - 2) + \ldots$

$2 + 1 = \frac{1}{2} n (n + 1)$ multiplications.

Theorem: If a polynomial $f(x)$ is divided by $x - d$ successively (i.e., first $f(x)$, then its quotient, then quotient's quotient ...), then the successive remainders are

$$f(\alpha),\ f'(\alpha),\ \frac{f'(\alpha)}{2!},\ \ldots.$$

Proof: We know that the Taylor's series expansion of $f(x)$ about the point $x = \alpha$ is a terminating series.

$$f(x) = f(\alpha) + (x - \alpha)\,\frac{f'(\alpha)}{1!} + (x - \alpha)^2\,\frac{f'(\alpha)}{2!} + \ldots \tag{1}$$

Hence, if $f(x)$ is divided by $x - \alpha$, the remainder is $f(\alpha)$ and the quotient is

$$\frac{f'(\alpha)}{1!} + (x - \alpha)\,\frac{f''(\alpha)}{2!} + (x - \alpha)^2\,\frac{f'''(\alpha)}{3!} + \ldots \tag{2}$$

If this quotient is divided by $x - \alpha$, we get the remainder $\dfrac{f'(\alpha)}{1!}$ and the quotient

$$\frac{f''(\alpha)}{2!} + (x - d)\,\frac{f'''(\alpha)}{3!} + \ldots \tag{3}$$

If (3) is divided by $x - \alpha$, then the remainder is $\dfrac{f''(\alpha)}{2!}$. Hence, the theorem.

Example 11. Find the quotient and remainder on dividing $x^4 - 5x^3 + 6x^2 + 7x - 9$ by $x - 1$.

Solution:

$x = 1$	$1 - 5 + 6 + 7 - 9$
	$+1 - 4 + 2 + 9$
	$1 - 4 + 2 + 9$ (+0)

$\therefore$ Quotient $= x^3 - 4x^2 + 2x + 9$,
Remainder $= 0$.

Example 12. Divide $2x^4 - 3x^3 + 5x^2 - 7x - 10$ by $(x - 2)$

Solution:

	2	−3	5	−7	−10
2		4	2	14	14
	2	1	7	7	4 = remainder = $f(2)$

Hence, the quotient $= 2x^3 + x^2 + 7x + 7$ and the remainder $= 4$.

Example 13. Find $f(-3)$, $f'(-3)$, $f''(-3)$ if $f(x) = x^5 + 6x^3 - 2x^2 + 18x + 25$

Solution:

	1	0	6	−2	18	25	
−3		−3	9	−45	141	−477	
−3	1	−3	15	−47	159		$-452 = f(-3)$
		−3	18	−99	438		
−3	1	−6	33	−146			$597 = f'(-3)$
		−3	27	−180			
	1	−9	60				$-326 = \dfrac{f''(-3)}{2!}$

Hence, $f(-3) = -452$, $f'(-3) = 597$, $f''(-3) = -652$

(j) **Synthetic division of a polynomial by a quadratic expression.** The division of a polynomial $f(x)$ by the quadratic $x^2 - \alpha x - \beta$ is carried out by the following synthetic scheme:
If a polynomial of degree n is divided by $x^2 - \alpha x - \beta$, then the quotient is a polynomial of degree $n - 2$ and the remainder is a linear expression.

$$f(x) = (x^2 - \alpha x - \beta)\ g(x) + \text{remainder}$$

or
$$a_0\, x^n + a_1 x^{n-1} + \ldots + a_n$$
$$= (x^2 - dx - \beta)\ (b_0 x^{n-2} + b_1\, x^{n-3} + \ldots + b^{n-2}) + (x - d)\ b_{n-1} + b_n.$$

On multiplying and equating the like powers of x, we get

$$\begin{aligned} a_0 &= b_0 & \qquad b_0 &= a_0 \\ a_1 &= b_1 - \alpha b_0 & b_1 &= a_1 + \alpha b_0 \\ a_2 &= b_2 - \alpha b_1 - \alpha b_0 \quad \text{or} & b_2 &= a_2 + \alpha b_1 + \alpha b_0 \\ &\ldots\ldots\ldots\ldots & &\ldots\ldots\ldots\ldots \\ &\ldots\ldots\ldots\ldots & &\ldots\ldots\ldots\ldots \\ a_n &= b_n - \alpha b_{n-1} - \beta b_{n-2} & b_n &= a_n + \alpha b_{n-1} + \beta b_{n-2} \end{aligned}$$

These equations can be shown in a tabular form:

	a_0	a_1	a_2	$\ldots$	a_{n-1}	a_n
α		αb_0	αb_1	$\ldots$	αb_{n-2}	αb_{n-1}
β			βb_0	$\ldots$	b_{n-x}	βb_{n-2}
	b_0	b_1	b_2	$\ldots$	b_{n-1}	b_n

Example 14. Divide $2x^5 - 3x^4 + 4x^3 - 5x^2 + 6x + 2$ by $2x^2 - x + 1$.

Solution: Since $2x^2 - x + 1 \equiv 2\left(x^2 - \frac{1}{2}x + \frac{1}{2}\right)$,

$\therefore\ \alpha = \frac{1}{2};\ \beta = -\frac{1}{2}$

2	–3	4	–5	6	2	
	1	–1	1	$-\frac{3}{2}$		$\frac{1}{2}$
		–1	1	–1	$\frac{3}{2}$	$-\frac{1}{2}$
2	–2	2	–3	$\frac{7}{2}$	$\frac{7}{2}$	

Hence, the quotient $= \frac{1}{2}(2x^4 - 2x^3 + 2x - 3)$

$$= x^4 - x^3 + x - \frac{3}{2}$$

and the remainder $= \frac{7}{2}x + \frac{7}{2} = \frac{7}{2}(x + 1)$

Example 15. Divide $x^4 - x^3 + 7x^2 + x - 10$ by $x^2 - 2x + 3$.

Solution:

1	–1	7	1	–10	
	2	2	12	20	
					2
		–3	–3	–18	–3
1	1	6	10	–8	

Hence, the quotient $= x^2 + x + 6$ and the remainder $= 10(x - 2) - 8 = 10x - 28$.

(k) **To diminish the roots of an equation $f(x) = 0$ by h.**

Let the given equation be

$$a_0x^n + a_1x^{n-1} + \ldots + a_{n-1}x + a_n = 0 \qquad (1)$$

If $\alpha_1, \alpha_2, \ldots, \alpha_n$ are the roots of $f(x) = 0$ then the equation whose roots are

$\alpha_1 - h, \alpha_2 - h, \ldots, \alpha_3 - h$ is given by $f(x + h) = 0$

i.e. $(x+h)^n + a_1(x+h)^{n-1} + \ldots + a_n = 0$

i.e. $x^n + b_1x^{n-1} + b_2x^{n-2} + \ldots + b_n = 0$

where $b_1, b_2, \ldots, b_n$ are constants to be determined, the coefficients b_n, $b_{n-1}, \ldots b_1$ (in this order) are the remainders obtained by dividing $f(x)$ by $x - h$ in succession n times.

Note: To increase the roots of $f(x) = 0$ by h we diminish the roots of $f(x) = 0$ by $-h$.

Example 16. Diminish the roots of the equation $x^3 + x^2 + x - 100$ by 4.

Solution:

4	1	1	1	–100
		4	20	84
	1	5	21	–16
		4	36	
	1	9	57	
		4		
	1	13		

$\therefore$ Transformed equation in $x^3 + 13x^2 + 57x - 3 = 0$

Example 17. Diminish the roots of $2x^5 - x^3 + 10x - 8 = 0$ by 5.

Solution:

5	2	0	–1	0	10	–8
		10	50	245	1225	6175
	2	10	49	245	1235	6167
		10	100	745	4950	
	2	20	149	990	6185	
		10	150	1495		
	2	30	299	2485		
		10	200			
	2	40	499			
		10				
	2	50				
	2					

The transformed equation is

$$2x^5 + 50x^4 + 499x^3 + 2485x^2 + 6185x + 6167 = 0$$

Example 18. Increase the roots of the equation $4x^5 - 2x^3 + 7x - 3 = 0$ by 2.

Solution: Here, we diminish the roots by – 2.

–2	4	0	–2	0	7	–3
		–8	16	–28	56	–126
	4	–8	14	–28	63	–129
		–8	32	–92	240	
	4	–16	46	–120	303	
		–8	48	–188		
	4	–24	94	–308		
		–8	64			
	4	–32	158			
		–8				
	4	–40				

Hence, the transformed equation is $4x^5 - 40x^4 + 158x^3 - 308x^2 + 303x - 129 = 0$.

EXERCISES

1. Divide $x^3 - x - 1$ by $x - 1$ (Ans: $x^2 + x - 1$)
2. Divide $x^4 - 3x^2 + 5x - 7$ by $x + 4$ (Ans: $x^3 - 4x^2 + 13x - 47$, 181)
3. Evaluate f, f^1, f^{11} at $x = 5$, if $f(x) = x^4 - x^3 + 7x^2 + 14$. (Ans: 689, 495, 284)
4. Evaluate f and f^1 at $x = -2$ if $f(x) = x^5 - 7x^3 + x + 5$ (Ans: 27, – 3)

5. Divide $x^4 - 10$ by $x^2 - 2x - 1$, find the quotient and the remainder.
(Ans. $x^2 + 2x + 5$, $12x - 5$)

6. Divide $x^3 - 5x + 7$ by $x^2 + 1.1x + 1.2$. (Ans: $x - 1.1$, $- 4.99x + 8.32$)

ITERATIVE METHODS

The following sections of the chapter deal with iterative methods.

Suppose we have to find a root α of the equation $f(x) = 0$. Let x_0 be an approximation to d. Using x_0 we generate a sequence of numbers $x_1, x_2 \ldots$. Under certain conditions this sequence converges to the root α. The method of generating better and better approximations from an initial guess is called an iteration method.

Order of Convergence

Let $e_i = x_1 - d$ be the error in the ith stage. If the sequence $\{x_1\}$ converges to d, then the sequence $(\in_1\}$ converges to 0. If the error $\in_1$ is related to $\in_{i+1} = x_{i+1} - d$ by a formula

$$|\in_{i+1}| \leq K\ |\in_i|^p,$$

where K and p are constants, $K > 0$, $p \geq 1$, then we say that the convergence is of order p.

If $p = 1$, the convergence is said to be linear.

If $p = 2$, the convergence is said to be quadratic.

If $p = 3$, the convergence is said to be cubic.

Obviously the convergence is faster if K is smaller and p is larger.

Initial Approximation

In order to obtain successive approximations of a root, we have to start with some number x_0 such that $f(x_0)$ is near to zero. The number x_0 is called the initial approximation of the root. Almost all the numerical methods of the solution of an equation require an initial rough approximation to its root. The choice of this initial approximation depends on the desired degree of accuracy of the root being found and the method employed for iteration. Graphical methods are also used to find an approximate value of the root which is explained below.

Graphical solution of equations

The real root of the equation $f(x) = 0$ can be determined approximately as the abscissa of the points of intersection of the graph of the function $y = f(x)$ with the x–axis.

(i) Let $f(x) = 0$ (1)

Find a small interval (a, b) between which the root of the equation lies.

$f(a) = -\text{ve}$ and $f(b) = +\text{ve}$

then the root of the eqaution (1) lies between a and b.

For example, let

$$f(x) = e^{x-1} + x - 3 = 0 \qquad (1)$$

$$f(1) = 1 + 1 - 3 = -\text{ve}$$

$$f(2) = e + 2 - 3 = 2.718 - 1 = +\text{ve}$$

∴ The root of the equation lies between 1 and 2.

(ii) Write the equation $f(x) = 0$ as $\phi(x) = \psi(x)$

For example

$$e^{x-1} + x - 3 = 0$$

or

$$e^{x-1} = 3 - x \qquad (2)$$

(iii) Prepare two tables for $y = \phi(x)$ and $y = \psi(x)$ taking values of x between a and b.

(iv) Plot these points and join them to get the smooth curves.

(v) Note down the abscissa of the point of intersection of the curves $y = \phi(x)$ and $y = \psi(x)$.

This is the required root of the equation.

Note: Sometimes we do not write $f(x) = 0$ as $\phi(x) = \psi(x)$. We adopt the following method.

(a) Form a table for the value of x and $y = f(x)$ directly.

(b) Plot these points and join them to get a smooth curve.

(c) Read the abscissa of the points where this curve cuts the x-axis, these are rough approximations to the roots of $f(x) = 0$.

Now let us recall equation (2)

i.e.

$$e^{x-1} = 3 - x$$

then, the abscissa of the points of intersection of the curves

$$y = e^{x-1} \qquad (3)$$

and

$$y = 3 - x \qquad (4)$$

will give the required root.

We plot (2), and prepare the following table

$x =$	1.1	1.2	1.3	1.4	1.5	1.6	1.7	1.8	1.9	2.0
$y = e^{x-1}$	1.11	1.22	1.35	1.49	1.65	1.82	2.01	2.23	2.46	2.72

Next, we plot the points (1, 2) and (2, 1) as shown below.

From the figure, we get the required root to be $x = 1.44$ nearly.

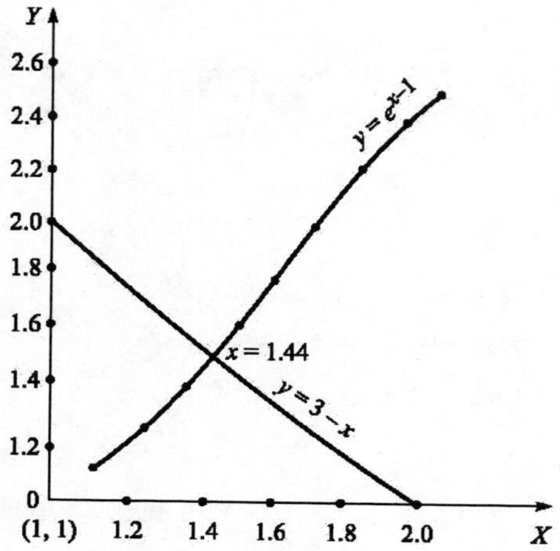

Fig. 2

Example 19. Solve graphically the equation

$$x = \sin x + \frac{\pi}{2}$$

Solution:

$$x = \sin x + \frac{\pi}{2}$$

or

$$\sin x - \frac{\pi}{2}$$

We take two equations $y = \sin x$ and $y = x - \frac{\pi}{2}$. Let us find out the abscissa of the point of intersection of the line $y = x - \frac{\pi}{2}$ and the curve $y = \sin x$ and give a rough estimate of the root. For the straight line $y = x - \frac{\pi}{2}$, we have the table.

x	0	$\frac{\pi}{2}$	π
$y = x - \frac{\pi}{2}$	$-\frac{\pi}{2}$	0	$\frac{\pi}{2}$

For the sine curve, we have the following table.

x	0	$\pi/4$	$\frac{\pi}{2}$	$\frac{3\pi}{4}$	π
$y = \sin x$	0	0.71	1.00	0.71	0

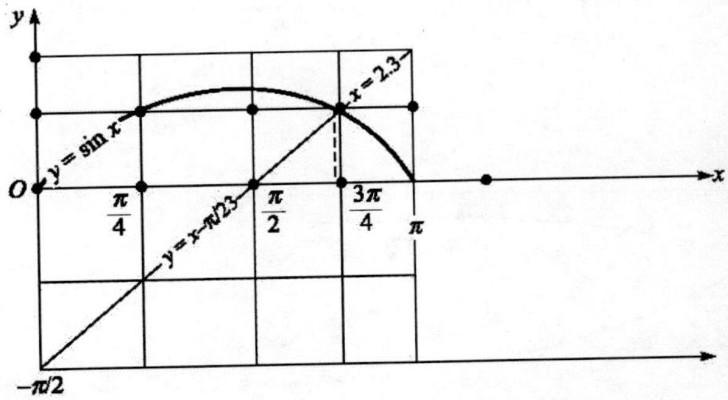

On the same axis and with the same scale construct the graphs of $y = x - \frac{x}{2}$ and $y = \sin x$

From the graph, we get, $x = 2.3$ rad approx.

Fig. 3

on the same axis and with the same scale constants the graphs of $y = x - \dfrac{\pi}{2}$ and

$y = \sin x$

From the graph, we set $x = 2.3$ rad approx.

Example 20. Solve graphically $\cos x \cosh x = -1$

Solution:

Here,
$$f(x) = \cos x \cosh x + 1 = 0 \qquad (1)$$
$$f(0) = +\text{ve},$$
$$f\left(\frac{\pi}{2}\right) = +\text{ve}$$
$$f(\pi) = -\text{ve}$$

The root of (1) lies between $\dfrac{\pi}{2}$ and π as $f\left(\dfrac{\pi}{2}\right)$ and $f(\pi)$ are opposite in sign.

[1] is written as $\cos x = -\text{sech } x$. We take two equations $y \cos x$ and $y = -\text{sech } x$. Let us find out the abscissa of the point of intersection of the curve $y = \cos x$

[2] and the curve $y = -\text{sech } x$

[3] and give a rough estimate of the root.

For the curve $y = \cos x$, we have the table

$x =$	$\frac{\pi}{2} = 1.57$	$\frac{3\pi}{4} = 2.36$	$\pi = 3.14$
$y = \cos x$	0	– 0.71	– 1

For the curve $y = -\text{sech } x$, we have the table

$x =$	1.57	2.36	3.14
$\cosh x =$	2.51	5.34	11.57
$y = -\text{sech } x$	– 0.4	– 0.19	– 0.09

on the same axes, and with the same scale construct the graphs of $y = \cos x$ and $y = -\operatorname{sech} x$.

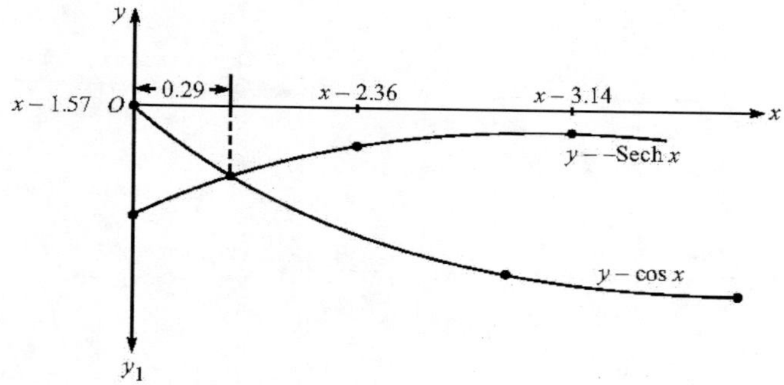

Fig. 4

From the graph, we get $x = 1.57 + 0.29 = 1.86$ radians approximately.

Example 21. Solve graphically $e^{-x} - \sin x = 0$

Solution:

Let
$$f(x) = e^{-x} - \sin x = 0$$
$$\Rightarrow \qquad e^{-x} = \sin x$$

We take two equations $y = e^{-x}$ and $y = \sin x$

Let us find out the abscissa of the point of intersection of the curve $y = e^{-x}$ and the curve $y = \sin x$ and give a rough estimate of the root. For the curve, we have the values of the abscissa as:

$$x = 1.57, 2.36, 3.14$$

On the same axes and with the same scale construct the graph of $y = e^{-x}$ and $y = \sin x$.

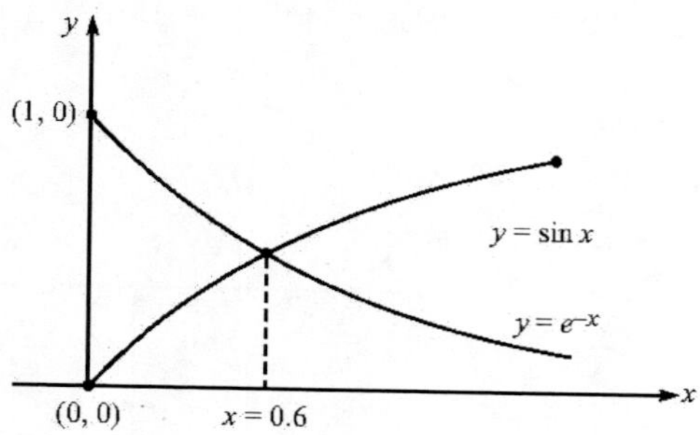

Fig. 5

From the graph, we get $x = 0.6$ radians approximately.

Example 22. Solve graphically $(x - 1)^2 = \log x$

Solution:

$$(x - 1)^2 = \log x$$

We take two equations $y = (x - 1)^2$ and $y = \log x$. Let us find out the abscissa of the point of intersection of the curve $y = (x - 1)^2$ and the curve $y = \log x$ and give a rough estimate of the root.

For the curve $y = (x - 1)^2$, we have the table:

x	0.5	1	1.5	2	2.5
$y = (x - 1)^2$	– 0.7	0	0.4	0.7	1.0

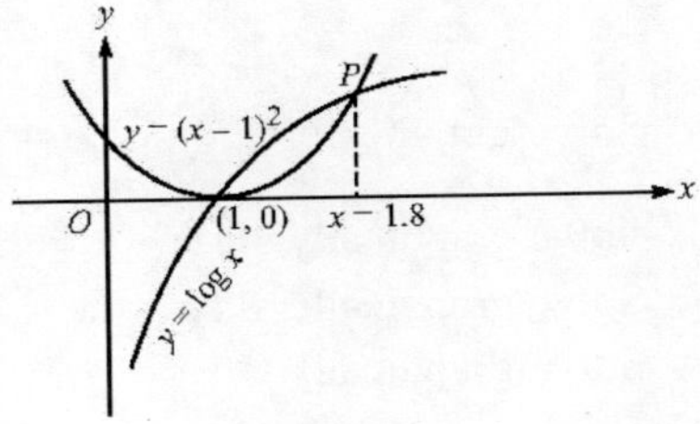

Fig. 6

on the same axes, and with the same scale construct the graphs of $y = (x - 1)^2$ and $y = \log x$.

From the graph, we get $x = 1$ and 1.8 approx.

$\because$ $(x - 1)^2 = \log x$ are 1 and 1.8 approx.

Example 23. Solve graphically $x^3 + x + 1 = 0$

Solution:

Let $$f(x) = x^3 + x + 1 = 0$$

This may be written as $x^3 = -(x + 1) = -x - 1$.

We take two equations $y = x^3$ and $y = -x - 1$. Let us find out the abscissa of the point of intersection of the curve $y = x^3$ and the curve $y = -x - 1$ and give a rough estimate of the root.

For the curve $y = x^3$, we have the table:

x	1	1.4	1.6	1.7
y	1	2.7	4.1	4.9

For the straight line, $y = -x - 1$, passes through $(0, -1)$ and $(-1, 0)$.

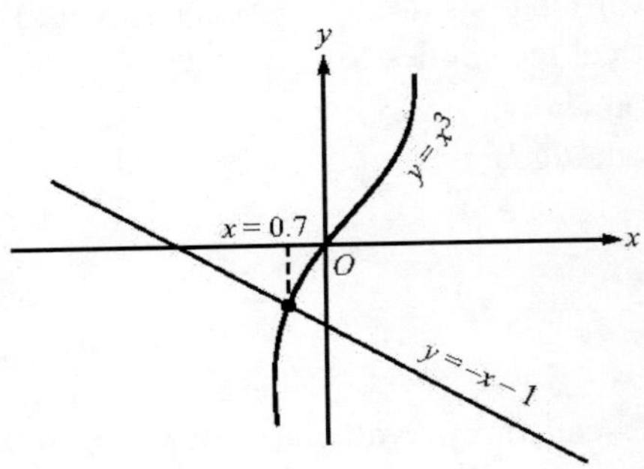

Fig. 7

From the graph, we get $x = +0.7$ approx.

EXERCISES

1. Solve graphically $x^3 - 2x - 5 = 0$. (Ans: 2.099)
2. Solve graphically $e^{3x} - 5x^2 - 17 = 0$. (Ans: 1.04)
3. Solve graphically $\tan x = 1.2\,x$. (Ans: 0.71 rad.)
4. Solve graphically $x = 3\cos\left(x - \frac{\pi}{4}\right)$. (Ans: 1.81 rad.)
5. Solve graphically $2.5\log_{10} x + \frac{x^2}{100} = 6.35$ (Ans: 17.93)
6. Solve graphically $\sin x = x^3 + 1$ (Ans: – 1.2 rad.)
7. Solve graphically $\sin x = (x - 1)^2$. (Ans: 0.4 and 1.9 rad.)
8. Solve graphically $\tan x = -3x + 1$. (Ans: – 1.8, 0.25, 1.8 rad. approx.)
9. Solve graphically $\sin = \frac{1-x}{1+x}$. (Ans: 0.4, 3.8, ... rad. approx.)
10. Solve graphically $x^3 - 6x^2 + 9x - 3 = 0$ (Ans: 0.45)
11. Solve graphically $8x^3 + 12x^2 + 10x - 17 = 0$ (Ans: 0.74)
12. Find graphically the approximate value of the smallest real root of $e^{-x} - \sin x = 0$. (Ans: 0.6)

NUMERICAL SOLUTION OF ALGEBRAIC AND TRANSCENDENTAL EQUATIONS

We are familiar with the solution of quadratic eqution $ax^2 + bx + c = 0$, which is analytic in nature.

Analytical methods of solving cubic and quadratic equations are also available. However, polynomial equations of degree greater than 4 are not solvable in analytical form and one has to resort to numerial methods to reach at a solution as correct as possible. On many occasions we come across transcendental equations in engineering. Analytical methods do not exist for solving such equations. We have to use only numerical techniques to solve them.

In this chapter, we deal with the most important numerical methods for solving non-linear equations, both algebraic and transcendental.

Algebraic Equation

An expression of the form

$f(x) = a_0x^n + a_1x^{n-1} + \ldots + a_{n-1}x + a_n$ where a's are constants ($a_0 \neq 0$) and n is a positive integer, is called a polynomial in x of degree n. The polynomial $f(x) = 0$ is called an algebraic equation of degree n in x.

Transcendental Equation

Equations which are not purely algebraic are called transcendental equations, that is, if $f(x)$ contains some other functions such as trigonometric, logarithmic, exponential etc. then $f(x) = 0$ is called a transcendental equation.

Examples (i) $2x + \log_{10}x - 5 = 0$

(ii) $x + \sin x + 2 = 0$

(iii) $e^x + \tan x + 3x = 10$

BISECTION (OR BOLZANO) METHOD

This method is based on the repeated application of the intermediate-value property. It is a lengthy and laborious method.

Let the given equation whose solution is to be found be

$$f(x) = 0 \tag{1}$$

where the function $f(x)$ is continuous on (a, b) and $f(a)\,f(b) < 0$. In order to find a root of (1) lying in the interval (a, b). We shall determine a very small interval (a_0, b_0) (by graphical method) in which $f(a_0)\,f(b_0) < 0$ and $f'(x)$ maintain the same sign in (a_0, b_0) so that there is only one real root of the equation $f(x) = 0$. Divide the interval in half and let

$$x_1 = \frac{a_0 + b_0}{2}$$

If $f(x) = 0$ then x_1 is a root of the equation. If $f(x_1) \neq 0$ then either $f(a_0)\,f(x_1) < 0$ of $f(b_0)\,f(x_1) <0$. If $f(a_0)\,f(x_1) < 0$ then the root of the equation lies in (a_0, x_1) otherwise the root of the equation lies in (x_1, b_1). We rename the interval in which the root lies as (a_1, b_1) so that

$$b_1 - a_1 = \frac{1}{2}\,(b_0 - a_0)$$

now we take $x_2 = \dfrac{a_1 + b_1}{2}$

If $f(x_2) = 0$ then x_2 is the root of $f(x) = 0$. If $f(x_2) \neq 0$ and $f(x_2)\,f(a_1) < 0$, then the root lies in (a_1, x_2). In which case we rename the interval as (a_2, b_2), otherwise (x_2, b_1) it is renamed as (a_2, b_2) where

$$a_2 - b_2 = \frac{1}{2^2}\,(b_0 - a_0)$$

Proceeding in this manner, we find

$$x_{n+1} = \frac{a_n + b_n}{2}$$

which gives us the $(n + 1)^{th}$ approximation of the root of $f(x) = 0$, and the root lies (a_n, b_n) where

$$b_n - a_n = \frac{1}{2^n} (b_0 - a_0)$$

Since the left hand points $a_1, a_2, ..., a_n$, form a monotonic non-decreasing bonded sequence, and the right hand points $b_1, b_2, ... b_n$, form a monotonic non-increasing bounded sequence, then there is a common limit

$$c = \lim_{n\to\infty} a_n = \lim_{n\to\infty} b_n$$

such that $f(c) = 0$ which means that c is a root of equation (1).

Example 24. Find a root of the equation $x^3 - 4x - 9 = 0$, using the bisection method in four stages.

Solution: We have

$$f(x) = x^3 - 4x - 9$$

Since $f(2) = 8 - 8 - 9 = - 9 < 0$: (negative)

and $f(3) = 27 - 12 - 9 = 6 > 0$; (positive)

$$\therefore \qquad f(2)\, f(3) < 0$$

$\Rightarrow$ a root lies between 2 and 3.

Let $$a_0 = 2,\ b_0 = 3$$

First approximation to the root is

$$x_1 = \frac{1}{2}(a_0 + b_0) = \frac{1}{2}(2 + 3) = 2.5$$

Thus, $f(x_1) = (2.5)^3 - 4\ (2.5) - 9 = - 3.375 < 0$ (negative)

$\Rightarrow$ the root lies between x_1 and 3

$\therefore$ Second approximation is:

$$x_2 = \frac{1}{2}\ (x_1 + 3) = 2.75$$

Thus, $f(x_2) = (2.75)^3 - 4\ (2.75) - 9 = 0.7969 > 0$ (positive)

⇒ the root lies between x_1 and x_2

∴ Third approximation is:

$$x_3 = \frac{1}{2}(x_1 + x_2) = 2.625$$

Then $$f(x_3) = (2.625)^3 - 4\,(2.625) - 9$$
$$= -\,1.4121 < 0 \text{ (negative)}$$

⇒ the root lies between x_2 and x_3

∴ Fourth approximation is

$$x_4 = \frac{1}{2}(x_2 + x_3) = 2.6875$$

Hence, the required approximate root is 2.6875.

Example 25. Using bisection method find a real root of the equation:

$$x^3 - 5x + 1 = 0.$$

Solution:

Let $$f(x) = x^3 - 5x + 1$$

Since $$f(0) = 1 > 0 \text{ (positive)}$$

and $$f(1) = -\,3 < 0 \text{ (negative)}$$

∴ $$f(0)\,f(1) < 0$$

⇒ a root lies between 0 and 1.

$$a_0 = 0, \quad b_0 = 1$$

∴ The first approximation of this root is

$$x_1 = \frac{1}{2}(a_0 + b_0) = \frac{1}{2}(0 + 1) = 0.5$$

Thus $f(x_1) = -\,1.375 < 0$. (negative)

∴ a root lies between a_0 and x_1

∴ The second approximation of the root is

$$x_2 = \frac{1}{2}(a_0 + x_1) = \frac{1}{2}(0 + 0.5) = 0.25$$

Thus $f(x_2) = -\,0.234375 < 0$ (negative)

∴ The root lies between $a_0 = 0$ and $x_2 = 0.25$

∴ The third approximation of the root is

$$x_3 = \frac{1}{2}(a_0 + x_2) = \frac{1}{2}(0 + 0.25) = 0.125.$$

thus $f(x_3) = 0.37695 > 0$.

Since $f(x_3) > 0$ and $f(x_2) < 0$

∴ The root lies between $x_3 = 0.125$ and $x_2 = 0.25$

∴ The fourth approximation of the root is

$$x_4 = \frac{1}{2}(x_3 + x_2) = \frac{1}{2}(0.125 + 0.25) = 0.1875$$

Thus $f(x_4) = 0.06910 > 0$. Since $f(x_4) > 0$ and $f(x_2) < 0$, the root lies between $x_4 = 0.1875$ and $x_2 = 0.25$.

Hence the fifth approximation of the root is $x_5 = \frac{1}{2}(x_4 + x_2) = \frac{1}{2}(0.1875 + 0.25) = 0.21875$.

∴ 0.21875 is taken as an approximate value of the root of the given equation that lies between 0 and 1.

Example 26. Using bisection method compute one root of $e^x - 3x = 0$ correct to two decimal places, in the interval [1.5, 1.6]

Solution:

Let $$f(x) = e^x - 3x$$

Since $$f(1.5) = -0.02 < 0,$$

and $$f(1.6) = 0.15 > 0,$$

here $$a_0 = 1.5,\ b_0 = 1.6$$

∴ first approximation of the root is

$$x_1 = \frac{1}{2}(a_0 + b_0) = \frac{1}{2}(1.5 + 1.6) = 1.55$$

Thus $$f(x_1) = f(1.55) = 0.06 > 0$$

∴ a root lies between a_0 and x_1

∴ second approximation of the root is

$$x_2 = \frac{1}{2}(a_0 + x_1) = \frac{1}{2}(1.5 + 1.55) = 1.525$$

Thus, $f(x_2) = 0.02 > 0$

∴ The root lies between $a_0 = 1.5$ and $x_2 = 1.525$

∴ Third approximation of the root is

$$x_3 = \frac{1}{2}(a_0 + x_2) = \frac{1}{2}(1.5 + 1.525) = 1.5125$$

Thus $f(x_3) = 0.00056 > 0$

∴ a root lies between a_0 and x_3.

∴ Fourth approximation of the root is

$$x_4 = \frac{1}{2}(a_0 + x_3) = \frac{1}{2}(1.5 + 1.5125) = 1.5062$$

Thus $f(x_4) = -\ 0.0090 < 0$

∴ The root lies between $x_3 = 1.51125$ and $x_4 = 1.5062$

∴ The fifth approximation of the root is

$$x_5 = \frac{1}{2}(x_3 + x_4) = \frac{1}{2}(1.5125 + 1.5062)$$
$$= 1.50935$$

Thus $f(x_5) = f(1.50935) = -\ 0.00426 < 0$

Check: ∴	$x_4 = 1.5062,\ x_5 = 1.50935$
	$x_6 = 1.51092,\ f(x_6) = -\ 0.00184 < 0.$

In the 4th step a_n, b_n and x^{n+1} are equal upto ten decimal places. Thus, $x = 1.51$ is the root of $f(x) = 0$, correct to two decimal places.

Example 27. By using the bisection method, find an approximate root of the equation $\sin x = \frac{1}{x}$ that lies between $x = 1$ and $x = 1.5$ (measured in radians). Carry out computation upto 7th stage.

Solution: $f(x) = x \sin x - 1 = 0$

Since $f(1) = -\ 0.158529 < 0$

and $f(1.5) = 0.496242 > 0$

∴ There is a root x_1 between $a_0 = 1$ and $b_0 = 1.5$

∴ The first approximation is

$$x_1 = \frac{1}{2}(a_0 + b_0) = \frac{1}{2}(1 + 1.25) = 1.25$$

Thus, $f(x_1) = 0.186231 > 0.$

Since $f(1) < 0$ and $f(x_1) > 0$

$\therefore$ The root lies between $a_0 = 1$ and $x_1 = 1.25$

$\therefore$ The second approximation is

$$x_2 = \frac{1}{2}(a_0 + x_1) = \frac{1}{2}(1 + 1.25) = 1.125$$

Thus, $f(x_2) = 0.015051 > 0$. Since $f(a_0) < 0$
and $f(x_2) > 0$, $\therefore$ the root lies between $a_0 = 1$ and $x_2 = 1.125$.

$\therefore$ The third approximation is

$$x_3 = \frac{1}{2}(a_0 + x_2) = \frac{1}{2}(1 + 1.125) = 1.0625$$

Thus, $$f(x_3) = -0.071827 < 0,$$
since $f(x_3) < 0$ and $f(x_2) > 0$,

$\therefore$ The root lies between $x_3 = 1.0625$ and $x_2 = 1.125$

$\therefore$ The fourth approximation is

$$x_4 = \frac{1}{2}(x_2 + x_3) = \frac{1}{2}(1.0625 + 1.125) = 1.09375$$

Thus, $$f(x_4) = -\ 0.028362 < 0,$$
since $f(x_4) < 0$ and $f(x_2) > 0$,

$\therefore$ The root lies between $x_4 = 1.09375$ and $x_2 = 1.125$.

$\therefore$ The fifth aproximation of the root is

$$x_5 = \frac{1}{2}(x_4 + x_2) = \frac{1}{2}(1.09375 + 1.125) = 1.109375$$

Thus, $f(x_5) = -\ 0.006643 < 0$.
Since $f(x_5) < 0$ and $f(x) > 0$, the root lies between $x_5 = 1.109375$ and $x_2 = 1.125$

$\therefore$ The sixth approximation of the root is

$$x_6 = \frac{1}{2}(x_5 + x_2) = \frac{1}{2}(1.109375 + 1.125) = 1.1171875$$

Thus, $f(x_6) = 0.004208 > 0$.
Since $f(x_5) < 0$ and $f(x_6) > 0$,
$\therefore$ The root lies between $x_5 = 1.109375$ and $x_6 = 1.1171875$.

$\therefore$ The seventh approximation of the root is

$$x_7 = \frac{1}{2}(x_5 + x_6) = \frac{1}{2}(1.109375 + 1.1171875) = 1.11328125$$

Which is the required approximate root of the desired order of the root of the given equation that lies between 1 and 1.5.

Example 28. Find the root of $\tan x + x = 0$ upto two decimal places, which lies between 2 and 2.1.

Solution:

Let $f(x) = \tan x + x$

Since $f(2) = -0.18 < 0$

and $f(2.1) = 0.39 > 0$

Thus, the root lies between 2.0 and 2.1

$\therefore$ $a_0 = 2,\ b_0 = 2.1$

$\therefore$ First approximation of the root is

$$x_1 = \frac{1}{2}(a_0 + b_0) = \frac{1}{2}(2 + 2.1) = 2.05$$

Thus, $f(x_1) = -0.12 > 0$

Similarly, we get the following successive approximation after computation,

$$\begin{aligned}
x_2 &= 2.025, & f(x_2) &< 0\\
x_3 &= 2.0375, & f(x_3) &> 0\\
x_4 &= 2.03125, & f(x_4) &< 0\\
x_5 &= 2.02812, & f(x_5) &< 0\\
x_6 &= 2.02968, & f(x_6) &> 0\\
x_7 &= 2.02890, & f(x_7) &> 0
\end{aligned}$$

In the fifth step, a_n, b_n and x_{n+1} are equal upto two decimal places.

$\therefore$ $x = 2.03$ is a root of $f(x) = 0$, correct to two decimal places.

Example 29. Find an approximate value of $\sqrt{3}$ correct to two decimal places, using bisection method.

Solution:

Let $x = \sqrt{3}$

$\Rightarrow$ $x_2 = 3$

$\Rightarrow$ $x_2 - 3 = 0$

Let $f(x) = x^2 - 3$

Since $f(1) = -2 < 0$

and $f(2) = 1 > 0$

$\therefore$ The root lies between $a_0 = 1,\ b_0 = 2$

$\therefore$ The first approximation of the root is

$$x_1 = \frac{1}{2}(a_0 + b_0) = \frac{1}{2}(1 + 2) = 1.5$$

Thus $f(x_1) = -0.75 < 0$, since $f(2) > 0$

$\therefore$ The root lies between $x_1 = 1.5$ and $b_0 = 2$

$$\therefore \quad x_2 = \frac{1}{2}(x_1 + b_0) = \frac{1}{2}(1.5 + 2) = 1.75$$

Similarly we get the following approximation

$$\left.\begin{aligned} x_3 &= 1.625 \\ x_4 &= 1.6875 \\ x_5 &= 1.71875 \\ x_6 &= 1.734375 \end{aligned}\right\}$$

Since $\sqrt{3} = 1.7320508$, we take that x_6 is an approx value of $\sqrt{3}$ correct to two decimal places.

EXERCISES

1. Find a root of the following equations, using bisection method correct to three decimal places.
   - (i) $x^3 - 4x - 9 = 0$ (Ans: 2.707)
   - (ii) $x^3 - x - 11 = 0$ in (2, 3) which lies between 2 and 3 (Ans: 2.375)
   - (iii) $x^4 - x - 10 = 0$ (Ans: 1.8125).
   - (iv) $x^3 - 3x - 5 = 0$ (Ans: 2.28)
   - (v) $x^3 - x^2 + x - 7 = 0$ in (2, 2.25) (Ans: 2.1049)
   - (vi) $e^x = x + 2$ in (1, 1.4) (Ans: 1.146)
   - (vii) $\tan x + \tan hx = 0$ in (1, 2) (Ans: 1.5707)
   - (ix) $x + \log x = 5$ in (3.2, 4) (Ans: 3.692)
   - (x) $\cos x - 1.3x = 0$ in (0, 1) (Ans: 0.625)
   - (xi) $x^4 + 2x^2 - 16x + 5 = 0$ in (0, 0.5) (Ans: 0.3261)
   - (xii) $x^3 - 5x - 4 = 0$ in (2, 3) (Ans: 2.5625)
   - (xiii) $x^3 - x - 4 = 0$ in (1.5, 2) (Ans: 1.7964)
2. Find the roots correct to one decimal place using bisection method.
   - (a) $x^2 - 4\sin x = 0$ near 2 (Ans: 1.9)
   - (b) $x - \frac{\pi}{2} - \sin x = 0$ near 2 (Ans: 2.3)
   - (c) $x - 1 - \tan hx = 0$ near 2 (Ans: 2.0)

METHOD OF FALSE POSITION OR REGULA-FALSI METHOD

This method is a great improvement of the bisection method and is the oldest method of finding the real root of an equation $f(x) = 0$. For the equation $f(x) = 0$, having found x_0 and x_1 such that $y_0 = f(x_0)$ and $y_1 = f(x_1)$ are of opposite sign, that is, the graph of $y = f(x)$ crosses the x-axis between these points (Fig. 8). This indicates that a root lies between x_0 and x_1 and consequently $f(x_0)\, f(x_1) < 0$.

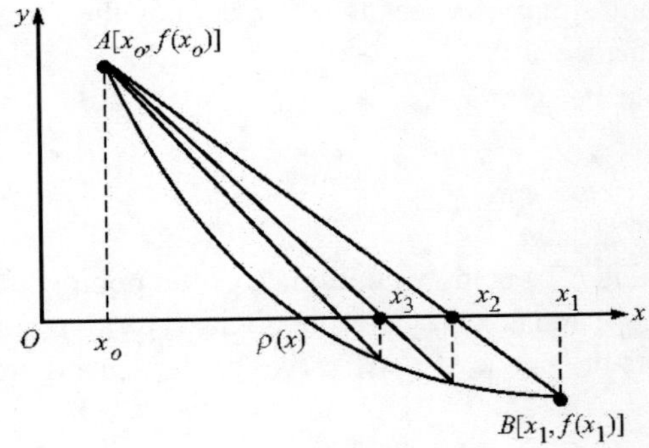

Fig. 8

Equation of the chord joining the points $A\ (x_0, f(x_0)]$ and $B[x_1, f(x_1)]$ is

$$y - f(x_0) = \frac{f(x_1) - f(x_0)}{x_1 - x_0}\,(x - x_0)$$

The method consists in replacing the curve AB by means of the chord AB and taking the point of intersection of the chord with the x-axis as an approximation to the root. So, the abscissa of the point where the chord cuts the x-axis ($y = 0$) is given by

$$x_2 = x_0 - \frac{x_1 - x_0}{f(x_1) - f(x)_0}\, f(x_0) \tag{1}$$

which is an approximation to the root. If we denote the first approximation by m_1 then, we may write (1) as

$m_1 = x_2 = x_0 - \left[\dfrac{x_1 - x_0}{f_1 - f_0}\right] f_0$ (2a) when $f_1 = f(x_1),\ f_0 = f(x_0)$
We can also write [2a] as:- $m_1 = x_2 = \left[\dfrac{x_0 f_1 - x_1 f_0}{f_1 - f_0}\right]$ gives the first approximation (2b)

Find $f_2 = f(x_2)$, then either f_0 and f_2 are of opposite signs or f_2 and f_1 are of opposite signs. Suppose f_0 and f_1 are of opposite signs. Then employ the formula (2) with x_0 and x_2 to find the value x_3 of x. Thus, the second approximation of the root is

$$m_2 = x_3 = x_0 - \left(\frac{x_2 - x_0}{f_2 - f_0}\right) f_0 \tag{3a}$$

on the other hand, if f_2 and f_1 are of opposite signs, employ the formula (2) with x_2 and x_1 to find the value x_3 of x.

Thus, in this case, can be written as

$$m_2 = x_3 = x_2 - \left(\frac{x_1 - x_2}{f_1 - f_2}\right) f_2 \tag{3b}$$

The x_3 given by (3a) and (3b) is the x-coordinate of the point where the chord joining the points $(x_0, f(x_0))$ and $(x_2, f(x_2))$ or the points $(x_2, f(x_2))$ and $(x_1, f(x_1))$, as the case may be, meets the x-axis. This x_3 serves as the second order approximation m_2 of m.

Repeat the process to determine the third-order, fourth order, ..., approximations of m_i stop the process when the desired degree of approximation is reached. When $f_k = f(x_k)$; $f_{k-1} = f(x_{k-1})$ are of opposite signs, the kth order approximation m_k of the root m is given by

$$\boxed{m_k = x_{k+1} = x_k - \frac{x_{k-1} - x_k}{f_{k-1} - f_k} f_k}$$

To find an approximation of m correct to p decimal places, we carry out successive approximations of m until two successive approximate values of m coincide upto p decimal places.

Example 30. By using the regula-falsi method find an approximate root of the equation $x^4 - x - 10 = 0$ that lies between 1.8 and 2. Carry out three approximations.

Solution: Let $f(x) = x^4 - x - 10$

since $f(1.8) = -1.3 < 0$

and $f(2) = 4 > 0$

Thus, the root lies between $x_0 = 1.8$ and $x_1 = 2$

∴ The first approximation of the root is

$$m_1 = x_2 = x_0 - \left(\frac{x_1 - x_0}{f_1 - f_0}\right) f_0$$

$$= 1.8 - \frac{2 - 1.8}{4 + 1.3} \times (-1.3)$$

$$= 1.849$$

Thus, $f_2 = f(x_2) = f(1.849) = -0.159 < 0$

and $$f_1 = f(x_1) = 4 > 0$$

$\therefore$ The root lies between $x_2 = 1.849$ and $x_1 = 2$

$\therefore$ The second approximation of the root is

$$m_2 = x_3 = x_2 - \left(\frac{x_1 - x_2}{f_1 - f_2}\right) f_2$$

$$= 1.8490 - \frac{2 - 1.849}{4 + 0.159} \times (-0.159)$$

$$= 1.8548$$

Thus, $$f_3 = f(x_3) = f(1.8548) = -0.019 < 0$$

and $$f_2 = f(x_2) = -0.159 < 0$$

Hence, f_3 and f_2 are of the same sign. Hence, the root m does not lie between x_2 and x_3. But f_3 and f_1 are of opposite signs. Hence m lies between x_3 and x_1.

$\therefore$ The third approximation of the root is

$$m_3 = x_4 = x_3 - \left(\frac{x_1 - x_3}{f_1 - f_3}\right) f_3$$

$$= 1.8548 - \frac{2 - 1.8548}{4 + 0.019} \times (-0.019)$$

$$= 1.8555$$

Example 31. Find the root of the equations $xe^x = \cos x$ using the regula-falsi method correct to four decimal places.

Solution:

Let us take $$f(x) = \cos x - xe^x$$

Since $$f(0) = 1,\ f(1) = \cos 1 - e = -2.17798$$

Taking $$x_0 = 0,\ x_1 = 1$$

We find that $$f(0) = 1 > 0,\ f(1) = -2.17798 < 0, \text{ or } f_0 > 0,\ f_1 < 0$$

Since f_0 and f_1 are of opposite signs, the first-order approximation of the root is

$$m_1 = x_2 = x_0 - \left(\frac{x_1 - x_0}{f_1 - f_0}\right) f_0$$

$$= 0 + \frac{1}{3.17798} \times 1$$

$$= 0.31647$$

We find that $f_2 = f(x_2) = f(0.31467) = 0.51987 > 0$ so that f_2 and f_1 are of opposite signs. Hence, the root m lies between x_2 and x_1 and the second order approximation of the root is

$$m_2 = x_3 = x_2 - \left(\frac{x_1 - x_2}{f_1 - f_2}\right) f_2$$

$$= 0.31467 + \frac{0.68533}{2.69785} \times 0.51987$$

$$= 0.44673$$

We find $f_3 = f(x_3) = f(0.44673) = 0.20356 > 0$
So that f_3 and f_2 are of the same sign.

Hence, the root m does not lie between x_2 and x_3. But f_3 and f_1 are of opposite signs. Hence, m lies between x_3 and x_1 and the third-order approximate value of the root is

$$m_3 = x_4 = x_3 - \left(\frac{x_1 - x_3}{f_1 - f_3}\right) f_3$$

$$= 0.44673 + \frac{0.55327}{2.38154} \times 0.20356$$

$$= 0.49402$$

Repeating the process, the successive approximations are:

$$m_4 = x_5 = 0.50995,\ m_7 = x_8 = 0.51748,$$
$$m_5 = x_6 = 0.51520,\ m_8 = x_9 = 0.51767,$$
$$m_6 = x_7 = 0.51692,\ m_9 = x_{10} = 0.51775 \text{ etc.}$$

Hence the root is 0.5177 correct to 4 decimal places.

Example 32. Find a real root of the equation $x^6 - x^4 - x^3 - 3 = 0$ by regula-falsi method in (1.5, 1.6) correct to three places of decimal.

Solution:

Let us take $f(x) = x^6 - x^4 - x^3 - 3$, and $x_0 = 1.5$, $x_1 = 1.6$

Since $f_0 = f(x_0) = f(1.5) = -0.047 < 0$

and $f_1 = f(x_1) = f(1.6) = 3.128 > 0$.

$\therefore$ f_0 and f_1 are of opposite signs, the equation $f(x) = 0$ has a root m between x_0 and x_1. The first-order approximation of this root is

$$m_1 = x_2 = x_0 - \left(\frac{x_1 - x_0}{f_1 - f_0}\right) f_0$$

$$= \frac{x_0 f_1 - x_1 f_0}{f_1 - f_0}$$

$$= \frac{1.5 \times 3.128 - 1.6 \times (-0.047)}{3.128 - (-0.047)}$$

$$m_1 = x_2 = \frac{4.7672}{3.175} = 1.501$$

We find that $\quad f_2 = f(x_2) = f(1.501) = -0.0215 < 0$

$$f_1 = f(x_1) = f(1.6) = 3.128 > 0$$

$\therefore$ $$m_2 = x_3 = x_2 - \left(\frac{x_1 - x_2}{f_1 - f_2}\right) f_2$$

$$= \frac{x_2 f_1 - x_1 f_2}{f_1 - f_2} = \frac{1.501 \times 3.128 - 1.6 \times (-0.215)}{3.128 - (-0.0215)}$$

$$m_2 = x_3 = \frac{4.729528}{3.1495}$$

$$= 1.5017$$

$$f(1.5017) = 0.0037$$

$\therefore$ Required root is 1.502.

Example 33. Using regula-falsi method, compute the real root of the equation $xe^x = 2$, in (0.8, 0.9) correct to four decimal places.

Solution:

Let $\quad f(x) = xe^x - 2,$

then $\quad f_0 = f(x_0) = f(0.8) = -0.2196 < 0$

and $\quad f_1 = f(x_1) = f(0.9) = 0.2136 > 0$

and $\quad x_0 = 0.8, \; x_1 = 0.9$

Since f_0 and f_1 are of opposite signs, the equation $f(x) = 0$ has a root m between x_0 and x_1.

The first order approximation of this root is

$$m_1 = x_2 = x_0 - \left(\frac{x_1 - x_0}{f_1 - f_0}\right) f_0$$

$$= \frac{x_0 f_1 - x_1 f_0}{f_1 - f_0}$$

$$= \frac{(0.8)(0.2136) - (0.9)(-0.2196)}{0.2136 - (-0.2196)}$$

$$= \frac{0.36812}{0.4332} = 0.8506$$

We find that $f_2 = f(x_2) = f(0.8506) = -0.00697$

$$m_2 = x_3 = x_2 - \left(\frac{x_1 - x_2}{f_1 - f_2}\right) f_2$$

$$= \frac{x_2 f_1 - x_1 f_2}{f_1 - f_2}$$

$$= \frac{(0.8506)(0.2136) - 0.9(-0.2196)}{0.2136 - (-0.2196)}$$

$$= \frac{0.36812}{0.4332} = 0.8506$$

We find that $f_3 = f(x_3) = f(0.8506) = -0.00697 < 0$

$$f_1 = f(x_1) = f(0.9) = 0.2136 > 0$$

$$m_3 = x_4 = x_3 - \left(\frac{x_1 - x_3}{f_1 - f_3}\right) f_3$$

$$= \frac{x_3 f_1 - x_1 f_3}{f_1 - f_2} = \frac{0.8506 \times 0.2136 - 0.9(-0.00697)}{0.2136 - (-0.00697)}$$

$$= \frac{0.187972}{0.22057} = 0.8523.$$

We find that $f_4 = f(x_4) = f(0.8523) = -0.00166 < 0$

$$f_1 = f(x_1) = f(0.9) = 0.2136 > 0$$

$$m_4 = x_5 = x_4 - \left(\frac{x_1 - x_4}{f_1 - f_4}\right) f_4$$

$$= \frac{x_4 f_1 - x_1 f_4}{f_1 - f_4}$$

$$= \frac{0.8523 \times (0.2136) - (0.9)(-0.00166)}{0.2136 - (-0.00166)}$$

$$= \frac{0.183545}{0.21526} = 0.8526.$$

$$f(0.8526) = -0.000023.$$

Hence, the required root of the given equation is 0.8526, correct to four significant figures.

Example 34. By using the regula-falsi method, find the root, correct to 3 decimal places of the equation $x \log_{10}^{x} = 1.2$ that lies between 2 and 3.

Solution:

Given: $f(x) = 0$, where

$f(x)\ x\log^{x}_{10} - 1.2$

and $x_0 = 2$, $x_1 = 3$. Then $f_0 = f(x_0) = f(2) = -0.59794 < 0$ and $f_1 = f(x_1)\ f(3) = 0.23136 > 0$.

Since f_0 and f_1 are of opposite signs, the equation $f(x) = 0$ has a root m between x_0 and x_1. The first approximation of this root is

$$m_1 = x_2 = x_0 - \left(\frac{x_1 - x_0}{f_1 - f_0}\right) f_0$$

$$= 2 - \frac{3-2}{0.23136 + 0.59794} \times (-0.59794)$$

$$= 2.721$$

Thus $f_2 = f(x_2) = f(2.721) = (2.721) = (2.721)\log_{10}(2.721) - 1.2 = -0.0171 < 0$

$\therefore$ f_1 and f_2 are of opposite signs.

$\Rightarrow$ there is a root m between x_2 and x_1.

Hence, the second approximation of the root is

$$m_2 = x_3 = x_2 - \left(\frac{x_1 - x_2}{f_1 - f_2}\right) f_2$$

$$= 2.721 - \frac{3-2.721}{0.23136 + 0.0171} \times (-0.0171)$$

$$= 2.74020$$

$\therefore$ $$f_3 = f(x_3) = 2.74020 \times \log_{10} 2.74020 - 1.2$$

$$= -0.00039 < 0$$

and $$f_2 = -0.0171 < 0$$

$\Rightarrow f_2$ and f_3 are of the same sign.

$\Rightarrow$ the root 'm' does not lie between x_2 and x_3. But f_3 and f_1 are of opposite signs; hence 'm' lies between x_3 and x_1.

$\therefore$ the third approximation of the root is $m_3 = x_4 = x_3 - \left(\frac{x_1 - x_3}{f_1 - f_3}\right) f_3$

$$= 2.74020 - \frac{3-2.74020}{0.23136 + 0.00039} \times (-0.00039)$$

$$= 2.74064$$

We find that $m_2 = 2.74020$ and $m_3 = 2.74064$ coincide upto 3 decimal places.

Hence $m_2 \approx m_6 \approx 2.740$ is the required root correct to three decimal places.

Example 35. Solve the equation $x^3 - 9x + 1 = 0$ by regula-falsi method.

Solution:

Let $$f(x) = x^3 - 9x + 1$$

Let us take $x_0 = 2$ and $x_1 = 3$

$\therefore$ $$f_0 = f(x_0) = f(2) = -9 < 0;$$

and $$f_1 = f(x_1) = f(3) = 1 > 0;$$

Hence, the root lies between 2 and 3.

$$m_1 = x_2 = x_0 - \left(\frac{x_1 - x_0}{f_1 - f_0}\right) f_0 \quad (1)$$

$$= 2 - \left(\frac{1}{1-(-9)}\right)(-9)$$

$$= 2 + \frac{9}{10} = 2.9$$

$\therefore$ $$f_2 = f(x_2) = f(2.9) = -0.711 < 0$$

Also,

$$f(x_2),\ f(x_1) = (-0.711 \times 1) = -0.711 < 0$$

Choosing x_2 and x_1,

$$m_2 = x_3 = x_2 - \left(\frac{x_1 - x_2}{f_1 - f_2}\right) f_2 = \frac{x_2 f_1 - x_1 f_2}{f_1 - f_2}$$

$$= \left[\frac{(2.9)(1) - 3(-0.711)}{1-(-0.711)}\right] = 2.9415$$

The process is continued unitil the value converges to 2.9428.

Example 36. Using regula-falsi method find a root of $f(x) = x^6 - x^4 - x^3 - 1 = 0$ in (1, 2). Carry out 3 iterations.

Solution:

Let $$f(x) = x^6 - x^4 - x^3 - 1; \begin{cases} f(1) = -2 < 0 \\ f(2) = 39 > 0 \end{cases}$$

$\therefore$ take $x_0 = 1.5$ (between the approximates)
and $x_1 = 1$

First approximation: (m_1)

$$m_1 = x_2 = x_0 - \left(\frac{x_1 - x_0}{f_1 - f_0}\right) f_0$$

$$= 1.5 - \frac{1.9532}{25.3126}$$

$$= 1.5 - 0.07716 = 1.4228$$

$$\boxed{m_1 = x_2 = 1.4228}$$

Second approximation: (m_2)

$$m_2 = x_3 = x_2 - \left(\frac{x_1 - x_2}{f_1 - f_2}\right) f_2$$

$$= 1.4228 - \frac{0.3177}{17.3902} = 1.4228 - 0.0182$$

$$= 1.4046$$

$\therefore$ $$\boxed{m_2 = x_3 = 1.4046}$$

Third approximation: (m_3)

$$m_3 = x_4 = x_3 - \left(\frac{x_1 - x_3}{f_1 - f_3}\right) f_3 = 1.4046 - \frac{0.0139}{15.7909} = 1.4038$$

$\therefore$ The root correct to 3 places of decimal is 1.404 as $x_3 \cong x_4$.

EXERCISES

1. By using the regula-falsi method, find the roots, correct to 4 decimal places, of the following equations over the indicated intervals:

(1) $x^3 + x^2 + x + 7 = 0$ over $(-2.5, 2.0)$ (Ans: -2.1049)
(2) $x^3 + 7x^2 + 9 = 0$ over $(-7.2, -7)$ (Ans: -7.1748)
(3) $x^4 + 2x^2 - 16x + 5$ over $(0,1)$ (Ans: 0.3265)
(4) $x^6 - x^4 - x^3 - 1 = 0$ over $(1.4, 1.5)$ (Ans: 1.4036)
(5) $x^3 - 2x - 5 = 0$ (Ans: 2.0946)
(6) $x^3 - 10x^2 + 40x - 35 = 0$ (Ans: 1.1975)
(7) $x^3 + x^2 + 3x + 4 = 0$ (Ans: -1.22248)
(8) $x^6 - x^4 - x^3 - 1 = 0$ (Ans: 1.4036)
(9) $x^3 - 9x + 1 = 0$ (Root between 2 & 3) (Ans: 2.9416)
(10) $x^3 - 5x - 7 = 0$ (Ans: 2.746)
(11) $3x^3 - 5x^2 + 3x - 5 = 0$ (Ans: 1.6629)

2. By using the regula-falsi method find the roots, correct to 3 decimal places, of the following equations over the indicated intervals.

(1) $x^3 - x^2 - 2 = 0$ over $(1, 2)$ (Ans: 1.69)
(2) $x^3 - 2x - 5 = 0$ over $(2, 2.5)$ (Ans: 2.094)

(3) $x^3 - 4x - 9 = 0$ over (2.5, 3.0) (Ans: 2.706)
(4) $x^3 - 9x + 1 = 0$ over (2.5, 3.0) (Ans: 2.942)
(5) $x^6 - x^4 - x^3 - 1 = 0$ over (1.5, 2.0) (Ans: 1.736)
(6) $e^x = 3x$ over (0.5, 1.0) (Ans: 0.619)
(7) $x = e^{-x}$ over (0.5, 1.0) (Ans: 0.567)
(8) $xe^x = 3$ over (1, 1.5) (Ans: 1.050)
(9) $x^2 - \log_e x = 12$ over (3, 4) (Ans: 3.646)
(10) $2x - \log_{10} x = 7$ over (3.5, 4) (Ans: 3.789)
(11) $\sin x = 1 + x^3$ over (–2, –1) (Ans: – 1.249)
(12) $x = 2 \sin x$ over ($\pi/2$, π) (Ans: 1.895)

3. By using regula-falsi method, find the roots, correct to 4 decimal places of the following equations over the indicated intervals:

(1) $xe^x = 2$ over (0.5, 1.0) (Ans: 0.583)
(2) $xe^x = \cos x$ over (0, 1) (Ans: 0.5177)
(3) $\cos x = 3x - 1$ over (0.5, 1.0) (Ans: 0.6071)
(4) $\tan x + \tan hx = 0$ over (2, 3) (Ans: 2.3650)

4. By using the regula-falsi method, find the approximate value of $\sqrt{3}$ correct to four decimal places. (Ans: 1.7320)

SECANT METHOD

This method being an improvement over the regula-falsi method is based on omitting the condition $f(a) \cdot f(b) < 0$ of that method and hence the method is less restrictive than the method of false position.

Let $f(x) = 0$ be the given equation and m be a root of $f(x) = 0$

We draw the graph of the function $y = f(x)$ as shown in figure.

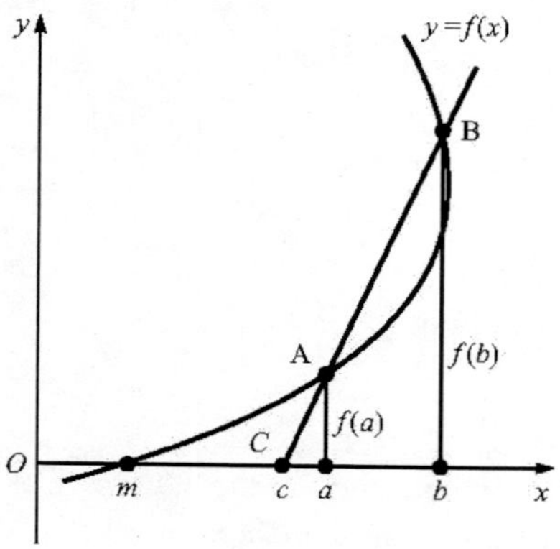

Fig. 9

The curve $y = f(x)$ meets the x-axis at the point $x = m$.

Let A $(a, f(a)$ and $B(b, f(b))$ be two neighbouring points on the curve $y = f(x)$. The equation of the chord

AB is
$$\frac{y - f(a)}{x - a} = \frac{f(b) - f(a)}{b - a} \tag{1}$$

At C, we have $y = 0$ and equation (1) yields

$$x = a - \frac{b - a}{f(b) - f(a)} f(a) \tag{2}$$

This gives the x-coordinate c of the point C. It the points A and B are sufficiently near to the x-axis on the curve $y = f(x)$, the point $x = c$ is sufficiently close to the point $x = m$ (where) the curve meets the x-axis and it serves as an approximation for m. The method of finding an approximate value of m on the basis of this hypothesis is known as the secant method or the chord method.

Working Rule for the Secant Method

Step 1. Choose $[x_0, x_1]$ such that $f(x) = 0$ has exactly one root between x_0 and x_1, $(x_0 < x_1)$.

Step 2. From equation (2) with $a = x_0$ and $b = x_1$ to find the value x_2 of x. We have

$$x_2 = x_0 - \frac{x_1 - x_0}{f(x_1) - f(x_0)} f(x_0)$$

where x_2 is the x-coordinate of the point where the chord through the points $(x_0, f(x_0))$ and $(x_1, f(x_1))$ meets the x-axis; it serves as the first approximation m_1 of m.

Let $f(x_0) = f_0$, $f(x_1) = f_1$.

Thus,
$$m_1 = x_2 = x_0 - \frac{x_1 - x_0}{f_1 - f_0} f_0 \tag{3}$$

Step 3. Equation (2) with $9 = x_1$ and $b = x_2$ to get the value x_3 of x.

Thus,
$$x_3 = x_1 - \frac{x_2 - x_1}{f(x_2) - f(x_1)} f(x_1)$$

writing f_2 for $f(x_2)$, we have,

$$m_2 = x_3 = x_1 - \frac{x_2 - x_1}{f_2 - f_1} f_1 \tag{4}$$

Proceeding like this to find successive approximation $m_3 = x_4$, $m_4 = x_5$, ..., of m by taking $a = x_2$, $b = x_3$, $a = x_4$, $b = x_5$, ..., successively in the equation (2). Stop the process when the desired order of accuracy of the root m is reached.

Thus, the formula for kth approximation m_k of the root m is given by

$$\boxed{m_k = x_{k+1} = x_{k-1} - \frac{x_k - x_{k-1}}{f_k - f_{k-1}} f_{k-1}, k = 1, 2, \ldots} \qquad (5)$$

where $f_{k-1} = f(x_{k-1})$; $k = 1, 2, \ldots$.

Example 37. By the secant method, find the root, correct to four decimal places, of the equation $x^3 - 2x - 1 = 0$ that lies between 1 and 2.

Solution:

Let
$$f(x) = x^3 - 2x - 1$$
$$x_0 = 1, f(x_0) = f_0 = f(1) = 1 - 2 - 1 = -2 < 0$$
$$x_1 = 2, f(x_1) = f_1 = f(2) = 8 - 4 - 1 = 3 > 0$$

since f_0 and f_1 are of opposite signs, the given equation does not have a root m between 1 and 2.

$$m_1 = x_2 = x_0 - \frac{x_1 - x_0}{f_1 - f_0} f_0$$

$$= 1 - \frac{2-1}{3+2}(-2) = 1 + \frac{2}{5} = \frac{7}{5} = 1.4$$

∴
$$f_2 = f(x_2) = f(1.4) = (1.4)^3 - 2(1.4) - 1$$
$$= -1.056 < 0$$

Thus,
$$m_2 = x_3 = x_1 - \frac{x_2 - x_1}{f_2 - f_1} f_1$$

$$= 2 - \frac{1.4 - 2}{1.056 - 3} \qquad (1)$$

$$= 2 - \frac{(-0.6)}{-4.056} = 2 - \frac{0.6}{4.056} = 1.5562$$

$$f(x_3) = (1.5562)^3 - 2(1.5562) - 1$$
$$= -0.3436 < 0$$

$$m_3 = x_4 = x_2 - \frac{x_3 - x_2}{f_3 - f_2} f_2$$

$$= \frac{1.4(-0.3436) - 1.5562(-1.056)}{-0.3436 + 1.056}$$

$$= 1.63153$$

$$f(m_3) = f(x_4) = (1.63153)^3 - 2(1.63153) - 1$$
$$= 0.07987$$

$$m_4 = x_5 = x_3 - \frac{x_4 - x_3}{f_4 - f_3}$$

$$= \frac{x_3 f_4 - x_4 f_3}{f_4 - f_3}$$

$$= \frac{(1.5562)(.07987) - (1.63153)(-0.3436)}{0.07987 + 0.3436}$$

$$= 1.61730465$$

∴ Approximate value of the root correct to one decimal place is 1.6.

Example 38. By the secant method, find the root, correct to four decimal places, of the equation $x^3 - 5x^2 - 29 = 0$ that lies between 5 and 6.

Solution: Let

$$f(x) = x^3 - 5x^2 - 29$$

$$x_0 = 5,\ f_0 = f(x_0) = f(5) = -29 < 0$$

$$x_1 = 6,\ f_1 = f(x_1) = f(6) = 7 > 0$$

∴ $f(x_0)\, f(x_1) < 0$

Thus,

$$m_1 = x_2 = x_0 - \frac{x_1 - x_0}{f_1 - f_0} f_0$$

$$= 5 - \frac{6 - 5}{7 + 29}(-29)$$

$$= 5.81$$

∴

$$f_2 = f(x_2) = f(5.81) = -1.66 < 0$$

Hence,

$$m_2 = x_3 = x_1 - \frac{x_2 - x_1}{f_2 - f_1} f_1$$

$$= 6 - \frac{5.81 - 6}{-1.66 - 7} \times 7 = 5.846$$

We find that

$$f_3 = f(x_3) = f(5.846) = -0.087 < 0$$

Hence,

$$m_3 = x_4 = x_2 - \frac{x_3 - x_1}{f_3 - f_2} f_2$$

$$= 5.81 - \frac{5.846 - 5.81}{-0.087 + 1.66} \times (-1.66) = 5.84799$$

We find that

$$f_4 = f(x_4) = f(5.84799) = 0.00039$$

Hence,

$$m_4 = x_5 = x_3 - \frac{x_4 - x_3}{f_4 - f_3} f_3$$

$$= 5.846 - \frac{5.84799 - 5.846}{0.00039 + 0.087} \times (-0.087)$$

$$= 5.84798$$

We note that $m_3 = m_4$

Hence, the approximate value of m correct to four places of decimal is $m = 5.8480$.

Example 39. By using the secant method, find the root, correct to four places of decimal, of the equation $\cos x = xe^x$ that lies between 0 and 1.

Solution: Let

$$f(x) = \cos x - xe^x.$$

$$x_0 = 0,\ f_0 = f(x_0) = f(0) = 1$$

$$x_1 = 1,\ f_1 = f(x_1) = f(1) = -\ 2.17798$$

Since f_0 and f_1 are of opposite signs, $f(x) = 0$, does have a root m between 0 and 1. The first approximate m_1 of m is

$$m_1 = x_2 = x_0 - \frac{x_1 - x_0}{f_1 - f_0} f_0 = 0 - \frac{1-0}{-2.17798-1} \times 1 = 0.31467$$

Example 40. Show that the order (rate) of convergence of the secant method is 1.618 approximately.

Solution: We know that the iteration formula for the secant method is given by

$$x_{k+1} = x_{k-1} - \frac{(x_k - x_{k-1})}{f(x_k) - f(x_{k-1})} f(x_{k-1}) \qquad (1)$$

Put $e_k = x_{k-m}$, where m is a root of $f(x) = 0$ for which x_{k+1} is the kth order approximation.

Then (1) becomes

$$e_{k+1} = e_{k-1} - \frac{(e_k - e_{k-1})}{f(m+e_k) - f(m+e_{k-1})} f(m + e_{k-1}) \qquad (2)$$

By using the Taylor's expansion, we get

$$f(m+e_k) = f(m) + e_k f'(m) + \frac{1}{2} e_k^2 f''(m) + \ldots$$

$$= e_k f'(m) + \frac{1}{2} e_k^2 f''(m) + \ldots \qquad (3)$$

because m is a root of $f(x) = 0$; that is, $f(m) = 0$

similarly, $f(m + e_{k-1}) = e_{k-1} f'(m) + \dfrac{1}{2} e_{k-1}^2 f''(m) + \ldots$ (4)

From (3) and (4), we get

$$f(m + e_k) - f(m + e_{k-1}) = (e_k - e_{k-1})\, f'(m) + \frac{1}{2}(e_k^2 - e_{k-1}^2)\, f''(m) + \ldots$$

$$= (e_k - e_{k-1})\, f'(m) \left\{1 + \frac{1}{2}(e_k + e_{k-1}) \frac{f''(m)}{f'(m)} + \ldots\right\}$$ so that

$$\frac{(e_k - e_{k-1})}{f(m+e_k) - f(m+e_{k-1})} = \frac{1}{f'(m)}\left\{1+\frac{1}{2}(e_k + e_{k-1})\frac{f''(m)}{f'(m)}+\ldots\right\}^{-1}$$

$$= \frac{1}{f'(m)}\left\{1-\frac{1}{2}(e_k + e_{k-1})\frac{f''(m)}{f'(m)}-\ldots\right\} \quad (5)$$

on the assumption that $f'(m) \neq 0$.

Using (4) and (5) in (2), we get

$$e_{k+1} = e_{k-1} - e_{k-1}\left\{1-\frac{1}{2}e_{k-1}\frac{f''(m)}{f'(m)}+\ldots\right\}\times\left\{1-\frac{1}{2}(e_k + e_{k-1})\frac{f''(m)}{f'(m)}-\ldots\right\}$$

$$= -\frac{1}{2}e_k e_{k-1}\frac{f''(m)}{f'(m)} \quad (6)$$

on neglecting third and higher degree terms in the errors.

Now, let p be the order of convergence of the iteration (1). Then

$$|e_{k+1}| \approx \alpha_k |e_k|^p \quad (7)$$

for some positive constant α_k, $k = 1, 2, \ldots$

The expression (7) may be rewritten as $|e_k| \approx \alpha_{k-1}\,|e_{k-1}|^p$,

or
$$|e_{k-1}| \approx \left\{\frac{1}{\alpha_{k-1}}|e_k|\right\}^{1/p}$$

With the aid of this, the expression (6) yields

$$|e_{k+1}| \approx \frac{1}{2}|e_k|\left\{\frac{1}{\alpha_{k-1}}|e_k|\right\}^{1/p}\frac{|f''(m)|}{|f'(m)|}$$

$$\approx \lambda_k\,|e_k|^{1+1/p} \quad (8)$$

where
$$\lambda_k = \frac{1}{2(\alpha_{k-1})^{1/p}}\frac{|f''(m)|}{|f'(m)|}$$

Comparing (7) and (8), we find that $\alpha_k = \lambda_k$ and

$$p = 1 + \frac{1}{p}$$

Thus, p satisfies the condition (9), which may be rewritten as $p^2 - p - 1 = 0$. Solving this quadratic equation and bearing in mind that p is positive, we get

$$p = \frac{1}{2}(1 + \sqrt{5}) = 1.618 \text{ (approx)}$$

Hence, the order of convergence of the secant method is 1.618 approximately. This processes the required result.

Example 41. Determine the order (rate) of convergence of the regula-falsi method.

Solution:

Let $f(x) = 0$ be the given equation and m is an exact root of $f(x) = 0$.

If the root m lies between two approximations x_{k-1} and x_k, then $f(x_{k-1})$ and $f(x_k)$ have opposite signs.

The iteration formula for regula-falsi method is given by

$$x_{k+1} = x_k - \frac{x_{k-1} - x_k}{f(x_{k-1}) - f(x_k)} f(x_k) \tag{1}$$

Putting $e_k = x_k - m$, this formula becomes

$$e_{k+1} = e_k - \frac{e_{k-1} - e_k}{f(m+e_{k-1}) - f(m+e_k)} f(m + e_k)$$

$$= \frac{e_k f(m+e_{k-1}) - e_{k-1} f(m+e_k)}{f(m+e_{k-1}) - f(m+e_k)} \tag{2}$$

Expanding $f(m + e_{k-1})$ and $f(m + e_k)$ by using the Taylor's expansion, substituting these expansions in (1) and simplifying the resulting expression, we obtain

$$e_{k+1} = \frac{1}{2} e_k \, e_{k-1} \frac{f''(m)}{f'(m)} \tag{3}$$

and terms of third and higher degree terms in the errors.

We note that for a change in sign in the r.h.s expression (3) is the same as expression (6) of, that is $\left(e_{k+1} = -\frac{1}{2} e_k e_{k-1} \frac{f''(m)}{f'(m)} \right)$. Now, taking p as the order (rate) of convergence and proceeding as in example 40, we find that $p = 1.618$ (approx) here also.

Thus, the rate (order) of convergence of the regula-falsi method is the same as the rate of convergence of the secant method, namely 1.618 (approx).

EXERCISES

1. By employing the secant method, find the roots, correct to 3 places of decimal, of the following equations in the specified intervals.

   1. $x^4 - x - 10 = 0$ in (1.5, 2.0) (Ans: 1.856)
   2. $x^3 + x^2 + x + 7 = 0$ in (– 2.5, – 2.0) (Ans: –2.1048)
   3. $x^3 - 5x + 1 = 0$ in (0.0.5) (Ans: 0.219)
   4. $x = e^{-x}$ in (0.5, 0.75) (Ans: 0.567)
   5. $x \sin x = 1$ in (1.0, 1.5) (Ans: 1.114)

MULLER'S METHOD

Muller's method is an iterative method. It requires three points $(x_0, f(x_0))$, $(x_1, f(x_1))$ and $(x_2, f(x_2))$. A parabola is constructed that passes through these points then the quadratic formula is used to find a root of the quadratic for the next approximation. Without loss of generality we assume that x_2 is the best approximation to the root and consider the parabola through the three starting values as shown in figure. We now suppose that x_{k-2}, x_{k-1}, x_k are three distinct approximations to the root of $f(x) = 0$.

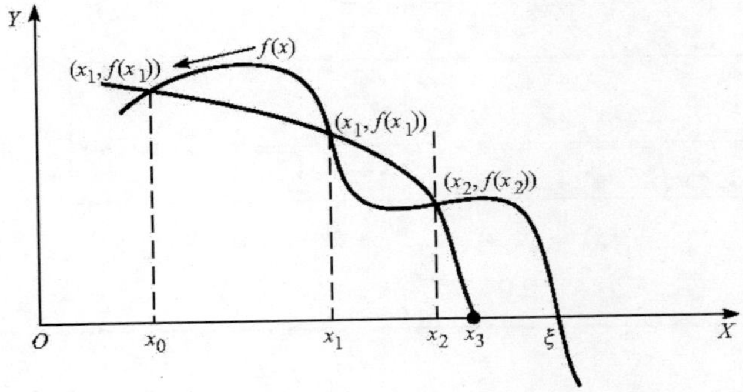

Let us start the method with the equations:

$$x_{k+1} = x_k + (x_k - x_{k-1})\,\lambda_{k+1},\; k = 1, 2, \ldots \tag{1}$$

where

$$h_k = X_{k-1} = x_k - x_{k-1},\; h_{k-1} = X_{k-2} = x_{k-1} - x_{k-2}$$

$$\lambda_k = \frac{h_k}{h_{k-1}}, \qquad \delta_k = 1 + \lambda_k$$

$$g_k = \lambda_k^2 f(x_{k-2}) - \delta_k^2 f(x_{k-1}) + (\lambda_k + \delta_k)\, f(x_k)$$

$$c_k = \lambda_k(\lambda_k f(x_{k-2}) - \delta_k f(x_{k-1}) + f(x_k)$$

$$\lambda_{k+1} = -\frac{2\delta_k f(x_k)}{g_k \pm \sqrt{g_k^2 - 4\delta_k c_k f(x_k)}}$$

The sign '±' in the denominator is chosen so that λ_{k+1} has the smallest absolute value. (i.e. sign of the sq. root in the denominator is that of g_k).

Note: Three initial approximations x_{k+1}, x_k, x_{k+2} are required to the root and one function evaluation per iteration for this method.

Example 42. Apply Muller's method to find the smaller positive root of the equation $x^3 - 5x + 1 = 0$ in (0, 1) (perform three iterations).

Solution:

Let $x_0 = 0$, $x_1 = 0.5$ and $x_2 = 1.0$ be the three initial approximations to the root.

$\therefore\; f_0 = 1, f_1 = -1.375, f_2 = -3$

1st Iteration

$$\begin{Bmatrix} h_2 = x_2 - x_1 = 0.5, \\ h_1 = x_1 - x_0 = 0.5, \end{Bmatrix}$$

$$\left\{\lambda_2 = \frac{h_2}{h_1} = 1,\ \delta_2 = 1 + \lambda_2 = 2\right\}$$

$$\begin{Bmatrix} g_2 = \lambda_2^2 f_0 - \delta_2^2 f_1 + (\lambda_2 + \delta_2) f_2 = -2.5 \\ c_2 = \lambda_2 (\lambda_2 f_0 - \delta_2 f_1 + f_2 = 0.75 \end{Bmatrix}$$

Since $g_2 < 0$, we get

$$\lambda_3 = \frac{-2\delta_2 f_2}{g_2 - \sqrt{g_2^2 - 4\delta_2 f_2 c_2}} = -\frac{12}{2.5 + \sqrt{24.25}} = -1.616286$$

$$x_3 = x_2 + (x_2 - x_1)\ \lambda_3 = 0.191857,$$

$$f_3 = 0.047777$$

2nd Iteration:

$$h_3 = x_3 - x_2 = -0.808143$$

$$\lambda_3 = \frac{h_3}{h_2} = -1.61628,\ \delta_3 = 1 + \lambda_3 = -0.616286$$

$$g_3 = \lambda_3^2 f_1 - \delta_3^2 f_2 + (\lambda_3 + \delta_3) f_3 = -2.559263$$

$$c_3 = \lambda_3 (\lambda_3 f_1 - \delta_3 f_2 + f_3) = -0.680961.$$

Since $g_3 < 0$, we get

$$\lambda_4 = \frac{-2\delta_3 f_3}{g_3 - \sqrt{g_3^2 - 4\delta_3 f_3 c_3}} = -\frac{0.058889}{2.559263 + \sqrt{6.469625}}$$

$$= -0.011541$$

$$x_4 = x_3 + (x_3 - x_2)\ \lambda_4 = 0.201184,$$

$$f_4 = 0.002223$$

3rd Iteration

$$h_4 = x_4 - x_3 = 0.009327$$

$$\lambda_4 = \frac{h_4}{h_3} = -0.0115441,\ \delta_4 = 1 + \lambda_4 = 0.988459$$

$$g_4 = \lambda_4^2 f_2 - \delta_4^2 f_3 + (\lambda_4 + \delta_4) f_4 = -0.044908$$

$$c_4 = \lambda_4 (\lambda_4 f_2 - \delta_4 f_3 + f_4) = 0.000120.$$

Since $g_4 < 0$, we get

$$\lambda_5 = -\frac{2\delta_4 f_4}{g_4 - \sqrt{g_4^2 - 4\delta_4 f_4 c_4}}$$

$$= \frac{0.004395}{0.044908 + \sqrt{0.002016}}$$

$$= 0.048940$$

$$x_5 = x_4 + (x_4 - x_3)\,\lambda_5 = 0.201640$$

$$\boxed{x_5 = 201640}$$

(Alternative Method)

Let $y = f(x)$ be a function of x. In this method, we approximate the graph of the function $y = f(x)$ in the neighbourhood of the root by a second degree curve and take one of its points of intersection with the x-axis as the next approximation.

Let $$f(x) = a_0\,(x - x_k)^2 + a_1(x - x_k) + a_2 = 0,\ a_0 \neq 0 \qquad (1)$$

by a polynomial of degree two in x.

We start with $$\boxed{\begin{aligned} x_{k+1} &= x_k + (x_k - x_{k+1})\,\lambda_{k+1} \\ k &= 1, 2, \ldots \end{aligned}} \qquad (2)$$

Substituting $x = x_k$, x_{k-1} and x_{k-2}, we determine a_0, a_1 and a_2 from the equations

$$\begin{aligned} f_k &= a_2 \\ f_{k-1} &= a_0(x_{k-1} - x_k)^2 + a_1(x_{k-1} - x_k) + a_2 \\ f_{k-2} &= a_0\,(x_{k-2} - x_k)^2 + a_1(x_{k-2} - x_k) + a_2 \end{aligned}$$

We obtain

$$a_2 = f_k$$

$$a_1 = \frac{1}{D}\,[(x_k - x_{k-2})^2\,(f_k - f_{k-1}) - (x_k - x_{k-1})^2\,(f_k - f_{k-2})] \qquad (3a)$$

$$a_0 = \frac{1}{D}\,[(x_k - x_{k-2})\,(f_k - f_{k-1}) - (x_k - x_{k-1})\,(f_k - f_{k-2})] \qquad (3b)$$

where

$$\boxed{\begin{aligned} D &= (x_{k-1} - x_k)^2\,(x_{k-2} - x_k) - (x_{k-2} - x_k)^2\,(x_{k-1} - x_k) \\ &= (x_{k-1} - x_k)\,(x_{k-2} - x_k)\,(x_{k-1} - x_k - x_{k-2} + x_k) \\ &= (x_k - x_{k-1})\,(x_k - x_{k-2})\,(x_{k-1} - x_{k-2}) \end{aligned}}$$

Solving the equation (1) for $(x - x_k)$ and replacing x by x_{k+1}, we obtain:

$$x_{k+1} = x_k + \frac{-a_1 \pm \sqrt{a_1^2 - 4a_0a_2}}{2a_0}$$

$$= x_k - \frac{2a_0}{a_1 \pm \sqrt{a_1^2 - 4a_0a_2}}, \quad k = 2, 3, \ldots \qquad (3b)$$

$$\boxed{x_{k+1} = x_k - \frac{2a_2}{D^v}, \text{ where } a_1 \pm \sqrt{a_1^2 - 4a_0a_2} = D^v} \qquad (4)$$

Note: (1) Since $D^v = a_1 \pm \sqrt{a_1^2 - 4a_0a_2}$, the sign in this expression which is in the denominator in equation (4) is chosen as that of a_1, so that $D^v = (a_1 \pm \sqrt{a_1^2 - 4a_0a_2})$, has the maximum absolute value i.e. $|D^v| = \left|a_1 \pm \sqrt{a_1^2 - 4a_0a_2}\right|$, that is, x_k changes by a smaller value.

Convergence

The Muller's method is an extension of the secant method. This method generally converges for all initial approximations. If no better approximations are known then we choose $x_0 = -1$, $x_1 = 0$, $x_2 = +1$. In this method, the next approximation x_{k+1} is found as the zero points (x_{k-2}, f_{k-2}), (x_{k-1}, f_{k-1}) and (x_k, t_k) where x_{k-2}, x_{k-1} and x_k are the initial approximations of the root and it requires one function evaluation per iteration. The order of convergence of the method is $p = 1.84$.

Solution for the example (1), by using the alternative Muller's method.

***Iteration* (1)**

$$\left\{\begin{aligned} a_2 &= f_2 = -3, D = (x_2 - x_1)(x_2 - x_0)(x_1 - x_0) = 0.25 \\ a_1 &= \frac{1}{D}[(f_2 - f_1)(x_2 - x_0)^2 - (f_2 - f_0)(x_2 - x_1)^2] = -2.5 \\ a_0 &= \frac{1}{D}[(f_2 - f_1)(x_2 - x_0) - (f_2 - f_0)(x_2 - x_1)] = 1.5 \end{aligned}\right\}$$

Since $a_1 < 0$, we get

$$x_3 = x_2 - \frac{2a_2}{a_1 - \sqrt{a_1^2 - 4a_0a_2}} = 1 - \frac{6}{2.5 + \sqrt{24.25}} = 0.191857.$$

***Iteration* (2)**

$$\begin{cases} a_2 = f_3 = 0.047777, D = (x_3 - x_1)(x_3 - x_2)(x_2 - x_1) = 0.124512 \\ a_1 = \frac{1}{D}[(f_3 - f_2)(x_2 - x_1)^2 - (f_3 - f_1)(x_3 - x_2)^2] = -5.138588 \\ a_0 = \frac{1}{D}[(f_3 - f_2)(x_3 - x_1) - (f_3 - f_1)(x_3 - x_2)] = 1.691854. \end{cases}$$

Since $a_1 < 0$, we get

$$x_4 = x_3 - \frac{2a_2}{a_1 - \sqrt{a_1^2 - 4a_0a_2}}$$

$$= 0.191857 + \frac{0.095554}{5.138588 + \sqrt{26.081760}} = 0.201183.$$

***Iteration* (3)**

$$\begin{cases} a_2 = f_4 = 0.002228, D = (x_4 - x_2)(x_4 - x_3)(x_3 - x_2) = 0.006020 \\ a_1 = \frac{1}{D}[(f_3 - f_2)(x_4 - x_2)^2 - (f_4 - f_2)(x_4 - x_3)^2] = -4.871483 \\ a_0 = \frac{1}{D}[(f_4 - f_3)(x_4 - x_2) - (f_4 - f_2)(x_4 - x_3)] = 1.393112. \end{cases}$$

Since $a_1 < 0$, we get

$$x_5 = x_4 - \frac{2a_2}{a_1 - \sqrt{a_1^2 - 4a_0a_2}}$$

$$= 0.201183 + \frac{0.004456}{4.871483 + \sqrt{23.718931}} = 0.201640$$

$$\boxed{x_5 = 0.201640}$$

Thus, we note that both the methods yield the same approximation for example (1).

HORNER'S METHOD

In this method, we determine both the commensurable (mensurable by the same standard) and the incommensurable real roots of a numerical polynomial equation. First, we find the integral part of the root and then by every iteration, we find each decimal place value in succession.

The principle underlying Horner's process is indicated in detail below:

If the root of an equation lies between a and $a + 1$, then the value of this root will be $a.\ bcd \ldots$, where $b, c, d \ldots$ are digits in its decimal part. To obtain these, we proceed as follows:

(a) Diminish the roots of the equation by 'a' so that the root of the new equation is $o.\ bcd \ldots$

(b) Multiply the roots of the transformed equation by 10 so that the root of the new equation is $b.cd \ldots.$

(c) Diminish the root by 'b' and multiply the roots of the resulting equation by 10 so that the root is $c.\ d \ldots.$

(d) Again diminish the root by 'c' and so on.

Repeating the process, the root may be evaluated to any desired degree of accuracy, digit by digit.

Note: The procedure in the solved example 43 below is given in a detailed manner, and rest of the solution for the other examples is identical and so in the solution of the other example (other than example 43) given below the equation obtained in the several stages and the corresponding roots are as shown.

Example 43. By Horner's method, find two places of decimal, the root of the equation $x^3 - 3x + 1 = 0$ which lies between 1 and 2.

Solution:

The given equation is $x^3 - 3x + 1 = 0$

$$f(x) = x^3 + 0.\ x^2 - 3x + 1 = 0 \quad \text{(i)}$$

$$f(1) = 1 - 3 + 1 = -ve;\ f(2) = 8 - 6 + 1 = +\ ve \quad \text{(ii)}$$

The positive root lies between 1 and 2.

Let it be $o.bcd \ldots$

Diminish the roots by 1.

By synthetic division,

1	1	0	–3	1
	0	1	1	–2
	1	1	–2	–1
1	0	1	2	
	1	2	0	
1	0	1		
	1	3		

$$x^3 + 3x^2 + 0.x - 1 = 0 \quad \text{(iii)}$$

There is a root of this equation between 0 and 1. So, we form the equation whose roots are 10 times the roots of (iii), that is,

$$\phi_1(x) = x^3 + 30x^2 - 1000 = 0 \qquad \text{(iv)}$$

(multiplying the roots of (iii) by 10)

The equation (iv) has a root *b.cd*

$$\phi_1(5) = 125 + 30(25) - 1000 = \text{negative};$$

$$\phi_1(6) = 216 + 1080 - 1000 = \text{positive} \quad \therefore\ b = 5$$

So, the equation (iv) has a root between 5 and 6.

Hence, the equation (i) has a root between 1.5 and 1.6. Diminish the roots of the equation $\phi_1(x) = 0$ by 5.

By synthetic division.

5	1	30	0	–1000
	0	5	175	875
	1	35	175	–125
5	0	5	200	
	1	40	375	
5	0	5		
	1	45		

So, we get, $\phi_2(x) = x^3 + 45x^2 + 375x - 125 = 0$ (v) has the root *o.cd*

Multiply the roots of this equation by 10.

Then, we get

$$\phi_3(x) = x^3 + 450x^2 + 37500x - 125000 = 0 \qquad \text{(vi)}$$

has a root *c.de* ...

For this equation, we have

$$\phi_3(3) = 27 + 4050 + 112500 - 125000 = \text{negative};$$

$$\phi_3(4) = 64 - 7200 + 150000 - 125000 = \text{positive}.$$

So, the equation (vi) has a real root between 3 and 4 $\therefore\ c = 3$.

$\therefore \qquad \phi_3(x) = 0$ has a root 3.*de* ...

Hence, the equation (i) has a root between 1.53 and 1.54.

Again, diminish the roots of the equation (vi) by 3.

3	1	450	37500	–125000
	0	3	1359	116577
	1	453	38859	–8423
3	0	3	1368	
	1	456	40227	
3	0	3		
	1	459		

So, we get $\phi_4(x) = x^3 + 459x^2 + 40227x - 8423 = 0$ (vii)
Multiplying the roots of this equation by 10, we get

$$\phi_4(x) = x^3 + 4590x^2 + 4022700x - 8423000 = 0 \quad \text{(viii)}$$

has a root *d.ef* ...
For this equation,

$$\phi_4(2) = 8 + 18360 + 8045400 - 8423000 = \text{negative};$$
$$\phi_4(3) = 27 + 41310 + 12068100 - 8423000 = \text{positive}$$

So, equation (vii) has a real root between 2 and 3. $\therefore d = 2$

Hence, equation (i) has a root between 1.532 and 1.533. As the root is required to be correct to two places of decimal, the root is ≡ (1.53).

Example 44. $x^3 - 3x - 4$ Roots

Solution:

	Roots
$x^3 - 3x - 4 = 0;\ 2 - 3,$	2
$x^3 + 6x^2 + 9x - 2 = 0$	
$x^3 + 60x^2 + 900x - 2000 = 0;\ 1 - 2,$	2.1
$x^3 + 63x^2 + 1023x - 1039 = 0$	
$x^3 + 630x^2 + 102300x - 1039000 = 0;\ 3 - 4$	2.13
$x^3 + 639x^2 + 106097x - 726403 = 0$	
$x^3 - 6390x^2 + 10609700x - 72643000 = 0;\ 6 - 7$	2.136

The root ≡ 2.136 = 214 (correct to two places)

Example 45. $x^3 + x^2 + x - 100 = 0$

Solution:

	Roots
$x^3 + x^2 + x - 100 = 0\ ;\ 4 - 5$	4
$x^3 + 130x^2 + 5700x - 16000 = 0;\ 2 - 3$	4.2
$x^3 + 1360x^2 + 623200x - 4072000 = 0;\ 6 - 7$	4.26
$x^3 + 13780x^2 + 63962800x - 283624000 = 0\ ;\ 4 - 5$	4.264

4.264 and so 4.26 to two places.

Example 46. Find the cube root of 20 to two places of decimals.

Solution: We have to solve the equation $x = (20)^{1/3}$; i.e. $x^3 = 20$
Adopting the Horner's process as usual we get:

	Roots
$x^3 - 20 = 0 \; ; 2 - 3$	2
$x^3 + 60x^2 + 1200x - 12000 = 0 \; ; 7 - 8$	2.7
$x^3 + 810x^2 + 218700x - 317000 = 0 \; ; 1 - 2$	2.71
$x^3 + 8130x^2 + 22032300x - 974898000 = 0 \; ; 4 - 5$	2.714

The root is 2.714 ≡ 2.71 (to two places)

Example 47. Find the cube root of 7 to three places of decimal

Solution: Here, we have to solve the equation $x = (7)^{1/3}$

	Roots
$x^3 - 7 = 0, \; 1 - 2$	1
$x^3 + 30x^2 + 300x - 6000 = 0, \; 9 - 10$	1.9
$x^3 + 570x^2 + 108300x - 141000 = 0, \; 1 - 2$	1.91
$x^3 + 5730x^2 + 10944300x - 32129000 = 0, \; 2 - 3$	1.912
$x^3 + 57360x^2 + 1096723200x - 10217472000 = 0, \; 9 - 10$	1.9129

For three places of decimal $(7)^{1/3} = 1.913$.

EXERCISES

Solve the following equation by Horner's method correct to 2 places of decimal.

1. $x^3 - 4x^2 + 5 = 0$ in (1, 2) (Ans: 1.38)
2. $x^3 - 9x^2 + 23x - 14 = 0$ in (4, 5) (Ans: 4.86)
3. $x^3 - x - 9 = 0$ (Ans: 2.24)
4. $x^3 + 3x - 1$ (Ans: 0.32)
5. $x^3 - 3x^2 + 2.5 = 0$ in (2, 3) (Ans: 2.64)
6. A negative root $2x^3 - 4x + 1$ (Ans: –1.52)
7. A negative root of $x^3 - 9x^2 + 18 = 0$ (Ans: –1.33)

LIN-BAIRSTOW'S METHOD

The Lin-Bairstow's method extracts a quadratic factor of the form $(x^2 + px + q)$ from the polynomial

$$f(x) = a_0x^n + a_1x^{n-1} + \dots + a_n \tag{1}$$

which may give a pair of complex roots or a pair of real roots. If we divide the polynomial (1) by the quadratic factor then we obtain a quotient polynomial of degree n–2 and a remainder term which is a polynomial of degree one, that is,

$$a_0x^n + a_1 x^{n-1} + \ldots + an$$
$$= (x^2 - px - q)(b_0 x^{n-2} + b_1 x^{n-3} + \ldots + b_{n-2}) + b_{n-1}(x - p) + b_n$$

Equating the like powers of x.

$$\begin{array}{lcl} a_0 = b_0 & & b_0 = a_0 \\ a_1 = b_1 - pb_0 & & b_1 = a_1 + pb_0 \\ a_2 = b_2 - pb_1 - qb_0 & \text{OR} & b_2 = a_2 + pb_1 + qb_0 \\ \ldots\ldots & & \ldots\ldots \\ \ldots\ldots & & \ldots\ldots \\ a_{n-1} = b_{n-1} - pb_{n-2} - qb_{n-3} & & b_{n-1} = a_{n-1} + pb_{n-2} + qb_{n-3} \\ a_n = b_n - pb_{n-1} - qb_{n-2} & & b_n = a_n + pb_{n-1} + qb_{n-2} \end{array} \quad (2)$$

The b's are functions of p and q.
If $x^2 - px - q$ is an exact factor of $f(x)$, then $b_n(p, q) = 0$, $b_{n-1}(p, q) = 0$.

The above equations are two simultaneous equations in two unknowns p and q. Suppose that (p_0, q_0) is an initial approximation and that $(p_1 = p_0 + h, q_1 = q_0 + k)$ is the true solution. Following the Newton-Raphson method, we obtain h and k from the equations.

$$\left\{\begin{array}{l} b_n + \dfrac{\partial b_n}{\partial p}h + \dfrac{\partial b_n}{\partial q}k = 0 \\ b_{n-1} + \dfrac{\partial b_{n-1}}{\partial p}h + \dfrac{\partial b_{n-1}}{\partial q}k = 0 \end{array}\right\}$$

Here b_n, b_{n-1} and the partial derivative are evaluated at $p = p_0$, $q = q_0$.
From (2), we get on differentiation

$$\frac{\partial b_n}{\partial p} = b_{n-1} + p\frac{\partial b_{n-1}}{\partial p} + q\frac{\partial b_{n-2}}{\partial p} \quad (3)$$

$$\frac{\partial b_n}{\partial q} = b_{n-2} + p\frac{\partial b_{n-1}}{\partial q} + q\frac{\partial b_{n-2}}{\partial q} \quad (4)$$

If we set $\dfrac{\partial b_n}{\partial p} = c_{n-1}$, $\dfrac{\partial b_n}{\partial q} = c_{n-2}$, the above equations take the form

$$\boxed{\begin{array}{l} c_{n-1} = b_{n-1} + pc_{n-2} + qc_{n-3}. \\ c_{n-2} = b_{n-2} + pc_{n-3} + qc_{n-4}. \end{array}}$$

Comparing these with (2), we find c's are obtained from b's the same way as b's are obtained from b's the same way as b's are obtained from a's. By applying Synthetic division twice, we compute c's from a's.
In terms of c's equations (3) and (4) take the form

$$b_n + c_{n-1}h + c_{n-2}k = 0 \tag{5}$$

$$b_{n-1} + c_{n-2}h + c_{n-3}k = 0 \tag{6}$$

Solving these equations, we get h and k and the improved values $p_1 = p_0 + h$, $q_1 = q_0 + k$.

The procedure is to be repeated till the desired accuracy is achieved by using the generalised results for finding a quadratic factor of the polynomial of degree n which are given by:

$$\left.\begin{array}{l} p_{k+1} = p_k + h, \quad q_{k+1} = q_k + k \quad k = 0,1,\ldots \\ \text{where} \quad h = -\dfrac{b_n c_{n-3} - b_{n-1} c_{n-2}}{c_{n-2}^2 - c_{n-3}(c_{n-1} - b_{n-1})} \\ k = -\dfrac{b_{n-1}(c_{n-1} - b_{n-1}) - b_n c_{n-2}}{c_{n-2}^2 - c_{n-3}(c_{n-1} - b_{n-1})} \end{array}\right\} \begin{array}{l} \text{XXX} \\ \text{XXX} \\ \text{XXX} \end{array} \tag{7}$$

We use the Synthetic division method to determine b_i's and c_i's.

Synthetic division method

$-p_n$	a_0	a_1	a_2	$\cdots$	a_{n-1}	a_n
		$-p_k b_0$	$-p_k b_1$	$\cdots$	$-p_k b_{n-2}$	$-p_k b_{n-1}$
$-q_k$			$-q_k b_0$	$\cdots$	$-q_k b_{n-3}$	$-q_k b_{n-2}$
	b_0	b_1	b_2	$\cdots$	b_{n-1}	b_n
	$= a_0$	$-p_k c_0$	$-p_k c_1$	$\cdots$	$-p_n c_{n-2}$	
			$-q_k c_0$	$\cdots$	$-q_k c_{n-3}$	
	c_0	c_1	c_2	$\cdots$	c_{n-1}	
	$= b_0$					

We have

$$\lim_{k\to\infty} b_n = 0; \quad \lim_{k\to\infty} b_{n-1} = 0$$

$$\lim_{k\to\infty} P_k = p; \quad \lim_{k\to\infty} q_k = q$$

The order of this method is 2.

Having found the values of p and q to the desired degree of accuracy, we obtain the next factor from the deflated polynomial $Q_{n-2}^{(n)} = b_0 x^{n-2} + b_1 x^{n-3} + \ldots +$

$b_{n-3}x + b_{n-2}$ which can be obtained from the first part of the above Synthetic division method.

Note: The equation (7) which yield the values of h and k on solving, can also be written as a particular case in the form: (for $n = 3$ in equation (7))

$$\begin{Bmatrix} h = \left\{ -\dfrac{b_3c_0 - b_2c_1}{c_1^2 - c_0(c_2 - b_2)} \right\} \\ k = \left\{ -\dfrac{b_2(c_2 - b_2) - b_3c_1}{c_1^2 - c_0(c_2 - b_2)} \right\} \end{Bmatrix} \tag{8}$$

> The values of h and k given by the equation (8) (as a particular case when $n = 3$) is more useful in solving for h and k in the examples under consideration.

Example 48. Find the quadratic factor of $x^3 - x - 1 = 0$ close to $x^2 + x + 1$ with $(p_0, q_0) = (1, 1)$

Solution:

Given: $f(x) = x^3 - x - 1 = 0$

Here, $p_0 = 1, q_0 = 1, \quad (a_0 = 1 = b_0)$

By Synthetic division twice by $x^2 + x + 1$, we get b_n, b_{n-1}, c_n, c_{n-1}, c_{n-2} as follows:

First iteration:

−1	1	0	−1	−1
−1		−1	1	1
			−1	1
	1	−1	$-1 = b_{n-1}$	$1 = b_n$
−1		−1	2	
−1			−1	
	$1 = c_{n-3}$	$-2 = c_{n-2}$	$0 = c_{n-1}$	

Hence, the equations $\begin{Bmatrix} b_n + c_{n-1}h + c_{n-2}k = 0 \\ b_{n-1} + c_{n-2}h + c_{n-3}k = 0 \end{Bmatrix}$

take the form $1 - 2k = 0$, $-1 - 2h + k = 0$ on solving for h and k, we get

$$h = -0.25, \; k = 0.5$$

The improved values are
$$p_1 = -1 + h = -1.25$$
$$q_1 = -1 + 0.5 = -0.5$$

Second iteration:

$$p_1 = -1.25,\ q_1 = -0.5$$

−1.25	1	0	−1	−1
−0.5		−1.25	1.5625	−0.078125
			−0.5	0.625
	1	−1.25	$0.0625 = b_{n-1}$	$-0.453125 = b_n$
		−1.25	3.125	
−1.25			−0.5	
−0.5				
	$1 = c_{n-3}$	$-2.5 = c_{n-2}$	$2.6875 = c_{n-1}$	

Hence, the equations
$$b_n + c_{n-1}h + c_{n-2}k = 0$$
$$b_{n-1} + c_{n-2}h + c_{n-3}k = 0$$

take the form
$$-0.453125 + 2.6875h - 2.5k = 0$$
$$0.0625 - 2.5h + k = 0.$$

Solving these equations, $h = -0.0833,\ k = -0.2708.$

Hence,
$$p_2 = p_1 + h = -1.3333$$
$$q_2 = q_1 + k = -0.7708.$$

The quadratic factor is $x^2 + 1.3333x + 0.7708$.
The approximate complex roots are:
$-0.6667 \pm 0.5713i$

Example 49. Apply Lin-Bairstow's method to $x^4 - 3x^3 + 20x^2 + 44x + 54 = 0$, to find a quadratic factor close to $x^2 + 2x + 2$. Do one step only with $(p_0, q_0) = (2, 2)$.

Solution:

Given: $f(x) \equiv x^4 - 3x^3 + 20x^2 + 44x + 54 = 0$

Here, $p_0 = 2,\ q_0 = 2,\ (a_0 = 1 = b_0)$

Using Synthetic division scheme as shown below:

−2	1	−3	20	44	54
		−2	10	−56	4
−2			−2	10	−56
	1	−5	28	$-2 = b_{n-1}$	$2 = b_n$
		−2	14	−80	
			−2	14	
	1	$-7 = c_{n-3}$	$40 = c_{n-2}$	$-68 = c_{n-1}$	

$\therefore$ $$\boxed{b_n = 2;\ b_{n-1} = -2;\ c_{n-1} = -68;\ c_{n-2} = 40;\ c_{n-3} = -7}$$

Hence, the equation
$$b_n + c_{n-1}h + c_{n-2}k = 0$$
$$b_{n-1} + c_{n-2}h + c_{n-3}k = 0$$

become
$$\left.\begin{matrix} 2 - 68h + 40k = 0 \\ -2 + 40h - 7k = 0 \end{matrix}\right\} \quad (1)$$

or,
$$40h - 7k = 2$$
$$-68h + 40k = -2$$

Solving (1) by cross-multiplication rule, we get

$$\frac{h}{\begin{vmatrix} -7 & -2 \\ 40 & 2 \end{vmatrix}} = \frac{-k}{\begin{vmatrix} 40 & -2 \\ -68 & 2 \end{vmatrix}} = \frac{1}{\begin{vmatrix} 40 & -7 \\ -68 & 40 \end{vmatrix}}$$

$\Rightarrow$ $$\frac{h}{-14 + 80} = \frac{-k}{80 + 136} = \frac{1}{1600 - 476}$$

$\Rightarrow$ $$\frac{h}{66} = \frac{-k}{-56} = \frac{1}{1124}$$

$\Rightarrow$ $$\boxed{\begin{aligned} h &= \frac{66}{1124} = \frac{33}{562};\ k = \frac{56}{1124} = \frac{14}{281} \\ &= 0.0580 \qquad\qquad = 0.0457 \end{aligned}}$$

$\therefore$ $$h = -0.0580, \qquad k = -0.0457$$

$\Rightarrow$ $$\left.\begin{matrix} p_1 = p_0 + h = 2 - 0.0580 = 1.9420, \\ q_1 = q_0 + k = 2 - 0.0457 = 1.9543, \end{matrix}\right\}$$

Hence, the required quadratic factor close to $x^2 + 2x + 2$ is $x^2 + p_1x + q_1$

i.e. $$\boxed{x^2 + 1.9420x + 1.9543}$$

Example 50. Using Lin-Bairstow's method obtain the quadratic factor of the equation $x^4 - x^3 + 6x^2 + 5x + 10 = 0$ with (perform two iterations) $(p_0, q_0) = (1.14, 1.42)$

Solution:

Given: $$f(x) = x^4 - x^3 + 6x^2 + 5x + 10 = 0$$

Here, $p_0 = 1.14,\ q_0 = 1.42,\ (a_0 = 1 = b_0)$

Using Synthetic division scheme, we have

First iteration:

–1.14	1	–1	6	5	10
–1.42		–1.14	2.4396	–8.0023	–0.0416
			–1.42	3.0388	–9.9678
	1	–2.14	7.0196	0.0365	–0.9678 = b_n
		–1.14	3.7392	10.6462	
			–1.14	4.6576	
	1	–3.28	9.3388	–5.9522 = c_{n-1}	
		= c_{n-3}	= c_{n-2}		

Equation (1) yields

$$\boxed{b_n + c_{n-1}h + c_{n-2}k = 0;\ b_{n-1} + c_{n-2}h + c_{n-3}k = 0} \tag{1}$$

$$\begin{Bmatrix} h = 0.0046, & k = 0.0019 \\ p_1 = 1.1446 & q_1 = 1.4219 \end{Bmatrix}$$

Second iteration:

$p_1 = 1.1446,\ q_1 = 1.4219$

–1.1446	1	–1	6	5	10
–1.4219		–1.1446	2.4547	–8.0498	–0.0005
			–1.4219	3.0494	–10.0
	1	–2.1446	7.0328	–0.0004 = b_{n-1}	–0.0005 = b_n
		–1.1446	3.7648	–10.7314	
			–1.4219	4.6769	
	1	–3.2892	9.3757	–6.3561 = c_{n-1}	
		= c_{n-3}	= c_{n-2}		

$$\boxed{\begin{aligned} b_n + c_{n-1}h + c_{n-2}k &= 0 \\ b_{n-1} + c_{n-2}h + c_{n-3}k &= 0 \end{aligned}}$$

Giving $h = 0,\ k = 0$

$p_2 = 1.1446,\ q_2 = 1.4219$

Example 51. Solve $x^4 - 8x^3 + 39x^2 - 62x + 50 = 0$ using Lin-Bairstow's method with $(p_0, q_0) = (0, 0)$

Solution: Given: $f(x) = x^4 - 8x^3 + 39x^2 - 62x + 50 = 0$

Here, $p_0 = 0, q_0 = 0, (a_0 = 1 = b_0)$

By Synthetic division twice, we get b_n, b_{n-1}, c_{n-1}, c_{n-2} as follows:

First iteration:

	1	–8	39	–62	50
0		0	0	0	0
0			0	0	0
	1	–8	39	$-62 = b_{n-1}$	$50 = b_n$
0		0	0		
0		0	0		
	1	$-8 = c_{n-3}$	$39 = c_{n-2}$	$-62 = c_{n-1}$	

The equations $\begin{Bmatrix} b_n + c_{n-1}h + c_{n-2}k = 0 \\ b_{n-1} + c_{n-2}h + c_{n-3}k = 0 \end{Bmatrix}$

take the form

$$50 - 62h + 39k = 0$$
$$-62 + 39h - 8k = 0$$

on solving, we get $h = 1.97$, $k = 1.85$

Solution by using Crammer's rule or cross-multiplication method,

$\begin{Bmatrix} -62h + 39k + 50 = 0 \\ 39h - 8k - 62 = 0 \end{Bmatrix}$ by cross-multiplication method, we have

$$\frac{h}{\begin{vmatrix} 39 & 50 \\ -8 & -62 \end{vmatrix}} = \frac{-k}{\begin{vmatrix} -62 & 50 \\ 39 & -62 \end{vmatrix}} = \frac{1}{\begin{vmatrix} -62 & 39 \\ 39 & -8 \end{vmatrix}}$$

$$\Rightarrow \left[\frac{h}{-2018} = \frac{k}{-1894} = \frac{1}{-1025}\right] \Rightarrow h = \frac{2018}{1025};\ k = \frac{1894}{1025}$$

$$\Rightarrow h = k =$$

and $p_1 = p_0 + h = 1.97$, $q_1 = q_0 + k = 1.85$.

Second iteration:

We proceed further and work with two decimal places

1.97	1	–8	39	–62	50
1.85		1.97	–11.88	57.07	–31.70
			1.85	–11.16	53.59
	1	–6.03	28.97	$-16.09 = b_{n-1}$	$71.90 = b_n$
1.97		1.97	–8.00	44.96	
1.85			1.85	–7.51	
	1	$-4.06 = c_{n-3}$	$22.82 = c_{n-2}$	$21.36 = c_{n-1}$	

Hence, equations for h and k are

$$\begin{Bmatrix} 71.90 + 21.36h + 22.82k = 0 \\ -16.09 + 22.82h - 4.06k = 0 \end{Bmatrix}$$

on solving the above equations, we get

$h = 0.12$, $k = -3.27$.

$\therefore \quad p_2 = 1.97 + 0.12 = 2.09$, $q_2 = 1.85 - 3.27 = -1.42$

Repeating the procedure once more, we get $p_3 = 2$, $q_3 = -2$ and the quadratic factor is $x^2 - 2x + 2$. The other quadratic factor is $x^2 - 6x + 25$. Hence, the roots are $1 \pm i$, $3 \pm 4i$.

Example 52. Solve $x^4 - 5x^3 + 20x^2 - 40x + 60 = 0$, given that all the roots of $f(x) = 0$ are complex, by using Lin-Bairstow's method with $(p_0, q_0) = (-4, 8)$.

Solution:

Given: $\quad f(x) = x^4 - 5x^3 + 20x^2 - 40x + 60 = 0$

with $\quad p_0 = -4$, $q_0 = 8$

First iteration:

4	1	–5	20	–40	60
–8		4	–4	32	0
			–8	8	–64
4	1	–1	8	$0\ (= b_{n-1})$	$-4\ (= b_n)$
		4	12	48	
–8			–8	–24	
	1	$3\ (= c_{n-3})$	$12\ (= c_{n-2})$	$24\ (= c_{n-1})$	

Hence, the equations for h and k are given by

$$12h + 3k = 0$$

$$24h + 12k = -4$$

on solving, we get

$$h = 0.1667,\ k = -0.6667$$

$$\therefore \quad p_1 = p_0 + h = -3.8333$$

$$q_1 = q_0 + k = 7.3333$$

Hence, the extracted quadratic factor is

$$(x^2 + P_1 x + q_1) \quad \text{or} \quad (x^2 - 3.8333x + 7.3333)$$

Second iteration:

$$P_1 = -3.8333,\ q_1 = 7.3333\ a_0 = b_0 = 1$$

3.8333	1	–5	20	–40	60
–7.3333		3.8333	–4.4723	31.4116	–.125
			–7.3333	8.5558	–60.092
	1	–1.1667	8.1944	$-0.0326 = b_{n-1}$	$-0.217 = b_n$
3.8333					
–7.3333		3.8333	10.2219	42.4845	
			–7.3333	–19.555	
	1	$2.6666 = c_{n-3}$	$11.083 = c_{n-2}$	$22.8969 = c_{n-1}$	

$$\therefore \quad c_{n-1} - b_{n-1} = 22.8969 + 0.0326 = 22.9295$$

Hence, the equations for corrections h and k are given by

$$11.083h + 2.6666k = -0.0326$$

$$22.9295h + 11.083k = -0.217$$

Solving, we get $h = 0.0033$ and $k = -0.0269$

$$p_2 = p_1 + h = -3.83,\ q_2 = q_1 + k = 7.3064$$

Hence, the extracted quadratic factor is

$$(x^2 + p_2 x + q_2) \quad \text{or} \quad (x^2 - 3.83x + 7.3064) \qquad (1)$$

If $\alpha \pm$ be the roots of (1), then
Sum of the roots $= 2\,\alpha = 3.82979 \cong 3.83$
Products of roots $= \alpha^2 + \beta^2 = 7.3064$
yielding $\alpha = 1.914895$ and $\beta = 1.907775$
We can take $\alpha = .9149$ and $\beta = 1.9077$
Hence, a pair of roots is $(1.9149 \pm 1.9077i)$

To find the remaining two roots of $f(x) = 0$, we divide $f(x)$ by (1) by using the Synthetic division method, as follows:

	1	–5	20	–40	60
3.83					
		3.83	–4.4811	31.4539	
–7.3064			–7.3064	8.5485	–60.0038
	1	–1.17	8.2125	0.0024	–0.0038
				≈ 0	≈ 0

∴ The extracted quadratic factor is $(x^2 - 1.17x + 8.2125)$ (2)

If $\gamma \pm i\delta$ be its roots (2) then $2\delta = 1.17$, $\gamma^2 + \delta^2 = 8.2125$

$\Rightarrow \{\gamma = 0.585$ and $\delta = 2.8054\}$

Hence, the pair of complex roots is $(0.585 \pm 2.8054\ i)$

> **Note:** Since p_0 and q_0 of P and q respectively are our choice, but the Lin-Bairstow's iteration will converge only if the starting values of p and q are sufficiently close to the true values.

Example 53. Perform one iteration of the Bairstow's method to extract a quadratic factor $x^2 + px + q$ from the polynomial $x^4 + x^3 + 2x^2 + x + 1 = 0$. Using initial approximation

$$(P_0, q_0) = (0.5, 0.5)$$

Solution: Given $\quad f(x) = x^4 + x^3 + 2x^2 + x + 1 = 0$

Here $\quad P_0 = 0.5,\ q_0 = 0.5$

By Synthetic division twice, we get b_n, b_{n-1}, c_{n-1}, c_{n-2}, c_{n-3} as follows:

–0.5	1	1	2	1	1
			–1.25	–0.625	–0.0625
–0.5		–0.5	–0.5	–0.25	–0.625
	1	0.5	$1.25 = b_2$	$0.125 = b_3$	$0.3125 = b_4$
		–0.5	0.0	–0.375	
			–0.5	0.0	
	1	$0.0 = c_1$	$0.75 = c_2$	$-0.25 = c_3$	

$$h = -\frac{b_4c_1 - b_3c_2}{c_2^2 - c_1(c_3 - b_3)} = 0.1667, \quad k = -\frac{b_3(c_3 - b_3) - b_4c_2}{c_2^2 - c_1(c_3 - b_3)} = 0.5$$

$$\left.\begin{array}{l} p_1 = p_0 + h = 0.6667 \\ q_1 = q_0 + k = 1.0 \end{array}\right\} \text{ The exact values of } p \text{ and } q \text{ are } 1.0.$$

Example 54. Using Lin-Bairstow's method obtain the quadratic factor of the equation: $x^3 - 3.7x^2 + 6.25x - 4.069 = 0$ (perform two iterations) with $(p_0, q_0) = (-2.5, 3)$

Solution:

Given: $f(x) = x^3 - 3.7x^2 + 6.25x - 4.069 = 0$

with $p_0 = -2.5,\ q_0 = 3\ (a_0 = 1 = b_0)$

By Synthetic division twice, we get $b_n,\ b_{n-1},\ c_{n-1},\ c_{n-2}$ as follows:

First iteration:

2.5	1	–3.7	6.25	–4.069
–3.0		2.5	–3.0	0.625
			–3.0	3.6
	1	–1.2	$0.25 = b_{n-1}$	$0.156 = b_n$
		2.5	3.25	
			–3.0	
	1	$1.3 = c_{n-2}$	$0.50 = c_{n-1}$	

Hence, by the equations: $b_n + c_{n-1}h + c_{n-2}k = 0;\ b_{n-1} + c_{n-2}h + c_{n-3}k = 0.$
Substituting the values of $b_n,\ b_{n-1},\ c_{n-1},\ c_{n-2}$
We obtain: $h = 0.1174,\ k = 0.0974$

$$\Rightarrow \quad \begin{Bmatrix} p_1 = p_0 + h = -2.5 - 0.1174 = -2.3826 \\ q_1 = q_0 + k = 3 + 0.0974 = 3.0974 \end{Bmatrix}$$

Second iteration:

$$h = 01174\ ,\ k = 0.0974$$
$$p_1 = -2.3826,\ q_1 = 3.0974$$

2.3826	1	–3.7	6.25	–4.069
–3.0974		2.3826	–3.1388	0.0329
			–3.0974	4.0805
	1	–1.3174	0.0138	$0.0444 = b_n$
		2.3826	2.5379	
			–3.0974	
	1	1.0652 $= c_{n-2}$	$-0.5454 = c_{n-1}$	

Hence, the equations $\begin{Bmatrix} b_n + c_{n-1}h + c_{n-2}k = 0 \\ b_{n-1} + c_{n-2}h + c_{n-3}k = 0 \end{Bmatrix}$ yield on solving for h and k with $b_n = 0.444,\ c_{n-1} = -0.5454,\ c_{n-2} = 1.0652$ etc., the values as $h = 0.0175,\ k = 0.0325,\ p_2 = 2.4001,\ q_2 = 3.1299.$

∴ The extracted quadratic factor is

$$(x^2 + p_2x + q_2)$$

or

$$(x_2 + 2.4001x + 3.1299)$$

Example 55. Perform two iterations of the Lin-Bairstow's method to extract a quadratic factor $x^2 + px + q$ from the polynomial $x^3 + x^2 - x + 2 = 0$. Using the initial approximations $(p_0, q_0) = (-0.9, 0.9)$

Solution:

Given: $$f(x) = x^3 + x^2 - x + 2 = 0 \quad (1)$$

and $$p_0 = -0.9,\ q_0 = 0.9$$

First iteration:

By using the Synthetic division method as follows, we obtain

0.9	1	1	–1	2
–0.9		0.9	1.71	–0.171
			–0.9	–1.71
	$1 = b_0$	1.9	$-0.19 = b_{n-1}$	$0.119 = b_n$
0.9		0.9	2.52	
–0.9			–0.9	
	$1 = c_{n-3}$	$2.8 = c_{n-2}$	$1.43 = c_{n-1}$	

Hence, the equations

$$b_n + c_{n-1}h + c_{n-2}k = 0$$
$$b_{n-1} + c_{n-2}h + c_{n-3}k = 0$$

become

$$\left.\begin{array}{l} 0.119 + 1.43h + 2.8k = 0 \\ -0.19 + 2.8h + k = 0 \end{array}\right\} \quad (2)$$

Solving equation (2) we obtain

$$h = -\frac{0.651}{6.22} = -0.1047$$

and

$$k = \frac{0.6410}{6.22} = 0.1031$$

$$p_1 = p_0 + h = -0.9 - 0.1047 = -1.0047$$
$$q_1 = q_0 + k = 0.9 + 0.1031 = 1.0031$$

Second iteration:

We have $p_1 = -1.0047;\ q_1 = 1.0031$

1.0047	1	1	–1	2
–1.0031		1.0047	2.0141	0.0111
			–1.0031	–2.0109
	$1 = b_0$	2.0047	$0.0110 = b_2$	$0.0002 = b_3$
		1.0047	3.0235	
			–1.0031	
	$1 = c_0$	$3.0094 = c_1$	$2.0314 = c_2$	

EXERCISES

1. Starting with the approximation $x^2 - 4x - 6$, find the quadratic factor $0 + x^3 - 3x^2 - x + 9$. [Ans: $x^2 - 4.25x + 5.90$]
2. Find the complex roots of $x^3 + x^2 - 2 = 0$ starting with $p_0 = q_0 = 0$. [Ans: $1 \pm i$]
3. Find the quadratic factors of
$$x^4 - 1.1x^3 + 2.3x^2 + 0.5x + 3.3 = 0$$
starting with the approximation $x^2 + x + 1$
[Ans: $(x^2 + 0.9x + 1.1)\ (x^2 - 2x + 3)$].
4. Use Lin-Bairstow's method to find a quadratic factor of $x^4 + 5x^3 + 3x^2 - 5x - 9 = 0$ after two iterations with $(p_0, q_0) = (3, -5)$
[Ans: $x^2 + 2.90255x - 4.91759$]
5. Determine the roots of the equation
$$x^4 + 9x^3 + 36x^2 + 51x + 27 = 0$$
to four places of decimal, using the iterative method of Lin-Bairstow.
[Ans: The extracted quadratic factors are $(x^2 - 2x + 2)$, $(x^2 - 6x + 25)$ and $(x^2 - 2x + 2)\ (x^2 - 6x + 25) = x^4 - 8x^3 + 39x^2 - 62x + 50 = 0$.]
6. Starting with $x^2 + 0.5x + 0.5$, improve the quadratic factor of $x^4 + x^3 + 2x^2 + x + 1$. [Ans: $x^2 + 0.67x + 1$ Exact : $x^2 + x + 1$]
7. Beginning with the final factor $x^2 - 4x + 5$ of $x^4 - 3.1x^3 + 2.1x^2 + 1.1x + 5.2$, improve it. [Ans: $x^2 - 4.1x + 5.2$]
8. By Lin-Bairstow's method, find a quadratic factor of the equation
$$5x^6 + 4x^5 + 7x^4 + 3x^3 + 20x^2 + 30x + 1 = 0.$$

DIRECT METHOD

Graeffe's Root Squaring Method

We give below Graeffe's root squaring method when all the roots of an algebraic equation may be real and distinct, real and equal or complex roots. This method has an advantage over the other methods as it does not require any prior information about the roots. But it is applicable to polynomial equations only and is capable of giving all the roots.

Working Rule for Finding the Roots

Step (i) From a given polynomial equation, construct an equation whose roots are the negative squares of the roots of the given equation

Step (ii) Repeat step (i), say m times

Step (iii) Estimate the roots λ_i of the final equation

Step (iv) Then, determine the roots of the given equation from λ_i

Take the given polynomial equation of degree n, as $f(x) = a_0 x^n + a_1 x^{n-1} + \ldots + a_n = 0$ (1)

If $\alpha_1, \alpha_2, \ldots \alpha_n$ are its roots, then

$$f(x) = a_0 (x - \alpha_1)(x - \alpha_2) \ldots (x - \alpha_n)$$

$$f(-x) = a_0 (-x - \alpha_1)(-x - \alpha_2) \ldots (-x - \alpha_n)$$

$$f(x) f(-x) = a_0^2 (-x^2 + \alpha_1^2)(-x^2 + \alpha_2^2) \ldots (-x^2 + \alpha_n^2)$$

$$= a_0^2 (y + \alpha_1^2)(y + \alpha_2^2) \ldots (y + \alpha_n^2)$$

(Setting $y = -x^2$)

Thus, the equation whose roots are $-\alpha_i^2$ is obtained by equating the product $f(x) f(-x)$ to zero and setting $y = -x^2$.

Now, let us consider $f(x) = a_0x^4 + a_1x^3 + a_2x^2 + a_3x + a_4 = 0$

$$f(-x) = a_0x^4 - a_1x^3 + a_2x^2 - a_3x + a_4 = 0$$

by actual multiplication, we obtain the equation

$$a_0^2 x^8 - (a_1^2 - 2a_0a_2)x^6 + (a_2^2 - 2a_1a_2 + 2a_0a_4)x^4 - (a_3^2 - 2a_2a_4)x^2 + a_4^2 = 0$$

on putting $x^2 = -y$, we get

$$b_0y^4 + b_1y^3 + b_2y^2 + b_3y + b_4 = 0 \qquad (2)$$

where $b_0 = a_0^2$;

$b_1 = a_1^2 - 2a_0a_1$;

$b_2 = a_2^2 - 2a_1a_3 + 2a_0a_4$;

$b_3 = a_3^2 - 2a_2a_4$; and $b_4 = a_4^2$.

In general,

$$b_i = a_i^2 - 2a_{i-1} + 2a_{i-2}\, a_{i+2} - \ldots \qquad (3)$$

In (3), the coefficients like $a_{-1}, a_{-2}, \ldots$, etc, which do not exist must be assumed zero.

Equation (2) is the desired equation whose roots are $-\alpha_i^2$, equation (3) can be put in a tabular form as shown below:

a_0	a_1	a_2	a_3	a_4
a_0^2	a_1^2	a_2^2	a_3^2	a_4^2
	$-2a_0a_2$	$-2a_1a_3$	$-2a_2a_4$	
		$2a_0a_4$		
b_0	b_1	b_2	b_3	b_4

The above process is repeated 'm' times to get an equation:

$$B_0u^4 + B_1u^3 + B_2u^2 + B_3u + B_4 = 0 \tag{4}$$

The roots λ_i of (4) are related to the roots of (1) by $\lambda_i = -\alpha_i^{2^m}$ (5)

Let α_i be real and $|\alpha_1| > |\alpha_2| > |\alpha_3| > |\alpha_4|$.

Then λ_i are widely separated that is, $|\lambda_1| >> |\lambda_2| >> |\lambda_3| >> |\lambda_4|$. (Here >> stands for "is much larger than".)

From equation (4), we have hence

$$-\frac{B_1}{B_0} = \lambda_1 + \lambda_2 + \lambda_3 + \lambda_4 = \lambda_1, \text{ approximately}$$

$$+\frac{B_2}{B_0} = \lambda_1\lambda_2 + \ldots = \lambda_1\lambda_2$$

$$-\frac{B_3}{B_0} = \lambda_1\lambda_2\lambda_3 + \ldots = \lambda_1\lambda_2\lambda_3$$

$$+\frac{B_4}{B_0} = \lambda_1\lambda_2\lambda_3\lambda_4$$

OR, $\quad \lambda_1 = -\dfrac{B_1}{B_0},\ \lambda_2 = -\dfrac{B_2}{B_1},\ \lambda_3 = -\dfrac{B_3}{B_2},\ \lambda_4 = -\dfrac{B_4}{B_3}$

The roots of the original equation are obtained from $|\alpha_i| = (-\lambda_i)^{1/2^m} = \left(\dfrac{B_i}{B_{i-1}}\right)^{1/2^m}$

The sign α_i can be determined by actual substitution in the given equation. The Descarte's rule of signs can be helpful in this respect.

Complex Roots

If α_k, α_{k+1} are two complex conjugate roots, then this would make the coefficients of x^{n-k} in the successive squarings to fluctuate both in magnitude and sign.

If α_k, $\alpha_{k+1} = \beta_k (\cos \phi_k + i \sin \phi_k)$ is the complex pair of roots, then the coefficient will fluctuate in magnitude and sign by the quantity $2\beta_k^m \cos m\, \phi_k$.

A complex pair is located only by such oscillation. If m is large, β_k can be got from

$$\beta_k^{2(2^m)} \approx \left| \frac{B_{k+1}}{B_{k-1}} \right| \text{ and } \phi \text{ is got from } 2\beta_k^m \cos m\, \phi_k = \frac{B_{k+1}}{B_{k-1}}$$

If the equation possesses only two complex roots $\xi \pm i\eta$, we have

$$\alpha_1 + \alpha_2 + \dots \alpha_{k-1} + 2\xi + \alpha_{k+2} + \dots + \alpha_n = -a_1$$

This gives the value ξ.

Since $|\beta_k|^2 = \xi^2 + \eta^2$ and $|\beta_k|$ is known already η is known from this relation.

Example 56. Find the roots of $x^2 + 4x - 5 = 0$ by two squarings.

Solution: The coefficients of the successive root squarings are given by the following table

m	2^m			
0	1	$1 = a_0$	$4 = a_1$	$-5 = a_2$
		1	16	25
		10		
1	2	1	26	25
		1	676	625
			–50	
		1	626	625

Note: $-2(1)(-5) = 10$

Note: $-50 = -2(1)(25)$

$B_0 = 1$, $B_1 = 626$, $B_2 = 625$.

$$|\alpha_1| = \left(\frac{B_1}{B_0}\right)^{1/4} = (626)^{1/4} = 5.0020.$$

$$|\alpha_2| = \left(\frac{B_2}{B_0}\right)^{1/4} = \left(\frac{625}{626}\right)^{1/4} = 0.9996.$$

By Descarte's rule of signs, we find one positive root and one negative root. By actual substitution we find the first root is negative.

So the second root must be positive and $\alpha_1 = -5.0020$, $\alpha_2 = 0.9996$. Exact roots: –5, 1.

Example 57. Find all the roots of the polynomial equation $x^3 - 2x + 2$. Using Graeffe's root squaring method.

Solution:

The coefficients of the successive root squarings are given by the following table

m	2^m				
0	1	$1 = a_0$	$0 = a_1$	$-2 = a_2$	$2 = a_3$
		1	0	4	4
			4	0	Note: $4 = (-2)(1)(-2)$ $0 = -2(0)(2) = 0$
1	2	1	4	4	4
		1	16	16	16
			−8	−32	Note: $-8 = (-2)(1)(4)$ $-32 = (-2)(4)(4)$
2	4	1	8	−16	16
		1	64	256	256
			32	−256	Note: $32 = (-2)(1)(-16)$ $-256 = (-2)(8)(16)$
3	8	1	96	0	256
		1	9216	0	65536
			0	−49152	Note: $0 = (-2)(1)(0)$ $-49152 = (-2)$
4	16	$1 = B_0$	$9216 = B_1$	$-49152 = B_2$	$65536 = B_3$

Since B_2 is alternatively positive and negative.

We have a pair of complex roots based on B_1, B_2, B_3. [$B_0 = 1, 9216 = B_1, -49152$

$= B_2$, $65536 = B_3$] are real roots $= |\alpha_1| = \left(\dfrac{B_1}{B_0}\right)^{\frac{1}{16}}$ $1 = (9216)^{1/16} = 1.7692$

On substituting into the given polynomial, we find that this root must be –ve.

Hence, one real root is $\alpha_1 = -1.7692$

To find the pair of complex roots.

$\rho_2\, e^{\pm i\phi_2} = \xi + i\eta$, we have

$$\sqrt{\xi^2 + \eta^2} = \beta$$

Where $|\beta|^{32} = \left|\frac{B_3}{B_1}\right|$

or $$\beta = 1.0632 = \sqrt{\xi^2 + \eta^2}$$

Also $$\alpha_1 + 2\xi = 0$$

or $$\xi = 0.8846$$

$$\eta^2 = \beta^2 - \xi^2 \quad \text{or} \quad \eta = 0.5898$$

Hence, roots are $0.8846 \pm 0.5898\ i$.

Example 58. Apply the Graeffe's root squaring method to find the roots of the equation $x^3 + 3x^2 - 4 = 0$ correct to two decimal places.

Solution: The coefficients of the successive root squarings are given by the following table.

m	2^m					
0	1	1	3	0	–4	
		1	9	0	16	
			0	24		$0 = (-2)(0)$ $24 = (-2)(3)(-4)$
1	2	1	9	24	16	
		1	81	576	256	
			–48	–288		$-48 = (-2)(1)(24)$ $-288 = (-2)(9)(16)$
2	4	1	33	288	256	
		1	1089	82944	65536	
			–576	–16896		$-576 = (-2)(1)(288)$ $-16896 = (-2)(33)(256)$
3	8	1	513	66048	65536	
		1	263169	4362338304	16^8	
			–132096	–67239936		$-132096 = (-2)(1)(66048)$ $-67239936 = (-2)(513)(65536)$
4	16	1	131073 almost half	4295098368	16^8	
		$= B_0$	$= B_1$	$= B_2$	$= B_3$	

Since B_1 is almost half of the corresponding value in the previous squaring, it indicates that there may be double root based on B_0, B_1 and B_2. Then we obtain one double root $| \alpha_1 |^{32} = | \alpha_2 |^{32} = | B_2 |$ which yields $| \alpha_1 | = 2.0000$. Substituting in the given equation we find that this root is negative. Hence $\alpha_1 = -2.0$

one simple real root $| \alpha_3 |^{16} = \left| \dfrac{B_3}{B_2} \right|$

which gives $\alpha_3 = 1.0000$. Substituting in the given equation, we find that the root is positive.

Hence, the roots are 1.0000, –2.0000, –2.0000

Example 59. Find the roots of the equation:

$x^3 - 12x^2 + 39.5x - 33 = 0$ using Graeffe's root squaring method.

Solution: The coefficients of the successive root squarings are given by the following table.

m	2^m				
0	1	1	–12	39.5	–33
		1	144	1560.25	1089
			0		(–2)(1)(39.5)
		1	144		etc

EXERCISES

I. Using Graeffe's method find the roots of the equation

$$x^3 - 3.7x^2 + 6.25x - 4.009 = 0$$

II. Solve by Graeffe's method by squaring the roots thrice:

1. $x^2 - 3x + 2 = 0.$ (Ans: 2.001, 0.9995)
2. $x^3 - 2x - 1 = 0.$ (Ans: 1.618, –1, –0.618)
3. $x^3 - 2x^2 - x + 1 = 0.$ (Ans: 2.247, –0.802, 0.555)
4. $x^3 - 2x^2 - 5x + 6 = 0.$ (Ans: 3.0144, –1.9914, 0.9995)
5. $x^3 - 5x^2 - 17x + 20 = 0.$ (Ans: 7.017, –2.974, 0.958)
6. $x^3 - 8x^2 + 17x - 10 = 0.$ (Ans: 5.0004, 2.0008, 0.9995)
7. $x^3 - 6x^2 + 11x - 6 = 0.$ (Ans: 3.0144, 1.9914, 0.9995)
8. $x^3 - 5x^2 - 17x + 21 = 0.$ (Ans: 7.001, 2.999, 0.9998)
9. $x^3 + 9x^2 + 23x + 14 = 0.$ (Ans: –4.8844, –3.2381, –0.8551)
10. $x^4 - 15x^2 - 10x + 24 = 0.$ (Ans: 4, –3, –2, 1)
11. $x^3 - 2x + 2 = 0.$ (Ans: 0.8846 ± 0.5898i, 1.7692)
12. $x^3 + 3x^2 - 4 = 0.$ (Ans: –2.0000, –2.0000, 1.0000)

NEWTON-RAPHSON METHOD

The most used iteration method is Newton-Raphson method also called Newton's method. By this method, we get closer approximation of the root of an equation if we already know its approximate root.

If m is a real root of an equation $f(x) = 0$, which lies near a point $x = a$, then, we have

(i) $f(m) = 0$ ...(1)

(ii) $f(m) = f(a + \overline{m - a}) = f(a) + (m - a) f'(a)$ ...(2)

Now, by the first mean-value theorem of differential calculus, which gives the first-order approximate value of $f(m)$ near the point a.

From (1) and (2), we obtain

$$0 = f(a) + (m - a) f'(a)$$

$$\Rightarrow \qquad m = a - \frac{f(a)}{f'(a)} \qquad ...(3)$$

Since (2) gives only an approximate value of $f(m)$, (3) gives only one approximate value of m (in terms of a).

Taking $a = x_0$ and $m = x_1$, in (3), we get

$$x_1 = x_0 - \frac{f(x_0)}{f'(x_0)} \qquad (4)$$

which gives the approximate value x_1 of m in terms of $a = x_0$. Here x_0 is the initial approximation of m and x_1 in the first-order approximation.

Replacing x_0 by x_1 and x_1 by x_2 in (4), we get the following expression for x_2, which is the second-order approximation of m.

$$x_2 = x_1 - \frac{f(x_1)}{f'(x_1)} \qquad (5)$$

Proceeding like this, we obtain successive approximations $x_3, x_4, \ldots$ of m.

Thus, the following in the formula for the $(k + 1)^{th}$-order approximation of m in terms of the kth-order approximation:

$$\boxed{x_{k+1} = x_k - \frac{f(x_k)}{f'(x_k)}} \qquad (6)$$

Formula (6) is an iterative formula that gives successive approximation for the root m of $f(x) = 0$, that lies near $x_0 = a$.

This formula is known as the Newton-Raphson formula or Newton's iterative formula.

Geometrical Meaning

Let $x = m$ be an exact root of $f(x) = 0$. The curve cuts the x-axis at the point $x = m$ which lies near the point $x = a$ on the x-axis.

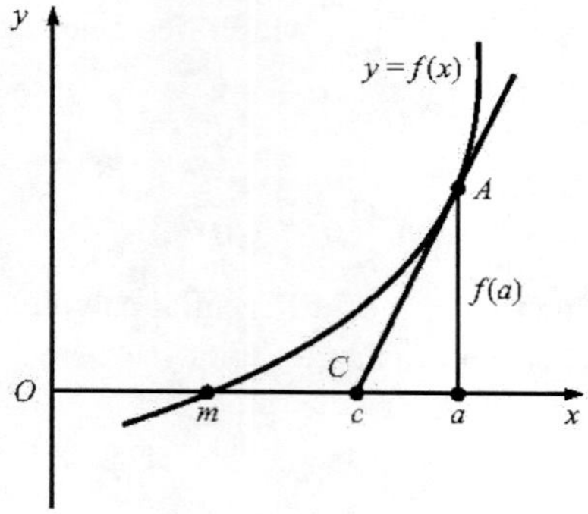

Fig. 10

Now, let 'a' be an approximate root of $f(x) = 0$. The ordinate of $x = a$ meets the curve at A $(a, f(a))$. The equation of the tangent of A to the curve is

$$y - f(a) = f'(a)\,(x - a) \tag{7}$$

This cuts the x-axis at the point C, where this tangent meets the x-axis, we have $y = 0$ and equation (7) yields

$$-f(a) = f'(a)\,(x - a)$$

$$\Rightarrow \qquad x - a = \frac{-f(a)}{f'(a)}$$

$$\therefore \qquad x = a - \frac{f(a)}{f'(a)} \tag{8}$$

This gives the x-coordinate 'c' of the point C.

Comparing the equations (3) and (8),

that is, $m = a - \dfrac{f(a)}{f'(a)}$ and $x = a - \dfrac{f(a)}{f'(a)}$ respectively

We find that 'm' determined by the equation (3) is the x-coordinate of the point where the tangent to the curve $y = f(x)$ at the point $(a, f(a))$ meets the x-axis.

Similarly, comparing equations (6) and (8),

that is, $x_{k+1} = x_k - \dfrac{f(x_k)}{f'(x_k)}$ and $x = a - \dfrac{f(a)}{f'(a)}$ respectively

We note that x_{k+1}, determined by the iterative formula (6), is the x-coordinate of the point where the tangent to the curve $y = f(x)$ at the point $(x_k, f(x_k))$ meets the x-axis. This is the geometrical meaning of the iterative formula (6).

Convergence

The Newton-Raphson iterative formula given by $x_{k+1} = x_k - \dfrac{f(x_k)}{f'(x_k)}$ is of the form

$$x_{n+1} = \phi(x_n), \text{ where } \phi(x) = x - \frac{f(x)}{f'(x)} \tag{9}$$

This shows that the Newton-Raphson iteration is a single point iteration.

Sequence of approximations $x_1, x_2, x_3, \ldots$ converges to the exact value if $|\phi'(x)| < 1$

that is, if $\left|1 - \dfrac{\{f'(x)\}^2 - f(x)f''(x)}{\{f'(x)\}^2}\right|$

that is, if $\left|\dfrac{f(x)f''(x)}{\{f'(x)\}^2}\right| < 1$

that is, if $|f(x)\,f''(x)| < \{f'(x)\}^2$ (10)

This is the criterion for the convergence. The interval containing m should be selected in which (10) is satisfied.

Order (or rate) of convergence of Newton-Raphson method

Let m be the root of $f(x) = 0$

Let e_k be the error at the kth stage of iteration.

Put $e_k = x_k - m$, $k = 0, 1, 2, \ldots$ where x_k is the approximate root at the kth iteration and $e_{k+1} = x_{k+1} - m$

$$= \left\{x_k - \frac{f(x_k)}{f'(x_k)}\right\} - m, \qquad \text{(using (6))}$$

$$= e_k - \frac{f(m+e_k)}{f'(m+e_k)} \qquad (\because x_k = e_k + m)$$

$$= e_k - \frac{f(m) + e_k f'(m) + \frac{1}{2!} e_k^2 f''(m) + \ldots}{f'(m) + e_k f''(m) + \frac{1}{2!} e_k^2 f'''(m) + \ldots} \tag{11}$$

(by Taylor's expansion)

Since $f(m) = 0$; $(\because\ m$ is a root of $f(x) = 0)$

If 'm' is not a repeated root, then $f'(m) \neq 0, f''(m) \neq 0, \ldots$ and the equation (11) yields

$$e_{k+1} = \frac{e_k[f'(m) + e_k f''(m) + \ldots] - [f(m) + e_k f'(m) + \frac{e_k^2}{2} f''(m) + \ldots]}{f'(m) + e_k f''(m) + \ldots}$$

$$e_{k+1} = \frac{\left[e_k^2 - \frac{e_k^2}{2}\right] f''(m) + \ldots}{f'(m)} \tag{12}$$

(Neglecting the higher degree terms in e_k and retaining only the leading terms in r.h.s of (12))

$$= \frac{e_k^2}{2} \cdot \frac{f''(m)}{f'(m)} = e_k^2 \left[\frac{f''(m)}{2f'(m)}\right]$$

Setting $\frac{f''(m)}{2f'(m)} = k$, we get

$$e_{k+1} = k e_k^2$$

$\therefore\ p = 2$, thus, the order of convergence of Newton-Raphson method is 2.

Limitations of the method

It has been assumed that 'm' is not a repeated root of $f(x) = 0$, in the derivation of equation (12): $\boxed{e_{k+1} = \frac{1}{2} e_k^2 \frac{f''(m)}{f'(m)}}$.

If 'm' is a repeated root, then $f'(m) = 0$ and the equation (12) is not valid and $x = m$ is the inequality given by the equation (10): $| f(x) f''(x) | < \{f'(x)\}^2$ does not hold good. Therefore, the convergence of iteration is not guaranteed. Thus, the Newton-Raphson formula (6) i.e, $\boxed{x_{k+1} = x_k - \frac{f(x_k)}{f'(x_k)}}$ is not applicable to find approximate values of a repeated root. The following situation where the formula (6) cannot be applied with 'a' as the initial approximation is explained.

Let $f(x) = 0$ has two roots at $a \pm h$ where h is infinitesimally small. Then as $h \to 0$, the two roots nearly equal, and $f'(a)$ becomes zero in a nbd of a. So, in this case, the Newton-Raphson formula (6) cannot be used with $x_0 = a$ as the initial approximations.

Example 60. Find the positive root of

$$x^4 - x - 10 = 0$$

Solution:

Here, $f(x) = x^4 - x - 10 = 0$

and $f'(x) = 4x^3 - 1$

Therefore, the Newton-Raphson iterative formula

$$x_{k+1} = x_k - \frac{f(x_k)}{f'(x_k)}$$

$$= x_k - \frac{x_x^4 - x_k - 10}{4x_k^3 - 1}$$

$$= \frac{3x_x^4 + 10}{4x_k^3 - 1} \qquad k = 0, 1, 2, \ldots \qquad (1)$$

By trial, we find:

$f(1) < 0$ and $f(2) > 0$

Hence, there is a root between 1 and 2. We take as our starting approximation

$x_0 = 1.5$ and compute $x_1 = \dfrac{3(15)^4 + 10}{4(15)^3 - 1} = 2.015.$

Similarly, we compute $x_2 = 1.8741$,

$$x_3 = 1.8558,\ x_4 = 1.8556,\ x_5 = x_4.$$

We now stop the iteration. The root correct to four decimal places is 1.8556.

Example 61. Apply Newton-Raphson method to find an approximate root, correct to three decimal places, of the equation $x^3 - 3x - 5 = 0$, which lies near $x = 2$.

Solution:

Here $f(x) = x^3 - 3x - 5$

$$f'(x) = 3\ (x^2 - 1)$$

The iteration formula is

$$x_{k+1} = x_k - \frac{f(x_k)}{f'(x_k)}$$

$$= x_k - \frac{x_k^3 - 3x_k - 5}{3(x_k^2 - 1)}$$

$$= \frac{2x_k^3 + 5}{3(x_k^2 - 1)}, \qquad k = 0, 1, 2,... \qquad (1)$$

To find the root near $x = 2$.

We take $x_0 = 2$. Then (1) gives

$$x_1 = \frac{2x_0^3 + 5}{3(x_0^2 - 1)} = \frac{16 + 5}{3(4 - 1)} = \frac{21}{9} = 2.3333$$

$$x_2 = \frac{2x_1^3 + 5}{3(x_1^2 - 1)} = \frac{2 \times (2.3333)^3 + 5}{3\{(2.3333)^2 - 1\}} = 2.2806$$

$$x_3 = \frac{2x_2^3 + 5}{3(x_2^2 - 1)} = \frac{2 \times (2.2806)^3 + 5}{3\{(2.2806)^2 - 1\}} = 2.2790$$

$$x_4 = \frac{2 \times (2.2790)^3 + 5}{3\{(2.2790)^2 - 1\}} = 2.2790$$

Since $x_3 = x_4$ upto 3 decimals.

We take $x_4 = 2.279$ as the required approximate root correct to 3 places of decimal.

Example 62. Find the positive root of $x - 2 \sin x = 0$.

Solution:

Here, $\qquad f(x) = x - 2 \sin x;$

and $\qquad f'(x) = 1 - 2 \cos x$

$$\therefore \qquad x_{k+1} = x_k - \frac{f(x_k)}{f'(x_k)}, \quad k = 0, 1, 2 ...$$

$$= x_k - \frac{x_k - 2\sin x_k}{1 - 2\cos x_k}$$

$$= \frac{2(\sin x_k - x_k \cos x_k)}{1 - 2\cos x_k}$$

$$x_{k+1} = \frac{\sin x_k - x_k \cos x_k}{05 - \cos x_k} \qquad (1)$$

With $x_0 = \frac{\pi}{2}$, we get $x_1 = 2$, $x_2 = 1.9010$, $x_3 = 1.8955$, $x_4 = x_3$. The root is 1.8955.

Example 63. Using the Newton-Raphson method find the root of the equation $f(x) = e^x - 3x = 0$ that lies between 0 and 1.

Solution:

Now, $f(x) = e^x - 3x$

and $f'(x) = e^x - 3$

Therefore, the Newton-Raphson formula is given by $x_{k+1} = x_k - \dfrac{e^{x_k} - 3x_k}{e^{x_k} - 3}$

$$= \frac{(x_k - 1)e^{x_k}}{(e^{x_k} - 3)}, \qquad k = 0,\ 1,\ 2\ \ldots \tag{1}$$

Since the required root is supposed to lie between 0 and 1, we take x_0 to be the average of 0 and 1; that is $x_0 = 0.5$.
Then (1) yields

$$x_1 = \frac{(0.5 - 1)e^{0.5}}{(e^{0.5} - 3)} = 0.61006$$

$$x_2 = \frac{(0.61006 - 1)\,e^{0.61006}}{e^{0.61006} - 3} = 0.618996$$

$$x_3 = \frac{(0.618996 - 1)\,e^{0.618996}}{e^{0.618996} - 3} = 0.619061$$

$$x_4 = \frac{(0.619061 - 1)\,e^{0.619061}}{e^{0.619061} - 3} = 0.619061$$

We find $x_3 = x_4$. We therefore take $m \approx 0.619061$ as an approximate root of the given equation.

Example 64. Find the complex root of $2x^2 - 2x + 13 = 0$ starting with $x_0 = 3i$.

Solution:

Here, $f(x) = 2x^2 - 2x + 13,\ f'(x) = 4x - 2$

$$x_{k+1} = x_k - \frac{2x_k^2 - 2x_k + 13}{4x_k - 2}$$

$$= \frac{2x_k^2 - 13}{4x_k - 2}$$

Hence, $$x_{k+1} = \frac{x_k^2 - 6.5}{2x_k - 1}$$

with $x_0 = 3i$, we find

$$x_1 = \frac{(3i)^3 - 6.5}{2(3i) - 1} = \frac{15.5}{1 - 6i} = \frac{15.5}{37}(1 + 6i)$$

$$= 0.4189 + 2.5135i$$

Similarly, we get

$x_2 = 0.4995 + 2.5135i$ which is very close to the exact root $\frac{1}{2}(1 + 5i)$.

Example 65. Using the Newton-Raphson method find the root that lies near $x = 4.5$ of the equation $\tan x = x$, correct to four decimal places. (Here x is measured in radians.)

Solution:

Let $f(x) = \tan x - x$

$\therefore$ $f'(x) = \sec_x^2 - 1$

$\therefore$ The Newton-Raphson formula is:

$$x_{k+1} = x_k - \frac{(\tan x_k) - x_k}{(\sec^2 x_k) - 1}$$

$$= \frac{x_k \sec^2 x_k - \tan x_k}{(\sec^2 x_k) - 1}$$

$$= \frac{x_k - \frac{1}{2}\sin 2x_k}{\sin^2 x_k} \quad (1)$$

Since the root lies near $x = 4.5$,
We take $x_0 = 4.5$, then (1) yields

$$x_1 = \frac{x_0 - \frac{1}{2}\sin 2x_0}{\sin^2 x_0} = \frac{4.5 - \frac{1}{2}\sin 9}{\sin^2 4.5} = 4.49361$$

$$x_2 = \frac{4.49361 - \frac{1}{2}\sin(8.98722)}{\sin^2(4.49361)} = 4.49341$$

$$x_3 = \frac{4.49341 - \frac{1}{2}\sin(8.98682)}{\sin^2(4.49341)} = 4.49341$$

Since $x_3 = x_4$

So, we take $x_4 \approx 4.4934$ (correct to four decimals) as the required root.

Example 66. Find the roots of the equation $2e^{-x} = \frac{1}{x+2} + \frac{1}{x+1}$ which has two roots > -1. Find these roots correct to 5 decimal places.

Solution:

Let $$f(x) = 2e^{-x} - \frac{1}{x+2} - \frac{1}{x+1} \text{ we find that}$$

$$f(-0.8) = -1.38,\ f(0) = 0.5$$
$$f(1.0) = -0.0976$$

$\Rightarrow$ roots greater than -1 lie in (0.8, 0) and (0, 1)

$\therefore$ By Newton-Raphson formula, we have

$$\left\{x_{k+1} = x_k - \frac{f(x_k)}{f'(x_k)},\ k = 0,1,2\ldots\right\}\left\{\begin{aligned} f(x) &= 2e^{-x} - \frac{1}{x+2} - \frac{1}{x+1} \\ f'(x) &= -2e^{-x} + \frac{1}{(x+2)^2} + \frac{1}{(x+1)^2} \end{aligned}\right\}$$

First root

Let $x_0 = -0.6$, be an approximate root of $f(x) = 0$

yielding $$\left.\begin{aligned} x_1 &= -0.737984, \\ x_2 &= -0.699338, \\ x_3 &= -0.689753, \\ x_4 &= -0.689753 \end{aligned}\right\} \Rightarrow \begin{array}{l} -0.68975 \text{ is the required root} \\ \text{correct to 5 places} \end{array}$$

Second root

Let $x_0 = 0.8$, be an approximate root to $f(x) = 0$

yielding $$x_1 = 0.769641,$$
$$x_2 = 0.770092,$$
$$x_3 = 0.770091$$

$\therefore$ The required root correct to 5 decimals is 0.77009.

Example 67. Find the real positive root of $3x - \cos x - 1 = 0$ by Newton's method correct to 6 decimal places.

Solution: Let $$f(x) = 3x - \cos x - 1;\ f(0) = -2 = -\text{ve}$$
$$f(1) = 3 - \cos 1 - 1 = 2 - \cos 1 = +\text{ve}$$

$\therefore$ a root lies between 0 and 1

$$f'(x) = 3 + \sin x$$

Therefore, the Newton-Raphson formula is:

$$x_{k+1} = x_k - \frac{3x_k - \cos x_k - 1}{3 + \sin x_k}$$

$$= \frac{x_k \sin x_k + \cos x_k + 1}{3 + \sin x_k}, \quad k = 0, 1, 2, \ldots \qquad (1)$$

Now take $x_0 = 0.5$

$\therefore$
$$x_1 = \frac{0.5(0.5) + \cos(0.5) + 1}{3 + \sin(0.5)}$$

$$= \frac{2.11729533}{3.479425539}$$

First approximation: $x_1 = 0.608518649$

$$x_2 = \frac{0.608518649 \times \sin(0.608518649) + \cos(0.608518649) + 1}{3 + \sin(0.608518649)}$$

$$= \frac{2.16835703}{3.57165265} = 0.607101878$$

Second approximation: $x_2 = 0.607101878$

$$x_3 = \frac{0.607101878 \times \sin(0.607101878) + \cos(0.607101878) + 1}{3 + \sin(0.607101878)}$$

$$= \frac{2.16765013}{3.5748962} = 0.607101648$$

$\therefore$ The root is 0.607102 correct to six places of decimal.

Example 68. By using the Newton-Raphson method find a root of the equation $1 - \cosh x \cdot \cos x = 0$, correct to four decimal places. Take $x_0 = 0.15$.

Solution:

Let $f(x) = 1 - \cosh x \cdot \cos x$,

and $f'(x) = \cosh x \cdot \sin x - \sinh x \cdot \cos x$

$\therefore$ The Newton-Raphson formula is

$$x_{k+1} = x_k - \frac{1 - (\cosh x_k)(\cos x_k)}{(\cosh x_k)(\sin x_k) - (\sinh x_k)(\cos x_k)}$$

Take $x_0 = 0.15$, in this formula, we get

$$x_1 = x_0 - \frac{1 - (\cosh x_0)(\cos x_0)}{(\cosh x_0)(\sin x_0) - (\sinh x_0)(\cos x_0)}$$

$$= 0.15 - \frac{1-(\cosh 0.15)(\cos 0.15)}{(\cosh 0.15)(\sin 0.15)-(\sinh 0.15)(\cos 0.15)}$$

$$\boxed{x_1 = 0.07382358}$$

$$x_2 = x_1 - \frac{1-(\cosh x_1)(\cos x_1)}{(\cosh x_1)(\sin x_1)-(\sinh x_1)(\cos x_1)}$$

$$\boxed{x_2 = 0.03628339}$$

$$x_3 = x_2 - \frac{1-(\cosh x_2)(\cos x_2)}{(\cosh x_2)(\sin x_2)-(\sinh x_2)(\cos x_2)}$$

$$\boxed{x_3 = 0.0178269}$$

Proceeding like this, we obtain:

$$\boxed{x_4 = 0.08758104,} \qquad \boxed{x_5 = 0.00430265}$$

$$\boxed{x_6 = 0.0021137805,} \qquad \boxed{x_7 = 0.001038444}$$

$$\boxed{x_8 = 0.000510116,} \qquad \boxed{x_9 = 0.000250628}$$

$$\boxed{x_{10} = 0.00012514} \qquad \boxed{x_{11} = 0.00006147709}$$

We observe that $x_{11} = x_{10} = 0.0001$, correct to four decimal places. Hence 0.0001 is taken as the required root.

Note: zero is an exact root of the given equation.

Example 69. Solve $3x + \sin x - e^x = 0$ correct to 4 decimal places using Newton Raphson method.

Solution: Let $f(x) = 3x + \sin x - e^x$

such that $f'(x) = 3 + \cos x - e^x$

Hence,

$$x_{k+1} = x_k - \frac{f(x_k)}{f'(x_k)}, \qquad k = 0, 1, 2 \ldots$$

$$x_{k+1} = x_k - \frac{3x_k + \sin(x_k) - e^{x_k}}{3 + \cos(x_k) - e^{x_k}}$$

$$= \frac{x_k \cos x_k + (1-x)e^{x_k} - \sin x_k}{3 + \cos x_k - e^{x_k}}$$

With $k = 0$, i.e. $x_0 = 0$, we get in succession

$$x_1 = 0.3333$$
$$x_2 = 0.3602$$
$$x_3 = 0.3604$$
$$x_4 = x_3$$

The root is 0.3604.

Example 70. Apply Newton-Raphson method to solve the equation $\log x = \cos x$ to 3 decimal places.

Solution:

$$\left.\begin{array}{l} \text{Let } f(x) = \log x - \cos x, \\ \text{so that } f'(x) = \dfrac{1}{x} + \sin x \end{array}\right\} \begin{array}{l} \text{since } f(1.3) = -0.00513 < 0 \\ \text{and } f(1.4) = 0.1665 > 0 \end{array}$$

The Newton-Raphson formula becomes

$$x_{k+1} = x_k - \frac{\log x_k - \cos x_k}{\dfrac{1}{x_k} + \sin(x_k)}$$

Take $x_0 = 1.3$ in the formula, we get

$$x_1 = x_0 - \frac{\log(x_0) - \cos(x_0)}{\dfrac{1}{x_0} + \sin(x_0)}$$

$$= 1.3 - \frac{(1.3)\log(1.3) - (1.3)(\cos 1.3^c)}{1 + (1.3)(\sin(1.3^c))}$$

$$= 1.3 - \left(\frac{-0.00667}{2.2526}\right)$$

$$= 1.3 + 0.0029 = 1.3029$$

$$x_2 = 1.3029 - \frac{(1.3029)\log(1.3029) - (1.3029)\cos(1.3029^c)}{1 + (1.3029)\sin(1.3029^c)}$$

$$= 1.3029 - \frac{-0.0001444}{1.99606}$$

$$= 1.3029 + 0.000072 = 1.302972$$

Since $x_1 = x_3$ (correct to 3 places of decimals)

$\therefore$ The required root is 1.302

Example 71. Using the Newton-Raphson method, find the real root of the equation $x \sin x + \cos x = 0$ near $x = \pi$.
Carry out four iterations. (Here, x is in radians.)

Solution:

Let $f(x) = x \sin x + \cos x$;

so that $f'(x) = x \cos x + \sin x - \sin x = x \cos x$;

Take $x_0 = \pi$, in the Newton-Raphson formula,

$$x_{k+1} = x_k - \frac{f(x_k)}{f'(x_k)}, \quad k = 0, 1, 2, \ldots$$

$$\therefore \quad x_1 = x_0 - \frac{f(x_0)}{f'(x_0)} = \pi - \frac{f(\pi)}{f'(\pi)}$$

$$= \pi - \frac{(\pi \sin \pi + \cos \pi)}{\pi \cos \pi}$$

$$\boxed{x_1 = \pi - \frac{1}{\pi} = 2.8233.}$$

$$x_2 = 2.8233 - \frac{(2.8233)(\sin(2.8233)) + \cos(2.8233)}{(2.8233)(\cos(2.8233))}$$

$$\boxed{x_2 = 2.7986}$$

$$x_3 = 2.7986 - \frac{(2.7986)(\sin(2.7986)) + (\cos(2.7986))}{(2.7986)(\cos(2.7986))}$$

$$\boxed{x_3 = 2.7984}$$

$$x_4 = 2.7984 - \frac{(2.7984)(\sin(2.7984)) + (\cos(2.7984))}{(2.7984)(\cos(2.7984))}$$

$$\boxed{x_4 = 2.7984}$$

Thus, $x_3 = x_4$, the required root is 2.7984 to 4 places of decimal.

Example 72. Newton-Raphson method for solving the equation $f(x) = c$, where c is a real valued constant, is applied to the function

$$f(x) = \begin{cases} \cos x & \text{when } |x| \leq 1 \\ \cos x + (x^2 - 1)^2 & \text{when } |x| \geq 1 \end{cases}$$

For which c is $\quad x_1 = (-1)^i$,

when $x_0 = 1$ and the calculation is carried out with no error?

Solution:

Let $$f(x) - c = f(x)$$

$\therefore$ By Newton-Raphson formula, we have:

$$x_{i+1} = x_i - \frac{f(x_i) - c}{f'(x_i)}, \; i = 0, 1, \ldots$$

$$\left(i = 0;\ x_1 = 1 - \frac{\cos 1 - c}{-\sin 1} = -1\right) \quad (i = k, \text{ in the formula used here})$$

$\Rightarrow$ $c = \cos 1 + 2 \sin 1$, with this value of c,

we obtain $$x_2 = 1,\ x_3 = -1, \ldots$$

$\Rightarrow$ $$x_i = (-1)^i$$

Example 73. Verify that the equation

$$e^x = 1 + x + \frac{x^2}{2} + \frac{x^3}{6} e^{0.3x}$$

has a root between 2 and 3. Find this root correct to 3 decimal places.

Solution: Let $f(x) = e^x - 1 - x - \frac{x^2}{2} - \frac{x^3}{6} e^{0.3x}$ (1)

since $f(2) = -0.4040$, $f(3) = 0.5173$;

$\Rightarrow$ the root lies between 2 and 3

Hence, let us take $x_0 = 2.5$ (which is the average of 2 and 3)

We find that

$$f'(x) = e^x - 1 - x - \frac{e^{0.3x}}{6}(3x^2 + 0.3x^3) \qquad (2)$$

With $x_0 = 2.5$, in the Newton-Raphson formula,

$$x_{k+1} = x_k - \frac{f(x_k)}{f'(x_k)}, \; k = 0, 1, 2 \ldots$$

We obtain the following successive approximations for the desired root.

$$x_1 = 2.461326,\ x_2 = 2.379358,\ x_3 = 2.363884,$$

$$x_4 = 2.363377,\ x_5 = 2.363376$$

since, $x_4 = x_5$ correct to five decimal places,

$\therefore$ 2.36338 is the required root.

Or, the required root correct to 5 places is $2.363376 \pm 0.5 \times 10^{-6}$

Example 74. Apply Newton-Raphson method to find the root of $2x \tan x - 1 = 0$ applying between 0.6 and 0.72.

Solution:

Given: $f(x) = 2x \tan x - 1$

so that $f'(x) = 2 \tan x + 2x \sec^2 x$

$= 2[\tan x + x(1 + \tan^2 x)]$ etc

Example 75. Find a real root of $x^2 + 4 \sin x = 0$ correct to 4 places of decimal by using Newton-Raphson method.

Solution:

$$\left\{\begin{array}{ll}\text{Let} & f(x) = x^2 + 4 \sin x; \\ \therefore & f'(x) = 2x + 4 \cos x,\end{array}\right\}$$

Let $x_0 = -1.9$, be an approximation to the root (since the term x^2 is positive of all real values of x so it is obvious that the equation will be satisfied only by a negative value of x).

$\therefore$ By Newton-Raphson formula, we have

First approximation: x_1

$$x_1 = x_0 - \frac{f(x_0)}{f'(x_0)}$$

$$= x - \frac{x_0^2 + 4\sin x_0}{2x_0 + 4\cos x_0}$$

$$= -1.9 - \frac{(-1.9)^2 + 4\sin(-1.9)}{2(-1.9) + 4\cos(-1.9)}$$

$$= -1.9 - \frac{3.61 - 3.78}{-3.8 - 1.293} = -1.93$$

Second approximation: x_2

$$x_2 = x_1 - \frac{x_1^2 + 4\sin x_1}{2x_1 + 4\cos x_1} = -1.93 - \frac{(-1.93)^2 + 4\sin(-1.93)}{2(-1.93) + 4\cos(-1.93)}$$

$$= -1.93 - \frac{0.0198}{5.266} = -1.93 - 0.0038 = -1.9338$$

$$\boxed{x_2 = -1.9338}$$

EXERCISES

I. By using Newton-Raphson method, find the real roots of the following equations, correct to 3 decimal places.

(1) $4x^3 - 12.3x^2 - x + 16.2 = 0$ near 1 (Ans: 1.550)
(2) $x^4 - 3x^3 + 2x^2 + 2x - 7 = 0$ near 2 (Ans: 2.327)
(3) $x^3 - 8x - 4 = 0$ near 3 (Ans: 3.051)
(4) $x^3 - 5x + 3 = 0$ near 2 (Ans: 1.834)
(5) $x^3 - 2x - 5 = 0$ near 2 (Ans: 2.095)

II. Using Newton-Raphson method compute the roots of the following equations correct to four decimal places:

(1) $x = \cos x, x > 0.$ (Ans: 0.7391)
(2) $e^x + x - 2 = 0.$ (Ans : 0.4492)
(3) $2x = \tan x, x > 0.$ (Ans: 1.1656)
(4) $xe^x = 1.$ (Ans: 0.5671)
(5) $x = \ln (3 - x).$ (Ans: 0.7921)
(6) $\sin x = \ln x.$ (Ans. 2.2219)
(7) $e^x = 2x^2, x < 0.$ (Ans: –0.5398)
(8) $e^x = 3x^2, x > 0.$ (Ans: 0.9100)
(9) $e^x = 3x^2, x < 0.$ (Ans: –0.4590)
(10) $e^{-x} = \sin x.$ (Ans: 0.5885)
(11) $e^{-x} = \cos x.$ (Ans: 1.2927)
(12) $x \sin^2 x = 1.$ (Ans: 1.1748)
(13) $e^{\sin} x = 4 \ln x.$ (Ans: 1.9030)
(14) $x^3 - x - 1 = 0, x > 0.$ (Ans: 1.3247)
(15) $e^x = x^3 + \cos 2\pi x$ near 4.5. (Ans: 4.5037)
(16) $x^3 - x - 3 = 0, x > 0.$ (Ans: 1.6717)
(17) $2x^3 + 3x - 10 = 0.$ (Ans: 1.4209)
(18) $x^3 - x - 3.875 = 0, x > 0.$ (Ans: 1.7818)
(19) $x^3 - x^2 + 20x + 10 = 0, x < 0.$ (Ans: –0.4827)
(20) $4x^4 - 9x^3 + 1 = 0$ near 2.2. (Ans: 2.2274)
(21) $2x^6 - 3x^5 - 5x^3 + 6x^3 - 5x - 3 = 0$ near 2. (Ans: 1.9689)
(22) $x^3 - 3x^2 - x + 9 = 0, x < 0.$ (Ans: –1.5251)
(23) $x^3 - x - 1 = 0$ near 1.5 (Ans: 1.3258)
(24) $x^3 - 6x + 4 = 0$ near 0.5 (Ans: 0.7321)
(25) $x^4 - x - 13 = 0$ near 2 (Ans: 1.9670)
(26) $4x^4 - x - 8 = 0$ near 1 (Ans: 1.2326)
(27) $x \sin x = 1$ near 1 (Ans: 1.1142)
(28) $x^2 = 4 \sin x$ near 2 (Ans: 1.9338)

(29) $3x = \cos x + 1$ near 0.5 (Ans: 0.6071)

(30) $x = \sin x + \frac{1}{2}$ near 1.5 (Ans: 1.4973)

(31) $x^2 + x = \cos x$ near 0.5 (Ans: 0.5500)

(32) $x \tan x = \frac{1}{2}$ near 0.5 (Ans: 0.6533)

(33) $e^x = 5x$ near 0.4 (Ans: 0.2592)

(34) $e^x - 1 = 2x$ near 1.5 (Ans: 1.2564)

(35) $e^{-x} = \sin x$ near 1 (Ans: 0.5885)

(36) $x \log_{10} x = 4.7772393$ near 6 (Ans: 6.0891)

(37) $\log_{10} x = 0.05x$ near 30 (Ans: 29.3531)

(38) $\log x = \cos x$ near 1.5 (Ans: 1.3030)

(39) $\cos x = x\, e^x$ near 0.5 (Ans: 0.518)

(40) $x \sin x + \cos x = 0$ near π (Ans: 2.798)

(41) $\cos x = x^2$ near 1 (Ans: 0.824)

(42) $e^x = 4x$ near 2 (Ans: 2.153)

(43) $x = e^{-x}$ near 0.5 (Ans: 0.567)

(44) $e^x = x^3 + \cos 25x$ near 4 (Ans: 4.545)

(45) $x \log_{10} x = 1.2$ near 2.5 (Ans: 2.741)

(46) $\log x - x + 3 = 0$ near 0.1 (Ans: 0.052)

III. By Newton-Raphson method, find the real roots of the following equations, correct to 5 decimal places:

(1) $x \tan x = 1.28$ near 1 (Ans: 0.5885)

(2) $2x - 3 \sin x = 5$ near 3 (Ans: 2.88324)

(3) $e^{-x} = -1$ near 0 (Ans: 0.56714)

(4) $x - 0.2 \sin x = 0.5$ near 0.5 (Ans: 0.61547)

(5) $\cos x - x^2 - x = 0$ near -1.5 (Ans: –1.25115)

IV. Find the roots, correct to five decimal places, of equation

$$2e^{-x} = \frac{1}{x+2} + \frac{1}{x+1} \qquad \text{(Ans: } -0.68975,\ 0.77009)$$

that are near $x = -0.6$ and $x = 0.8$.

V. Verify that the equation $f(x) = 0$, where

$$f(x) = \cos\left\{\frac{\pi}{8}(x+1)\right\} + 0.148\,x - 0.9062,$$

has a root in the interval $(-1, 0)$. Find this root correct to 4 decimal places.

VI. Verify that the equation $xe^{-x^2} = 1/10$ has a root in the interval (0, 1). Find this root of 6 decimal places by taking $x_0 = 0.1$. (Ans: 0.101026)

$x^3 + x^2 - 2 = 0$, complex root. (Ans: $1 \pm i$)

$x^3 + 2x^2 + 3.61x + 7.22 = 0$, complex root. (Ans: $\pm 1.9i$)

VII (a) Find the double root near 3 of the equation

$x^3 - 6.1x^2 + 8.99x + 0.961 = 0$. (Ans: 3.1)

(b) There is a triple root near 1 for the equation

$x^4 - 2.2x^3 + 2.662x + 1.4641 = 0$. Find it. (Ans: 1.1)

(c) Find the double root of

$x^3 - 1.1x^2 - 1.21x + 1.331 = 0$. (Ans: 1.1)

(d) Find the double root of

$x^3 - 3.9x^2 + 3.99x - 0.361 = 0$. (Ans: 1.9)

(e) Find the point of intersection of the curves.

$y = e^x - 2$ and $y = \ln(x + 2)$. (Ans: 1.1462)

(f) Find abscissa of the inflection point of the curve

$y = e^{-x} \ln x$ to the four correct decimals. (Ans: 2.5524)

VIII (1) Apply Newton-Raphson method to find an approximate solution of the equation $e^x - 3^x = 0$ correct upto three significant figures (assume $x = 0.4$ as an approximate root of the equation). (Ans: 0.619)

(2) Determine approximately the root of the equation $x + \log_{10} x = 3.375$ correct to two significant figures. (Ans: 2.911)

(3) Determine approximately the smallest positive root of the equation $x^2 + 2x - 2 = 0$, correct to two significant figures using Newton-Raphson method. (Ans: 0.7482)

(4) Design a Newton-Raphson iteration to compute the cube-root of a positive number, C. Perform two iterations of this method to compute $(2)^{1/3}$ starting from $x_0 = 1$. (Ans: 1.264)

(5) A root of the equation $e^x = 1 + x + \dfrac{x^2}{2} + \dfrac{x^2 e^{0.3x}}{6}$ is close to 2.5. Find this root to three decimal places, using Newton-Raphson method. (Ans: 2.364)

A GENERALISED NEWTON'S METHOD

Multiple Roots

A root 'm' is said to be a multiple root of equation $f(x) = 0$ if 'm' occurs more than once among the list of roots of $f(x) = 0$. If 'm' is repeated 'r' times, then m is said to be a root of multiplicity 'v' and $f(x) = (n - m)^v \phi(x)$ where $\phi(m) \neq 0$.

Let us suppose that 'm' is a multiple root of order p of the equation $f(x) = 0$. Then m is a multiple root of order $(p - 1)$ of $f'(x) = 0$.

Hence, if
$$g(x) = \frac{f(x)}{f'(x)}$$

Then m is a simple root of $g(x) = 0$. Applying the Newton-Raphson formula to the function $g(x)$, we get

$$x_{k+1} = x_k - \frac{g(x_k)}{g'(x_k)}$$

Remembering that $g'(x) = \dfrac{\{f'(x)^2\} - f(x) f''(x)}{\{f'(x)\}^2}$

We obtain
$$x_{k+1} = x_k - \frac{f_k f'_k}{(f'_k)^2 - f_k f''_k} \tag{1}$$

This formula (1) known as the deflated Newton-Raphson formula is employed to find successive approximations of the multiple root m of $f(x) = 0$.

Note: Since formula (1) is free from 'r' the order of multiplicity of the root, so, this formula may be applied even when the order of multiplicity of the root is not known.

> If m is a root of multiplicity p, (i.e. m is a repeated root, repeated p times), then the N.R. formula:
>
> $$x_{n+1} = x_n - p\frac{f(x_n)}{f'(x_n)}$$
>
> $\Rightarrow \dfrac{1}{p}f'(x_n)$ is the slope of the line through (x_n, y_n) and meeting the x-axis at $(x_{n+1}, 0)$.
>
> We are given that m is a root of $f(x) = 0$ with multiplicity p.
>
> $\Rightarrow$ It is a root of $f'(x) = 0$, with multiplicity $(p - 1)$, also a root of $f''(x) = 0$ with multiplicity $(p - 2)$ and so on.
>
> $$\therefore \quad x_0 - p\frac{f(x)}{f'(x)},\ x_0 - (p-1)\frac{f'(x_0)}{f''(x_0)},\ x_0 - (p-2)\frac{f''(x_0)}{f'''(x_0)}$$
>
> will all have the same value provided the initial approximation 'x_0' is selected very close to the actual root m.

Use of Synthetic Division Scheme

The Newton-Raphson formula is given by

$$x_{k+1} = x_k - \frac{f(x_k)}{f'(x_k)}, \quad k = 0, 1, 2, \ldots$$

For the sake of brevity, we take $f_k = f(x_k)$; $f'_k = f'(x_k)$ and soon

to find $\frac{f_k}{f'_k}$, we use Synthetic division process as a labour saving device. When the Newton-Raphson method by use of synthetic division process, also called BIRGE-VEITA method.

1. Use the Birge-Vieta method to find a real root correct to three decimals of the following equation.

$$x^6 - x^4 - x^3 - 1 = 0, \; x_0 = 1.5$$

Perform 3 iterations

First Iteration: (x_1), take $x_0 = 1.5$

1.5	1	0	–1	–1	0	0	–1
		1.5	2.25	1.875	1.3125	1.9688	2.9532
	1	1.5	1.25	0.875	1.3125	1.9688	1.9532
		1.5	4.5	8.625	14.25	23.3438	
	1	3	5.75	9.5	15.5625	25.3126	

$$x_1 = 1.5 - \frac{1.9532}{25.3126} = 1.4228$$

Second Iteration: (x_2)

1.4228	1	0	–1	–1	0	0	–1
		1.4228	2.0244	1.4575	0.6509	0.9261	1.3177
	1	1.4228	1.0244	0.4575	0.6509	0.9261	0.3177
		1.4228	4.0487	7.2180	10.9207	16.4641	
	1	2.8456	5.0731	7.6755	11.5716	17.3902	

$\therefore$

$$x_2 = 1.4228 - \frac{0.3177}{17.3902} = 1.4045$$

Third Iteration: (x_3)

1.4045	1	0	−1	−1	0	0	−1
		1.4045	1.9726	1.3660	0.5140	0.7219	1.0139
	1	1.4045	0.9726	0.3660	0.5140	0.7219	0.0139
		1.4045	3.9452	6.9071	10.2151	15.0690	
	1	2.8090	4.9178	7.2731	10.7291	15.7909	

$$x_3 = 1.4045 - \frac{0.0139}{15.7909} = 1.4036$$

The root correct to three decimals is 1.404

2. Use Birge-Vieta method to solve correct to 3 decimal places

$$x^5 - x + 1 = 0, \; x_0 = -1.5$$

Solution: **First Iteration:** $x_0 = -1.5$

−1.5	1	0	0	0	−1	1
		−1.5	2.25	−3.375	5.026	−6.0938
	1	−1.5	2.25	−3.375	4.0625	−5.0938
	0	−1.5	4.5	−10.125	20.25	
	1	−3	6.75	−13.5	24.3125	

$$x_1 = -1.5 + \frac{5.0938}{24.3125} = -1.2905$$

Second Iteration: $x_1 = -1.2905$

−1.2905	1	0	0	0	−1	1
		−1.2905	1.6654	−2.1492	2.7735	−2.2887
	1	−1.2905	1.6654	−2.1492	1.7735	−1.2887
		−1.2905	3.3308	−6.4476	11.0941	
	1	−2.5810	4.9962	−8.5968	12.8676	

$$x_2 = -1.2905 + \frac{1.2887}{12.8676} = -1.1903.$$

Third Iteration

−1.1903	1	0	0	0	−1	1
		−1.1903	1.4168	−1.6864	2.0073	−1.1990
	1	−1.1903	1.4168	−1.6864	1.0073	−0.1990
		−1.1903	2.8336	−5.0593	8.0294	
	1	−2.3806	4.2504	−6.7457	9.0367	

$$x_3 = -1.1903 + \frac{0.1990}{9.0367} = -1.1683$$

Fourth Iteration

−1.1683	1	0	0	0	−1	1
		−1.1683	1.3649	−1.5946	1.8630	−1.0082
	1	−1.1683	1.3649	−1.5946	0.8630	−0.0082
		−1.1683	2.7298	−4.7838	7.4519	
	1	−2.3366	4.0947	−6.3784	8.3149	

$$x_4 = -1.1683 + \frac{0.0082}{8.3149}$$

$$= -1.1673$$

The root correct to three decimals is $\boxed{-1.167}$

Some deductions from Newton-Raphson formula:

***D*₁:** **Reciprocal of a number**

Let $x = \frac{1}{N}$, or $N = \frac{1}{x} (N \neq 0)$

Let $f(x) = N - \frac{1}{x}$

$\Rightarrow$ $f'(x) = -x^{-2}$

Hence, we take, $x_{n+1} = x_n - \frac{f(x_n)}{f'(x_n)}$

$$= x_n + \frac{x_n^{-1} - N}{x_n^{-2}}$$

$$\left\{\boxed{x_{n+1} = x_n(2 - Nx_n)}\right\} \tag{1}$$

***D*₂: Square root of a number**

Let $\quad x^2 = N$ or, $x^2 - N = 0$

$$f(x) = x^2 - N \quad \Rightarrow \quad f'(x) = 2x$$

$\therefore$ $$x_{n+1} = x_n - \frac{x_n^2 - N}{2x_n}$$

$$= \frac{1}{2}\left(x_n + \frac{N}{x_n}\right)$$

$\therefore$ $$\left\{\boxed{x_{n+1} = \frac{1}{2}\left[x_n + \frac{N}{x_n}\right]}\right\} \tag{2}$$

This is known as HERO's formula for square roots.

D*₃: The *p*th root of a number *N

It is the root of $\quad x^p - N = 0$

Let $\quad f(x) = x^p - N$

$\Rightarrow \quad f'(x) = px^{p-1}$

$$x_{n+1} = x_n - \frac{x_n^p - N}{px_n^{p-1}}$$

$$= \frac{(p-1)x_n^p + N}{px_n^{p-1}}$$

$$\left\{x_{n+1} = \frac{1}{p}\left[(p-1)\,x_n + \frac{N}{x_n^{p-1}}\right]\right\} \tag{3}$$

For $p = 2$, (sq. root)

D*$_4$: The reciprocal p^{th} root of *N

It is the root of the equation

$$f(x) = N - x^{-p} = 0$$

Then $$f'(x) = px^{-(p+1)}$$

$$x_{k+1} = x_k - \frac{f_k}{f'_k} \qquad \text{(Note the form)}$$

This yields $$x_{k+1} = x_k - \frac{N - x_k^{-p}}{px_k^{-(p+1)}}$$

$$\left\{x_{k+1} = \frac{x_k}{p}[p+1 - N x_k^p]\right\} \tag{4}$$

gives the reciprocal p^{th} root of N.

Example 76. Evaluate by Newton's method:

(a) $\sqrt{\pi}$ (6 places of decimal);

(b) $\sqrt{12}$ (4 places of decimal);

(c) $\sqrt{8}$ (4 places of decimal);

Solution: (a) We know that

$$x_{n+1} = \frac{1}{2}\left[x_n + \frac{N}{x_n}\right] \tag{1}$$

(Hero's formula for square root)

Here, $$N = 3.1416 = \pi$$

Assume $$x_0 = 1.5$$

$$x_1 = \frac{1}{2}\left[1.5 + \frac{3.1416}{1.5}\right] = 1.7972$$

$$x_2 = \frac{1}{2}\left[1.7972 + \frac{3.1416}{1.7972}\right] = 1.772626$$

$$x_3 = \frac{1}{2}\left[1.772626 + \frac{3.1416}{1.772626}\right] = 1.7724559$$

$$x_4 = \frac{1}{2}\left[1.7724559 + \frac{3.1416}{1.7724559}\right] = 1.7724559$$

$\Rightarrow$ $x_3 = x_4$ or the required sq. root is 1.772455 (to 6 places)

(b) We know that

$$x_{n+1} = \frac{1}{2}\left[x_n + \frac{N}{x_n}\right] \qquad (1)$$

Here, $N = 12$ and let $x = \sqrt{12}$

$\Rightarrow$ $x^2 - 12 = 0$

$f(x) = 0 \Rightarrow f(3) = -3 < 0$

$f(4) = 4 > 0$

Root lies between 3 and 4

Take $x_0 = 3.5$: (between approximation)

Using (1), we have,

$$x_1 = \frac{1}{2}\left[x_0 + \frac{N}{x_0}\right]$$

$$= \frac{1}{2}\left[3.5 + \frac{12}{3.5}\right] = 3.464$$

$$x_2 = \frac{1}{2}\left[3.464 + \frac{12}{3.464}\right] = 3.4641$$

$$x_3 = \frac{1}{2}\left[3.4641 + \frac{12}{3.4641}\right] = 3.4641$$

Thus, $x_2 = x_3$, correct to 4 places of decimal

Hence, $\sqrt{12} = 3.4641$

(c) Since $4 < 8 < 9$

$\Rightarrow$ $\sqrt{4} < \sqrt{8} < \sqrt{9}$

$\Rightarrow$ $2 < \sqrt{8} < 3$

Hence, $\sqrt{8}$ lies between 2 and 3.

Take $x_0 = 2.5; \quad N = 8$

$$x_{n+1} = \frac{1}{2}\left[x_n + \frac{N}{x_n}\right]$$

Hence, $x_1 = \frac{1}{2}\left[x_0 + \frac{8}{x_0}\right] = 2.85$

$$x_2 = \frac{1}{2}\left[2.85 + \frac{8}{2.85}\right] = 2.8285$$

$$x_3 = \frac{1}{2}\left[2.8285 + \frac{8}{2.8285}\right] = 2.8284$$

Since $x_2 \cong x_3$; $\sqrt{8} = 2.8284$ (to 4 places of decimal)

Newton-Raphson's method for simultaneous equations

Suppose we wish to solve simultaneously, the equations:

$$f(x, y) = 0, \; g(x, y) = 0 \tag{1}$$

Let (α, β) be a solution, i.e., $f(\alpha, \beta) = g(\alpha, \beta) = 0$. In fact (α, β) is a point of intersection of the two curves represented by (1).

If (x_0, y_0) be an approximation to the solution (α, β) and $\alpha = x_0 + h$, $\beta = y_0 + k$, then, by Taylor's series,

$$f(\alpha, \beta) = f(x_0 + h, y_0 + k) = f(x_0, y_0) + h f_x(x_0, y_0) + k f_y(x_0, y_0)$$
$$g(\alpha, \beta) = g(x_0 + h, y_0 + k) = g(x_0, y_0) + h g_x(x_0, y_0) + k g_y(x_0, y_0)$$

dropping the higher powers of h and k.

Noting that $f(\alpha, \beta) = g(\alpha, \beta) = 0$, we have

$$f(x_0, y_0) + h f_x(x_0, y_0) + k f_y(x_0, y_0) = 0 \tag{2}$$
$$g(x_0, y_0) + h g_x(x_0, y_0) + k g_y(x_0, y_0) = 0 \tag{3}$$

Solving (2) and (3) for h and k, we get a better approximation $x_1 = x_0 + h$, $y_1 = y_0 + k$.

This process can be repeated till the desired accuracy is achieved.

This is Newton-Raphson method for simultaneous equations.

Example 77. Solve: $x^2 + 3x - y - 1 = 0$, $xy + 3y + 9 = 0$ starting with the approximation $x_0 = -4$, $y_0 = 6$.

Solution: Here

$$f = x^2 + 3x - y - 1, \; f_x = 2x + 3, \; f_y = -1,$$
$$g = xy + 3y + 9, \; g_x = y, \; g_y = x + 3.$$

Since $\left.\begin{aligned} x_0 &= -4; \\ y_0 &= 6; \end{aligned}\right\}$ (given)

We find that

$$f = 16 - 12 - 6 - 1 = -3;$$
$$f_x = 2x + 3; \; f_y = -1;$$

$\therefore$ $f_x(x_0, y_0) = -5; \; f_y(x_0, y_0) = -1$

Similarly, $g = 3; \; g_x = 6, \; g_y = -1$

Hence, from equations (2) and (3), we have,

$$\left.\begin{cases}-3-5h-k=0\\ 3+6h-k=0\end{cases}\right\} \Rightarrow h = -\frac{6}{11} = -0.55;\ k = -\frac{3}{11} = -0.22$$

$$x_1 = -4 + b = -4.35,\ y_1 = 6 + k = 5.73.$$

Repeating the process once more, we get $x_2 = -4.53$, $y_2 = 5.90$.

ACCELERATION OF CONVERGENCE

Let us recall the formula $|x_n - m| \le k|x_{n-1} - m|$, $k < 1$, for the convergence of the iteration method. It is clear from this formula that the iteration method is linearly convergent. This means that the convergence is very slow in the sense that a large number of iterations are required to reach the root. It is the first order convergence. It is possible to accelerate the process and one such method is due to Aitken and is known as Aitken's Δ^2 process.

AITKEN'S Δ^2 METHOD

Let $x = m$ be a root of the equation

$$f(x) = 0 \tag{1}$$

and let 1 be an interval containing the point $x = m$. The equation (1) can be written as $x = \phi(x)$ such that $\phi(x)$ and $\phi'(x)$ are continuous in I and $|\phi'(x)| < 1$ for all x is I.

Let x_{k-1}, x_k and x_{k+1} be three successive approximations of the desired root m. Since $|\phi'(x)| < 1$ for $k = 0, 1, 2, \ldots$, we may write the error relation as: $e_{k+1} = \lambda e_k$, where λ is such that $|\phi'(x)| \le \lambda <|$ for $k > 0$.

Then, we know that

$$e_k = m - x_k = \lambda e_{k-1} = \lambda(m - x_{k-1}) \text{ and } e_{k+1} = m - x_{k+1} = \lambda e_k = \lambda(m - x_k)$$

(λ is a constant such that $|\phi'(x)| \le \lambda <|$ for all k)

This yields (on division)

$$\frac{m - x_k}{m - x_{k+1}} = \frac{m - x_{k-1}}{m - x_k}$$

Solving this equation for m, we get

$$m = x_{k+1} - \frac{(x_{k+1} - x_k)^2}{(x_{k+1} - 2x_k - x_{k-1})} \tag{2}$$

Since $\quad \Delta x_k = x_{k+1} - x_k$

and

$$\begin{aligned}\Delta^2 x_{k-1} &= (E - 1)^2 x_{k-1}\\ &= (E^2 - 2E + 1)\, x_{k-1}\\ &= x_{k+1} - 2x_k + x_{k-1}\end{aligned}$$

(2) can be written as:

$$\boxed{m = x_{k+1} - \frac{(\Delta x_k)^2}{\Delta^2 x_{k-1}}} \tag{3}$$

Which yields successive approximation to the root.

Example 78. By Aitken's Δ^2 method find the real root of the equation $\cos x - x\, e^x = 0$, correct to four places of decimal.

Solution: The given equation may be rewritten as

$$x = x + \frac{1}{2}(\cos x - xe^x)$$

Then
$$\varphi(x) = x + \frac{1}{2}(\cos x - x\, e^x)$$

$$\varphi'(x) = 1 - \frac{1}{2}(\sin x + x\, e^x + e^x)$$

We check that $|\varphi'(x)| < 1$ for $0 \le x < 1$. Hence, the iteration method applied to $\varphi(x)$ is convergent near 0.

Let us take $x_0 = 0$ as the initial approximation of the root.

Then
$$x_1 = \varphi(x_0) = \varphi(0) = 0.5$$
$$x_2 = \varphi(x_1) = \varphi(0.5) = 0.52661096$$

Now, using the Aitken's formula (3) with x_0, x_1, x_2, we first form the following difference table.

Difference Table

$x_0 = 0.0$		
	$\Delta x_0 = 0.5$	
$x_1 = 0.5$		$\Delta^2 x_0 = -0.473389$
	$\Delta x_1 = 0.02861096$	
$x_2 = 0.526651096$		

The formula (3) now gives

$$x_3 = 0.52661096 - \frac{(0.02661096)^2}{-0.473389} = 0.52810686$$

We find x_4 and x_5 by iteration. Thus,

$$x_4 = \varphi(x_3) = 0.51222771$$
$$x_5 = \varphi(x_4) = 0.52059979$$

To find x_6, we use Aitken's formula. For this, we have the following table.

Difference Table

$x_3 = 0.52810686$		
	$\Delta x_3 = -0.015881$	
$x_4 = 0.5122271$		$\Delta^2 x_3 = 0.02425308$
	$\Delta x_4 = 0.0083721$	
$x_5 = 0.52059979$		

Therefore,

$$x_6 = 0.52059979 - \frac{(0.0083721)^2}{0.02425308} = 0.517709$$

Similarly, we find x_7 and x_8 by iteration and x_9 by using Aitken's formula. We find that

$$\begin{Bmatrix} x_7 = 0.517782, \\ x_8 = 0.517744 \\ x_9 = 0.517757 \end{Bmatrix}$$

Hence, we take $m \approx 0.5177$ as the required root, correct to 4 decimal places.

Example 79. Perform two iterations of the linear iteration method followed by one iteration of the Aitken's Δ^2 method to find the root of the equation $x = e^{-x}$ that lies near 1.

Example 80. Find a real root of the equation $\cos x = 3x - 1$ correct to 3 decimal places using

(i) Iteration method
(ii) Aitken's Δ^2 method.

Solution: (i) Let $f(x) = \cos x - 3x + 1 = 0$

$f(0) = 2$ = positive;

$$f(\pi/2) = -\frac{-3\pi}{2} + 1 = \text{negative};$$

$\therefore$ a root lies between 0 and $\frac{\pi}{2}$.

Rewriting the given equation as

$$x = \frac{1}{3}(\cos x + 1) = \phi(x)$$

we have $\phi'(x) = -\frac{\sin x}{3}$

and $|\phi'(x)| = \frac{1}{3}\,|\sin x| < 1$ in $\left(0, \frac{\pi}{2}\right)$.

Hence we take $x_0 = 0$. Then the successive approximations are:

$$x_1 = \phi\,(x_0) = \frac{1}{3}[\cos 0 + 1] = [0.6667]$$

$$x_2 = \phi\,(x_1) = \frac{1}{3}[\cos (.6667) + 1] = [0.5953]$$

$$x_3 = \phi\,(x_2) = \frac{1}{3}[\cos (.5953) + 1] = [0.6093]$$

$$x_4 = \phi\,(x_3) = \frac{1}{3}[\cos (.6393) + 1] = [0.6067]$$

$$x_5 = \phi\,(x_4) = \frac{1}{3}[\cos (0.6067) + 1] = [0.6072]$$

$$x_6 = \phi\,(x_5) = \frac{1}{3}[\cos (0.6072) + 1] = [0.6071]$$

Since $x_5 \equiv x_6$, the root is 0.607 correct to 3 decimal places.

(ii) We first find x_1, x_2, x_3 as shown in (i). To use Aitken's method, we have

x	Δx	$\Delta^2 x$
$x_1 = 0.667$		
	-0.0714	
$x_2 = 0.5953$		0.0854
	$\underline{0.014}$	
$x_3 = 0.6093$		

$$\therefore \qquad x_4 = x_3 - \frac{(\Delta x_2)^2}{\Delta^2 x_1} = 0.6093 - \frac{(0.014)^2}{0.0854} = 0.607$$

This corresponds to six iterations in normal form. Thus, the root correct to 3 places is $\boxed{0.607}$.

EXERCISES

1. By Aitken's Δ^2 process, find the roots of the following equations near the given points x_0 and correct to n places of decimals.

   (1) $2x - \log_{10} x = 7$, $x_0 = 3.5$, $n = 4$ (Ans: 3.7898)

   (2) $\cos x = 3\,x - 1$, $x_0 = 0$, $n = 3$ (Ans: 0.607)

(3) $e_x = 5\,x$, $x_0 = 0.3$, $n = 3$ (Ans: 0.259)

(4) $x = \cos x$, $x_0 = 0.7$, $n = 3$ (Ans: 0.739)

(5) $2x = \cos x + 3$, $x_0 = \frac{\pi}{2}$, $n = 3$ (Ans: 1.524)

(6) $\log_e{}^x = x - 3$, $x_0 = 4$, $n = 3$ (Ans: 4.505)

2. By Aitken's Δ^2 method, find the value of $\sqrt{3}$ upto seven places of decimal. (Ans: 1.7320509)

3. By Aitken's Δ^2 process, find the root $x = 1.497300$, correct to 6 decimal places, of the equation $x = \frac{1}{2} + \sin x$. How many iterations are required to compute the same root by linear iteration? (Ans: Six iterations)

ITERATION METHOD

To find a root of an equation $f(x) = 0$, we transform it into the form

$$x = \phi(x) \tag{1}$$

where $\phi(x)$ satisfies the following conditions

(i) For two real numbers a and b, $a \le x \le b$ implies $a \le \phi(x) \le b$, and

(ii) For all x', x'' lying in the internal (a, b), we have

$$|\phi(x') - \phi(x'')| \le \alpha\,|x' - x''|$$

where $0 \le \alpha \le 1$.

It can be proved that the equation (1) has a unique root m in (a, b). To find the approximate value of this root, we start with an initial approximation x_0 of the root m and find $\phi(x_0)$ and generate further approximations:

We put

$$\begin{aligned} x_1 &= \varphi(x_0),\\ x_2 &= \varphi(x_1),\\ x_3 &= \varphi(x_2)\\ &\dots\ \dots\\ &\dots\ \dots \end{aligned}$$

or, in general,

$$\begin{aligned} x_{i+1} &= \phi(x_i),\ i = 0, 1, 2, \dots\\ x_n &= \phi(x_{n-1}),\ n \ge 1. \end{aligned}$$

Thus, the nth approximation is given by

$$\boxed{x_n = \varphi(x_{n-1}),\quad n \ge 1} \tag{2}$$

(2) is called the iteration formula and such a process is called iteration process or just iteration. The successive approximations $x_1, x_2, \dots,$ obtained by iteration are called the iterates, the nth approximation x_n is called the nth iterate.

Convergence of an Iteration

The convergence of the sequence is not guaranteed always unless the choice of x_0 is properly chosen. Under certain conditions, to be stated next, the sequence $x_0, x_1, x_2, \ldots$ converges to the desired root α.

Single point (or fixed point) iteration

The computation of x_n from the formula $x_n = \phi(x_{n-1})$, $n \geq 1$, involves the computation of the function $\phi(x)$ at a single point, x_{n-1}. For this reason, the iteration formula $x_n = \phi(x_{n-1})$, $n \geq 1$ is called a single point (or fixed point) iteration.

Multipoint iterations

In the bisection, regula-falsi and secant methods, the computation of x_n involves the computation of a function at more than one of the preceding approximations. The iterations involved in these methods are examples of what are called multipoint iterations. The geometrical grounds upon which the bisection, regula-falsi and secant methods are based indicate that the iterations involved in these methods are always convergent.

Order (or rate) of convergence of an iterative process

Let $x_0, x_1, \ldots$ be the successive approximations of the root α of $f(x) = 0$. Let e_i be the error in the root x_k, $k = 1, 2, 3, \ldots$

If α the exact root,

$$e_k = x_k - \alpha \qquad \text{and} \qquad e_{k+1} = x_{k+1} - \alpha$$

Thus, an iterative method is said to be of order (or rate) of convergence p, where 'p' is a positive real number, if $|e_{k+1}|$ is of the order $|e_k|^p$; that is, if $|e_{k+1}| \approx \alpha_k |e_k|^p$, where α_k is a constant.

For iteration given by the formula

$e_{k+1} = e_k \, \phi'(\xi_k)$, we have

$|e_{k+1}| = |\phi'(\xi)| \, |e_k|$; as such, here $p = 1$; the rate (or order) of convergence of this iteration is therefore 1.

The condition for the convergence of the method

Theorem:

Let $f(x) = 0$ be the given equation whose actual root is m. The equation $f(x) = 0$ can be written as $x = \phi(x)$. Let I be an interval containing the root m of the equation $x = \phi(x)$.

It $|\phi'(x)| < 1$ for all x in the interval I, then for any choice of x_0 in 1, the iteration given by the equation:

$$x_n = \phi(x_{n-1}),\ n \geq 1 \text{ is convergent.}$$

Proof: Since m is a root of $x = \phi(x)$, we have $m = \phi(m)$ (1)

Further,
$$x_1 = \phi(x_0)$$
$$x_2 = \phi(x_1)$$
$$\ldots \quad \ldots$$
$$\ldots \quad \ldots$$
$$x_n = \phi(x_{n-1}) \tag{2}$$

from which the sequence $x_0, x_1, x_2, \ldots x_n$ of approximations is obtained. Hence,
$$x_n - m = \phi(x_{n-1}) - \phi(m) \tag{3}$$

By mean value theorem of differential calculus,
$$\phi(x_{n-1}) - \phi(m) = (x_{n-1} - m)\ \phi'(\theta)$$

where $x_{n-1} < \theta < \alpha$.

Using in (3),
$$x_n - m = (x_{n-1} - m)\ \phi'(\theta) \tag{4}$$

Let $|\phi'(x)| \le k,\ \forall\ x$ in I which contains
$$x_0, x_1, x_2, \ldots x_n, m$$

Hence, (4) reduces to,
$$|x_n - m| \le |x_{n-1} - m|\ k$$

Similarly,
$$|x_{n-1} - m| \le |x_{n-2} - m|\ k$$
$$|x_{n-2} - m| \le |x_{n-3} - m|\ k$$
$$\ldots \ \ldots \ \ldots$$
$$\ldots \ \ldots \ \ldots$$
$$|x_1 - m| \le |x_0 - m|\ k$$

Multiplying vertically and cancelling the factors
$$|x_n - m| \le k_n\ |x_0 - m| \tag{5}$$

Let $\quad k < 1,\ k^n \to 0$ as $n \to \infty$

Hence, $\quad |x_n - m| \to 0$ as $n \to \infty$

is $\quad \lim_{n\to\infty} x_n = m$

Therefore, the sequence of approximations $x_0, x_1, x_2, \ldots x_n \ldots$ converges to the exact root m if $|\phi'(x)| < k < 1\ \ \forall\ \ x \in 1$.

The sequence will converge rapidly if $|\phi'(x)|$ is very small.

If $|\phi'(x)| > 1$, $|x_n - m|$ will become very great and the sequence will not converge.

Note 1. Since $|x_n - m| \le k\ |x_{n-1} - m|$ where k is a constant, the convergence is linear and the convergence is or order one.

Note 2. The sufficient condition for the convergence is $|\phi'(x)| < 1$ for all x in 1.

Example 81. Find the positive root of $x^4 - x - 10 = 0$ by iteration.

Solution: This equation can be written in the form $x = \phi(x)$ several ways, such as $x = x^4 - 10$, $x = \dfrac{10}{x^3 - 1}$, $x = (x + 10)^{1/4}$.

Only the last form satisfies the convergence criterion $|\phi'(x)| < 1$.
Hence, taking it, we write the iteration formula

$$x_{i+1} = (x_i + 10)^{1/4},\ i = 0, 1, 2, \ldots$$

We find $f(1) < 0$, $f(2) > 0$ from the given equation. Hence, there is a root between 1 and 2.

Choosing

$$\begin{aligned}
x_0 &= 1.5, \text{ we get} \\
x_1 &= (1.5 + 10)^{1/4} = (11.5)^{1/4} = 1.8415 \\
x_2 &= (1.8415 + 10)^{1/4} = (11.8415)^{1/4} = 1.8550 \\
x_3 &= (1.8550 + 10)^{1/4} = (11.8550)^{1/4} = 1.8556 \\
x_4 &= (1.8556 + 10)^{1/4} = (11.8556)^{1/4} = 1.8556
\end{aligned}$$

Hence the root is 1.8556 correct to four decimal places.

Example 82. Solve $x = 1 + \tan^{-1} x$ by the method of iteration.

Solution: Here, $\phi(x) = \tan^{-1} x - 1$

$$\phi'(x) = \frac{1}{1+x^2} < 1.$$

$$\therefore \qquad |\phi'(x)| < 1$$

Hence, the process converges and we take

$$x_{i+1} = 1 + \tan^{-1} x_i$$

$$\left\{\begin{matrix} \text{with } x_0 = 1, & \text{we compute } x_1 = 1.7854, \\ x_1 = 2.602, & x_3 = 2.1189,\ x_4 = 2.1298, \\ x_5 = 2.1318, & x_6 = 2.1322,\ x_7 = 2.1323, \end{matrix}\right\}$$

$x_8 = x_9$. Hence, the root is 2.1323.
Had we started with $x_0 = 2$, we would have got

$$\begin{aligned}
x_1 &= 2.1071,\ x_2 = 2.1277,\ x_3 = 2.1314, \\
x_4 &= 2.1321,\ x_5 = 2.1322,\ x_6 = 2.1323
\end{aligned}$$

Example 83. Solve $e^x - 3x = 0$ by the method of iteration.

Solution: Let

$$\begin{aligned}
f(x) &= e^x - 3x = 0 \\
f(0) &= 1 = \text{positive} \\
f(1) &= e - 3 = \text{negative}
\end{aligned}$$

$\therefore$ a root lies between 0 and 1.

Let $$x = \frac{1}{3}e^x = \phi(x)$$

$$\phi'(x) = \frac{1}{3}e^x \text{ and } \phi'(0) = \frac{1}{3},\ \phi'(1) < 1.$$

In the interval (0, 1), $|\phi'(x)| < 1$

Select $$x_0 = 0.6,\ x_1 = \frac{1}{3}e^{x_0} = \frac{1}{3}e^{0.6} = 0.60737$$

$$x_2 = \frac{1}{3}e^{0.60737} = 0.61187,\ x_3 = \frac{1}{3}e^{0.61187} = 0.61452$$

$$x_4 = \frac{1}{3}e^{0.61452} = 0.61626,\ x_5 = \frac{1}{3}e^{0.61626} = 0.61733$$

$$x_6 = \frac{1}{3}e^{0.61733} = 0.61799,\ x_7 = \frac{1}{3}e^{0.61799} = 0.61840$$

$$x_8 = \frac{1}{3}e^{0.61840} = 0.61865,\ x_9 = \frac{1}{3}e^{0.61865} = 0.61881$$

$$x_{10} = \frac{1}{3}e^{0.61881} = 0.61891,\ x_{11} = \frac{1}{3}e^{0.61891} = 0.61897.$$

$$x_{12} = \frac{1}{3}e^{0.61897} = 0.61900,\ x_{13} = \frac{1}{3}e^{0.61900} = 0.61902$$

We can take 0.6190 as the correct value of the root of the equation.

Example 84. By the fixed point iteration process, find the root, correct to 3-decimal places, of the equation $x = \cos x$ near ($x = \pi/4$.)

Solution:

Given: $x = \cos x$ near $x = \dfrac{\pi}{4}$.

This is of the form $x = \phi(x)$,
where $\phi(x) = \cos x$; $\therefore\ \phi'(x) = -\sin x$
Hence, $|\varphi'(x)| = |\sin x| < 1$ for all x.

Hence, the iteration process $x_n = \varphi\,(x_n - 1)$ is convergent in every interval. Since the root is required near $\pi/4$, let us take the initial approximation of the root as $x_0 = \pi/4 = 0.785398$ (in radians). Then, the iteration formula $x_n = \varphi\,(x_{n-1})$ gives

$$x_1 = \varphi(x_0) = \cos\ (\pi/4) = 0.7071068,$$
$$x_2 = \varphi(x_1) = \cos x_1 = 0.7602446,$$
$$x_3 = \cos x_2 = 0.7246675,$$

$$x_4 = 0.7487199, \qquad x_5 = 0.7325608,$$
$$x_6 = 0.7434642, \qquad x_7 = 0.7361282,$$
$$x_8 = 0.7410736, \qquad x_9 = 0.7377441,$$
$$x_{10} = 0.7399878, \qquad x_{11} = 0.7384768,$$
$$x_{12} = 0.7394947, \text{ and so on.}$$

Hence, we take 0.739 (3 places) as an approximation of the required root.

Example 85. The equation $4x = e^x$ has two roots, one near 0.3 and the other near 2.1. Find them.

Solution: Given: $x = \frac{1}{4} e^x$

or $x = \log (4x)$

$$\left\{ \begin{array}{l} \text{Near } 0.3, \ \frac{d}{dx}\left[\frac{1}{4}e^x\right] = \frac{1}{4}e^x < 1 \\ \text{Near } 2.1, \ \frac{d}{dx}[\log 4x] = \frac{1}{x} < 1 \end{array} \right\}$$

Hence, with $x_0 = 0.3$, $x_{i+1} = \frac{1}{4} e^{x_i}$, we get

$$x_1 = 0.3375,$$
$$x_2 = 0.3503,$$
$$x_3 = 0.3549,$$
$$x_4 = 0.3565,$$
$$x_5 = 0.3571,$$
$$x_6 = 0.3573,$$
$$x_7 = 0.3574,$$
$$x_8 = x_7.$$

With $x_0 = 2.1$, $x_{i+1} = \log 4x_i$, we get

$$x_1 = 2.1282,$$
$$x_2 = 2.1416,$$
$$x_3 = 2.1478,$$
$$x_4 = 2.1508,$$
$$x_5 = 2.1521,$$
$$x_6 = 2.1527,$$
$$x_7 = 2.1530,$$
$$x_8 = 2.1532,$$
$$x_9 = x_8.$$

Hence, the two roots are: 0.3574, 2.1532

Example 86. Solve the Keplar's equation iteratively for $m = 0.8$; $E = 0.2$ by writing in the form: $x = m + E \sin x$ and starting with $x_0 = m = 0.8$.

Solution:

$$\left\{\begin{array}{l} x_1 = \phi(x_0) = \phi(0.8) = 0.8 + 0.2\sin(0.8) = 0.94347 \\ x_2 = \phi(x_1) = \phi(0.94347) = 0.8 + 0.2\sin(0.94347) = 0.96192 \\ x_3 = \phi(x_2) = \phi(0.96192) = 0.8 + 0.2\sin(0.96192) = 0.96405 \\ x_4 = \phi(x_3) = \phi(0.96405) = 0.8 + 0.2\sin(0.96405) = 0.9643 \\ x_5 = \phi(x_4) = \phi(0.9643) = 0.8 + 0.2\sin(0.9643) = 0.96433 \\ \Rightarrow \text{required root} = 0.9643, \text{ as } x_4 = x_5 \text{ to 4 places of decimal.} \end{array}\right\}$$

Example 87. Solve the equation $x = 0.21 \sin(0.5 + x)$ by iteration method with the approximate root as 0.1.

Solution:

$$\left\{\begin{array}{l} x_1 = 0.21\sin(0.5 + 0.1) = 0.1129 \\ x_2 = 0.21\sin(0.5 + 0.1129) = 0.1208 \\ x_3 = 0.21\sin(0.5 + 0.1208) = 0.1221 \\ x_4 = 0.21\sin(0.5 + 0.1221) = 0.12238 \\ x_5 = 0.21\sin(0.5 + 0.12238) = 0.12242 \\ x_6 = 0.21\sin(0.5 + 0.12242) = 0.12243 \end{array}\right\}$$

required root = 0.12243

EXERCISES

1. By the fixed point iteration process find the roots of the following equations near the given points x_0 and correct to n places of decimals.

(1) $x^3 - 2x + 1 = 0$, $x_0 = 0$, $n = 4$ (Ans: 0.6180)
(2) $x^3 - x - 1 = 0$, $x_0 = -1$, $n = 5$ (Ans: –1.32472)
(3) $x^3 - 2x + 5 = 0$, $x_0 = -2$, $n = 5$ (Ans: –2.09455)
(4) $x^3 + 2x - 1 = 0$, $x_0 = 0.75$, $n = 4$ (Ans: 0.7549)
(5) $x^3 + x^2 - 100 = 0$, $x_0 = 4.2$, $n = 5$ (Ans: 4.33105)
(6) $\sin x - 2x + 1 = 0$, $x_0 = 1.5$, $n = 4$ (Ans: 1.4973)
(7) $\cos x - 3x + 1 = 0$, $x_0 = 0$, $n = 3$ (Ans: 0.607)
(8) $\cos x - 2x + 3 = 0$, $x_0 = \pi/2$, $n = 3$ (Ans: 1.524)
(9) $x - 0.2 \sin x - 0.5 = 0$, $x_0 = 0.6$, $n = 5$ (Ans: 0.61547)
(10) $x = 0.21 \sin(x + 0.5)$, $x_0 = 0.12$, $n = 5$ (Ans: 0.12242)

2. Find the approximate values of the following by using single point iterative method.

   (1) $\sqrt{30}$ (2) $(1/3)^{1/4}$ (3) $1/\sqrt{15}$

   (Ans: 5.4772) (Ans: 0.7598) (Ans: 0.2582)

3. By using the iterative method, find the smallest root of the equation

$$1 - x + \frac{x^2}{(2!)^2} - \frac{x^3}{(3!)^2} + \frac{x^4}{(4!)^2} + \ldots = 0.$$

   correct to two decimal places. (Ans: 1.44)

4. If α and β are two real roots of the equation $x^2 + ax + b = 0$, show that the iteration $x_{n+1} = -(x_n^2 + b)/a$ is convergent near $x = \alpha$ if $|\alpha| < (1/2)\,|\alpha + \beta|$.

5. If α is such that $|1 + 2\alpha| < 1$, show that the iteration given by the formula

   $x_{n+1} = x_n + \alpha\,(x_n^2 - 3)$ is linearly convergent near the point $x_0 = \sqrt{3}$.

Chapter 3

Simultaneous Linear Equations

INTRODUCTION

We are familiar with methods of solving 'n' simultaneous equations in 'n' unknowns by Crammer's rule or by matrix inversion. If the equations written in the matrix form are $AX = B$ and if $|A| \neq 0$ and B are not a null matrix, they admit of a unique solution $x = A^{-1} B$. If B, a zero matrix, then non-trivial set of solutions of the equations exist provided $|A| = 0$.

We shall reconsider the solution of linear simultaneous equations by methods which require less calculations. The methods are divided into two categories.

(i) The direct methods and
(ii) The iterative methods.

We are familiar with the method of solving linear simultaneous equation by the matrix inversion method.

If $|A| \neq 0$ and we have the matrix equation $AX = B$, then $|A| \neq 0$, so inverse of A exists and we have $X = A^{-1} B$ as the solution.

SYSTEM OF LINEAR EQUATIONS

Let the simultaneous equations be:

$$\begin{aligned} a_1x + b_1y + c_1z &= d_1 \\ a_2x + b_2y + c_2z &= d_2 \\ a_3x + b_3y + c_3z &= d_3 \end{aligned}$$

Then $$D = \begin{vmatrix} a_1 & b_1 & c_1 \\ a_2 & b_2 & c_2 \\ a_3 & b_3 & c_3 \end{vmatrix} = a_1 (b_2c_3 - c_2b_3) - b_1(a_2c_3 - c_2a_3) + c_1(a_2b_3 - b_2a_3)$$

$$D_1 = \begin{vmatrix} d_1 & b_1 & c_1 \\ d_2 & b_2 & c_2 \\ d_3 & b_3 & c_3 \end{vmatrix}$$

$$= d_1(b_2c_3 - c_2b_3) - b_1\,(d_2c_3 - c_2d_3) + c_1(d_2b_3 - b_2d_3)$$

$$D_2 = \begin{vmatrix} a_1 & d_1 & c_1 \\ a_2 & d_2 & c_2 \\ a_3 & d_3 & c_3 \end{vmatrix}$$

$$= a_1(d_2c_3 - c_2d_3) - d_1\,(a_2c_3 - c_2a_3) + c_1(a_2d_3 - d_2a_3)$$

$$D_3 = \begin{vmatrix} a_1 & b_1 & d_1 \\ a_2 & b_2 & d_2 \\ a_3 & b_3 & d_3 \end{vmatrix}$$

$$= a_1(b_2d_3 - d_2b_3) - b_1\,(a_2d_3 - d_2a_3) + d_1(a_2b_3 - b_2a_3)$$

$$x = \frac{D_1}{D}$$

$$y = \frac{D_2}{D}; \quad z = \frac{D_3}{D}$$

Example 1. Solve the following system of linear simultaneous algebraic equations:

$$4x_1 + 3x_2 + 6x_3 = 13 \quad \text{(i)}$$
$$2x_1 - 4x_2 + x_3 = 8 \quad \text{(ii)}$$
$$3x_1 - 2x_2 + 6x_3 = 17 \quad \text{(iii)}$$

Solution: The problem is solved by the method of determinants.

Step (i): Form the determinant from coefficients of x_1, x_2 and x_3 and solve as,

$$= \begin{vmatrix} 4 & 3 & 6 \\ 3 & -2 & 6 \\ 2 & -4 & 1 \end{vmatrix}$$

$$= 4\,(-24 + 2) - 3\,(12 - 3) + 6\,(-4 + 12)$$
$$= -88 - 27 + 48 = -67.$$

Step (ii): In the above determinant, the first column is changed with the constants and solve as,

$$A_1 = \begin{vmatrix} 13 & 3 & 6 \\ 8 & -4 & 1 \\ 17 & -2 & 6 \end{vmatrix}$$

$$= 13\ (-24 + 2) - 3\ (48 - 17) + 6\ (-16 + 68)$$
$$= -286 - 93 + 312 = -67$$

Step (iii): In this change the second column of step (i) with the constants and solve as

$$A_2 = \begin{vmatrix} 4 & 13 & 6 \\ 2 & 8 & 1 \\ 3 & 17 & 6 \end{vmatrix}$$

$$= 4\ (48 - 17) - 13\ (12 - 3) + 6\ (34 - 24)$$
$$= 124 - 117 + 60 = 67$$

Step (iv): In this change the column of step (i) determinant with constant and solve as

$$A_3 = \begin{vmatrix} 4 & 3 & 13 \\ 2 & -4 & 8 \\ 3 & -2 & 17 \end{vmatrix}$$

$$= 4\ (-68 + 16) - 3\ (34 - 24) + 13\ (-4 + 12)$$
$$= -208 - 30 + 10 = -134$$

Step (v): The values of variables are calculated as

$$x_1 = \frac{A_1}{A} = \frac{-67}{-67} = 1$$

$$x_2 = \frac{A_2}{A} = \frac{67}{-67} = -1$$

$$x_3 = \frac{A_3}{A} = \frac{-134}{-67} = 2$$

Methods of Inversion of Matrices

The methods of inversion of matrices are as follows:

(i) *Cofactor method:* This method is applicable for 3 × 3 matrix. The rectangular matrix cannot be solved by this method.
The cofactors are calculated as shown below:

$$\begin{bmatrix} A_{11} & A_{12} & A_{13} \\ A_{21} & A_{22} & A_{23} \\ A_{31} & A_{32} & A_{33} \end{bmatrix} \begin{bmatrix} x_1 \\ x_2 \\ x_3 \end{bmatrix} = \begin{bmatrix} c_1 \\ c_2 \\ c_3 \end{bmatrix}$$

$$[A]\ [X] = [C]$$

$$a_{11} = (-1)^{1+1} \begin{vmatrix} A_{22} & A_{23} \\ A_{32} & A_{33} \end{vmatrix} = \{A_{22} \times A_{33} - A_{23} \times A_{32}\}$$

$$a_{12} = (-1)^{2+1} \begin{vmatrix} A_{21} & A_{23} \\ A_{31} & A_{33} \end{vmatrix} = -\{A_{21} \times A_{33} - A_{23} \times A_{31}\}$$

$$a_{13} = (-1)^{1+3} \begin{vmatrix} A_{21} & A_{22} \\ A_{31} & A_{32} \end{vmatrix}$$

$$= + \{A_{21} \times A_{32} - A_{22} \times A_{31}\}$$

In the same way, the other cofactors are also calculated.

Then $$M = \begin{vmatrix} a_{11} & a_{12} & a_{13} \\ a_{21} & a_{22} & a_{23} \\ a_{31} & a_{32} & a_{33} \end{vmatrix}$$

$$M^T = \text{Transpose of matrix } M$$

$$M^T = \begin{vmatrix} a_{11} & a_{21} & a_{31} \\ a_{12} & a_{22} & a_{32} \\ a_{13} & a_{23} & a_{33} \end{vmatrix}$$

$$[A^{-1}] = \frac{M^T}{|A|}$$

After solving we will get the values of variables.

(ii) *Augmented Matrix Method:* This is also called eliminating method and used for 4×4 matrix.

The procedure of solving is as shown below:

$$A = \begin{vmatrix} A_{11} & A_{12} & A_{13} \\ A_{21} & A_{22} & A_{23} \\ A_{31} & A_{32} & A_{33} \end{vmatrix} \quad I = \begin{vmatrix} 1 & 0 & 0 \\ 0 & 1 & 0 \\ 0 & 0 & 1 \end{vmatrix}$$

$$\left|\begin{array}{ccc|c} A_{11} & A_{12} & A_{13} & 1 \\ A_{21} & A_{22} & A_{23} & 0 \\ A_{31} & A_{32} & A_{33} & 0 \end{array}\right|$$

By the row operation bring the last column into first column. Then delete the first column of matrix and introduce the second column of unit matrix into the last column of the obtained matrix. Repeat the procedure for third column also.

Example 2. If C is a square matrix as given below:

$$C = \begin{bmatrix} 2 & 1 & 1 \\ 1 & 2 & 1 \\ 1 & 1 & 2 \end{bmatrix}$$

Show that $CC^{-1} = I$ where C^{-1} is the inverse of C and I is unit matrix.

Solution:

$$C^{-1} = \frac{Adj.\,C}{|C|}$$

Let C_{ij} denote the cofactors of C_{ij} in the matrix C.
Then

$$A_{11} = (-1)^{1+1} \begin{vmatrix} 2 & 1 \\ 1 & 2 \end{vmatrix} = (-1)^2 \{4 - 1\} = 3$$

$$A_{12} = (-1)^{1+2} \begin{vmatrix} 1 & 1 \\ 1 & 2 \end{vmatrix} = (-1)^3 \{2 - 1\} = -1$$

$$A_{13} = (-1)^{1+3} \begin{vmatrix} 1 & 2 \\ 1 & 1 \end{vmatrix} = (-1)^4 \{1 - 2\} = -1$$

$$A_{21} = (-1)^{2+1} = \begin{vmatrix} 1 & 1 \\ 1 & 2 \end{vmatrix} = (-1)^3 \{2 - 1\} = -1$$

$$A_{22} = (-1)^{2+2} \begin{vmatrix} 2 & 1 \\ 1 & 2 \end{vmatrix} = (-1)^4 \{4 - 1\} = 3$$

$$A_{23} = (-1)^{2+3} \begin{vmatrix} 2 & 1 \\ 1 & 1 \end{vmatrix} = (-1)^5 \{2 - 1\} = -1$$

$$A_{31} = (-1)^{3+1} \begin{vmatrix} 1 & 1 \\ 2 & 1 \end{vmatrix} = (-1)^4 \{1 - 2\} = -1$$

$$A_{32} = (-1)^{3+2} \begin{vmatrix} 2 & 1 \\ 1 & 1 \end{vmatrix} = (-1)^5 \{2 - 1\} = -1$$

$$A_{33} = (-1)^{3+3} \begin{vmatrix} 2 & 1 \\ 1 & 2 \end{vmatrix} = (-1)^6 \{4 - 1\} = 3.$$

$$Adj.\ C = \begin{vmatrix} A_{11} & A_{21} & A_{31} \\ A_{12} & A_{22} & A_{32} \\ A_{13} & A_{23} & A_{33} \end{vmatrix} = \begin{vmatrix} 3 & -1 & -1 \\ -1 & 3 & -1 \\ -1 & -1 & 3 \end{vmatrix}$$

$$|C| = 2\,(4-1) - 1\,(2-1) + 1\,(1-2)$$
$$= 6 - 1 - 1 = 4$$

$$\therefore \quad C^{-1} = \frac{Adj.\,C}{|C|} = \frac{1}{4}\begin{vmatrix} 3 & -1 & -1 \\ -1 & 3 & -1 \\ -1 & -1 & 3 \end{vmatrix}$$

$$\therefore \quad CC^{-1} = \begin{bmatrix} 2 & 1 & 1 \\ 1 & 2 & 1 \\ 1 & 1 & 2 \end{bmatrix}\begin{bmatrix} 0.75 & -0.25 & -0.25 \\ -0.25 & 0.75 & -0.25 \\ -0.25 & -0.25 & 0.75 \end{bmatrix}$$

$$= \begin{vmatrix} 1.5-0.25-0.25 & -0.5+0.75-0.25 & -0.5-0.25+0.75 \\ 0.75-0.5-0.25 & -0.25+1.5-0.25 & -0.25-0.5+0.75 \\ 0.75-0.25-0.5 & -0.25+0.75-0.5 & -0.25-0.25+1.5 \end{vmatrix}$$

$$= \begin{vmatrix} 1 & 0 & 0 \\ 0 & 1 & 0 \\ 0 & 0 & 1 \end{vmatrix} = I$$

$\therefore$ $CC^{-1} = I$, where I is an unit matrix.

Example 3. Find the inverse of the following matrix

$$A = \begin{bmatrix} 3 & 3 & 3 \\ 2 & 3 & 2 \\ 2 & 2 & 3 \end{bmatrix}$$

Solution: $A^{-1} = \dfrac{Adj.\,A}{|A|}$, if $|A| \neq 0$.

Let A_{ij} denote the cofactor of a_{ij} in the matrix A,

Then
$$A_{11} = (-1)^{1+1}\begin{vmatrix} 3 & 2 \\ 2 & 3 \end{vmatrix} = (-1)^2\,\{9-4\} = 5$$

$$A_{12} = (-1)^{1+2}\begin{vmatrix} 2 & 2 \\ 2 & 3 \end{vmatrix} = (-1)^3\,\{6-4\} = -2$$

$$A_{13} = (-1)^{1+3}\begin{vmatrix} 2 & 3 \\ 2 & 2 \end{vmatrix} = (-1)^4\,\{4-6\} = -2$$

$$A_{21} = (-1)^{2+1}\begin{vmatrix} 3 & 3 \\ 2 & 3 \end{vmatrix} = (-1)^3\,\{9-6\} = -3$$

$$A_{22} = (-1)^{2+2} \begin{vmatrix} 3 & 3 \\ 2 & 3 \end{vmatrix} = (-1)^4 \{9 - 6\} = 3$$

$$A_{23} = (-1)^{2+3} \begin{vmatrix} 3 & 3 \\ 2 & 2 \end{vmatrix} = (-1)^5 \{6 - 6\} = 0$$

$$A_{31} = (-1)^{3+1} \begin{vmatrix} 3 & 3 \\ 3 & 2 \end{vmatrix} = (-1)^4 \{6 - 9\} = -3$$

$$A_{32} = (-1)^{3+2} \begin{vmatrix} 3 & 3 \\ 2 & 2 \end{vmatrix} = (-1)^5 \{6 - 6\} = 0$$

$$A_{33} = (-1)^{3+3} \begin{vmatrix} 3 & 3 \\ 2 & 3 \end{vmatrix} = (-1)^6 \{9 - 6\} = 3$$

$$Adj{:}\ A = \begin{vmatrix} A_{11} & A_{21} & A_{31} \\ A_{12} & A_{22} & A_{32} \\ A_{13} & A_{23} & A_{33} \end{vmatrix} = \begin{vmatrix} 5 & -3 & -3 \\ -2 & 3 & 0 \\ -2 & 0 & 3 \end{vmatrix}$$

$$|A| = \begin{vmatrix} 3 & 3 & 3 \\ 2 & 3 & 2 \\ 2 & 2 & 3 \end{vmatrix} = 3\ (9 - 4) - 3\ (6 - 4) + 3\ (4 - 6)$$

$$= 15 - 6 - 6 = 3$$

$$\therefore \quad A^{-1} = \frac{1}{3} \begin{vmatrix} 5 & -3 & -3 \\ -2 & 3 & 0 \\ -2 & 0 & 3 \end{vmatrix}$$

$$= \begin{vmatrix} 5/3 & -1 & -1 \\ -2/3 & 1 & 0 \\ -2/3 & 0 & 1 \end{vmatrix}$$

Example 4. Solve the set of simultaneous equations by matrix inversion method:

$$x_1 + 2x_2 - 3x_3 = -4$$
$$2x_1 - 3x_2 + x_3 = -1$$
$$3x_1 - 5x_2 + 4x_3 = 5$$

Solution: The above problem is solved by the 'cofactors method'.

$$\begin{bmatrix} 1 & 2 & -3 \\ 2 & -3 & 1 \\ 3 & -5 & 4 \end{bmatrix} \begin{bmatrix} x_1 \\ x_2 \\ x_3 \end{bmatrix} = \begin{bmatrix} -4 \\ -1 \\ 5 \end{bmatrix}$$

Cofactors:

$$A_{11} = (-1)^{1+1} \begin{vmatrix} -3 & 1 \\ -5 & 4 \end{vmatrix} = -12 + 5 = -7$$

$$A_{12} = (-1)^{1+2} \begin{vmatrix} 2 & 1 \\ 3 & 4 \end{vmatrix} = -(8 - 3) = -5$$

$$A_{13} = (-1)^{1+3} \begin{vmatrix} 2 & -3 \\ 3 & -5 \end{vmatrix} = -10 + 9 = -1$$

$$A_{21} = (-1)^{2+1} \begin{vmatrix} 2 & -3 \\ -5 & 4 \end{vmatrix} = -(8 - 15) = 1$$

$$A_{22} = (-1)^{2+2} \begin{vmatrix} 1 & -3 \\ 3 & 4 \end{vmatrix} = (4 + 9) = 13$$

$$A_{23} = (-1)^{2+3} \begin{vmatrix} 1 & 2 \\ 3 & -5 \end{vmatrix} = -(-5 - 6) = 11$$

$$A_{31} = (-1)^{3+1} \begin{vmatrix} 2 & -3 \\ -3 & 1 \end{vmatrix} = (2 - 9) = -7$$

$$A_{32} = (-1)^{3+2} \begin{vmatrix} 1 & -3 \\ 2 & 1 \end{vmatrix} = -(1 + 6) = -7$$

$$A_{33} = (-1)^{3+3} \begin{vmatrix} 1 & 2 \\ 2 & -3 \end{vmatrix} = (-3 - 4) = -7$$

$$M = \begin{vmatrix} A_{11} & A_{12} & A_{13} \\ A_{21} & A_{22} & A_{23} \\ A_{31} & A_{32} & A_{33} \end{vmatrix}$$

Hence,

$$A = \begin{vmatrix} -7 & -5 & -1 \\ 7 & 13 & 11 \\ -7 & -7 & -7 \end{vmatrix}$$

$$A^T = \begin{vmatrix} -7 & 7 & -7 \\ -5 & 13 & -7 \\ -1 & 11 & -7 \end{vmatrix}$$

$$|A| = |(-12 + 5) - 2(8 - 3) - 3(-10 + 9)|$$
$$= -7 - 10 + 3 = -14$$

$$[A^{-1}] = \frac{M^T}{|A|} = \frac{1}{-14}\begin{vmatrix} -7 & 7 & -7 \\ -5 & 13 & -7 \\ -1 & 11 & -7 \end{vmatrix}\begin{vmatrix} -4 \\ -1 \\ 5 \end{vmatrix}$$

$$= \frac{1}{-14}\begin{vmatrix} (28-7-35) \\ (20-13-35) \\ (4-11-35) \end{vmatrix} = \frac{1}{-14}\begin{vmatrix} -14 \\ -28 \\ -42 \end{vmatrix}$$

$$\therefore \qquad \begin{Bmatrix} x_1 \\ x_2 \\ x_3 \end{Bmatrix} = \begin{vmatrix} 1 \\ 2 \\ 3 \end{vmatrix}$$

Example 5. Find the inverse of the coefficient matrix for the system of equations given:

$$8x_1 + x_2 - x_3 = 8$$
$$2x_1 + x_2 + 9x_3 = 12$$
$$x_1 - 7x_2 + 3x_3 = 4$$

Solution:

$$A = \begin{vmatrix} 8 & 1 & -1 \\ 2 & 1 & 9 \\ 1 & -7 & 3 \end{vmatrix}$$

A^{-1} is the inverse of matrix 'A'

$$A^{-1} = \frac{Adj. A}{|A|}$$

Cofactors of a_{ij} of the matrix are A_{ij}

$$A_{11} = (-1)^{1+1}\begin{vmatrix} 1 & 9 \\ -7 & 3 \end{vmatrix} = (-1)^2 \{3 + 63\} = 66$$

$$A_{12} = (-1)^{1+2}\begin{vmatrix} 2 & 9 \\ 1 & 3 \end{vmatrix} = (-1)^3 \{6 - 9\} = 3.$$

$$A_{13} = (-1)^{1+3} \begin{vmatrix} 2 & 1 \\ 1 & -7 \end{vmatrix} = (-1)^4 \{-14 - 1\} = -1$$

$$A_{21} = (-1)^{2+1} \begin{vmatrix} 1 & -1 \\ -7 & 3 \end{vmatrix} = (-1)^3 \{3 - 7\} = 4$$

$$A_{22} = (-1)^{2+2} \begin{vmatrix} 8 & -1 \\ 1 & 3 \end{vmatrix} = (-1)^4 \{24 + 1\} = 25$$

$$A_{23} = (-1)^{2+3} \begin{vmatrix} 8 & 1 \\ 1 & -7 \end{vmatrix} = (-1)^5 \{-56 - 1\} = 57$$

$$A_{31} = (-1)^{3+1} \begin{vmatrix} 1 & -1 \\ 1 & 9 \end{vmatrix} = (-1)^4 \{9 + 9\} = 10$$

$$A_{32} = (-1)^{3+2} \begin{vmatrix} 8 & -1 \\ 2 & 9 \end{vmatrix} = (-1)^5 \{72 + 2\} = 74$$

$$A_{33} = (-1)^{3+3} \begin{vmatrix} 8 & 1 \\ 2 & 1 \end{vmatrix} = (-1)^6 \{8 + 2\} = 10$$

$$Adj: A = \begin{vmatrix} A_{11} & A_{21} & A_{31} \\ A_{12} & A_{22} & A_{32} \\ A_{13} & A_{23} & A_{33} \end{vmatrix}$$

$$= \begin{vmatrix} 66 & 4 & 10 \\ 3 & 25 & 74 \\ -15 & 57 & 10 \end{vmatrix}$$

$$|A| = \begin{vmatrix} 8 & 1 & -1 \\ 2 & 1 & 9 \\ 1 & -7 & 3 \end{vmatrix} = 8\ (3 + 63) - 1\ (6 - 9) - 1\ (14 - 1)$$

$$= 528 + 3 + 15 = 546$$

$$A^{-1} = \frac{Adj A}{|A|}$$

$$= \frac{1}{546} \begin{vmatrix} 66 & 4 & 10 \\ 3 & 25 & 74 \\ -15 & 57 & 10 \end{vmatrix}$$

GAUSS' ELIMINATION METHOD

To solve the system of equations

$$a_{11}x_1 + a_{12}x_2 + a_{13}x_3 = b_1$$
$$a_{21}x_1 + a_{22}x_2 + a_{23}x_3 = b_2$$
$$a_{31}x_1 + a_{32}x_2 + a_{33}x_3 = b_3$$

x_1 is eliminated first of all from equations (1) and (2) and then from (1) and (3). After eliminating x_1, unknown x_2 is eliminated and then value of x_3 is evaluated.

The augmented matrix of this system is

$$[A : B] = \begin{bmatrix} a_{11} & a_{12} & a_{13} & : & b_1 \\ a_{21} & a_{22} & a_{23} & : & b_2 \\ a_{31} & a_{32} & a_{33} & : & b_3 \end{bmatrix} \quad (1)$$

R_1 is multiplied by $r_1 = \left(\frac{a_{21}}{a_{11}}\right)$ and subtracted from R_2 and R_1 is multiplied by $r_2 = \frac{a_{31}}{a_{11}}$ and subtracted from R_3, we get

$$[A : B] = \begin{bmatrix} a_{11} & a_{12} & a_{13} & : & b_1 \\ 0 & \alpha_{22} & \alpha_{23} & : & \beta_2 \\ 0 & \alpha_{32} & \alpha_{33} & : & \beta_3 \end{bmatrix} \quad (2)$$

where $\alpha_{22} = a_{22} - \left(\frac{a_{21}}{a_{11}}\right) a_{12},$

$$\alpha_{23} = a_{23} - \left(\frac{a_{21}}{a_{11}}\right) a_{13}$$

$$\alpha_{32} = a_{32} - \left(\frac{a_{31}}{a_{11}}\right) a_{12},$$

$$\alpha_{33} = a_{33} - \left(\frac{a_{31}}{a_{11}}\right) a_{13}$$

and $\beta_2 = b_2 - \left(\frac{a_{21}}{a_{11}}\right) b_1, \beta_3 = b_3 - \left(\frac{a_{31}}{a_{11}}\right) b_1$

where $(a_{11} \neq 0) = b_3 - \left(\frac{a_{31}}{a_{11}}\right) b_1$

Next multiply R_2 by $\frac{\alpha_{32}}{\alpha_{22}}$ and subtracting it from R_3 in (2)

we get

$$[A:B] = \begin{bmatrix} a_{11} & a_{12} & a_{13} & : & b_1 \\ 0 & \alpha_{22} & \alpha_{23} & : & \beta_2 \\ 0 & 0 & r_{33} & : & \delta_3 \end{bmatrix} \tag{3}$$

where $$r_{33} = \alpha_{33} - \left(\frac{\alpha_{32}}{\alpha_{22}}\right) \alpha_{23}$$

$$\delta_3 = \beta_3 - \left(\frac{\alpha_{32}}{\alpha_{22}}\right) \beta_2$$

(when $\alpha_{22} \neq 0$)

The augmented matrix is an upper diagonal system which may be solved by backward substitution. The solution obtained will be the solution of (1).

Example 6. Solve the equations

$$2x_1 + x_2 + x_3 = 10$$
$$3x_1 + 2x_2 + 3x_3 = 18$$
$$x_1 + 4x_2 + 9x_3 = 16$$

using the Gauss' elimination method.

Solution: The augmented matrix is

$$[A:B] = \begin{bmatrix} 2 & 1 & 1 & : & 10 \\ 3 & 2 & 3 & : & 18 \\ 1 & 4 & 9 & : & 16 \end{bmatrix} \tag{1}$$

Apply,

$$R_2 - \frac{3}{2} R_1$$

$$R_3 - \frac{1}{2} R_1$$

$$[A:B] \sim \begin{bmatrix} 2 & 1 & 1 & : & 10 \\ 0 & \frac{1}{2} & \frac{3}{2} & : & 3 \\ 0 & \frac{7}{2} & \frac{17}{2} & : & 11 \end{bmatrix} \tag{2}$$

Computation

$$3 - \frac{3}{2} \cdot 2 = 0$$

$$2 - \frac{3}{2} \cdot 1 = \frac{1}{2}$$

$$3 - \frac{3}{2} = \frac{3}{2}$$

Now, using $R_3 - 7R_2$, we get

$$[A : B] = \begin{bmatrix} 2 & 1 & 1 & : & 10 \\ 0 & \frac{1}{2} & \frac{3}{2} & : & 3 \\ 0 & 0 & -2 & : & -10 \end{bmatrix}$$

This augmented matrix corresponds to the following upper triangular system

$$2x_1 + x_2 + x_3 = 10$$

$$\frac{1}{2} x_2 + \frac{3}{2} x_3 = 3$$

$$-2x_3 = 10$$

By backward substitution, we get

$$x_1 = 7, \quad x_2 = -9, \quad x_3 = 12$$

Example 7. Solve

$$4x_1 + x_2 + x_3 = 4$$
$$x_1 + 4x_2 - 2x_3 = 4$$
$$3x_1 + 2x_2 - 4x_3 = 6$$

by Gauss' elimination method.

Solution: Consider the augmented matrix

$$[A : B] = \begin{bmatrix} 4 & 1 & 1 & : & 4 \\ 1 & 4 & -2 & : & 4 \\ 3 & 2 & -4 & : & 6 \end{bmatrix} \quad (1)$$

Applying

$$R_2 - \frac{1}{4} R_1$$

and

$$R_3 - \frac{3}{4} R_1,$$

$$[A : B] = \begin{bmatrix} 4 & 1 & 1 & : & 4 \\ 0 & 15/4 & -9/4 & : & 3 \\ 0 & 5/4 & -19/4 & : & 3 \end{bmatrix} \quad (2)$$

Again apply, $R_3 - \frac{5}{4} R_2$, we get

$$[A : B] \sim \begin{bmatrix} 4 & 1 & 1 & : & 4 \\ 0 & \frac{15}{4} & -\frac{9}{4} & : & 3 \\ 0 & 0 & -4 & : & 2 \end{bmatrix}$$

Back substituting gives the

such $$x_3 = -\frac{1}{2}, \quad x_2 = \frac{1}{2}$$

and $$x_1 = 1$$

Example 8. Solve:

$$x + y + z = 6$$
$$3x + 3y + 4z = 20$$
$$2x + y + 3z = 13$$

using Gauss' elimination method.

Solution: The augmented matrix is

$$[A : B] = \begin{bmatrix} 1 & 1 & 1 & : & 6 \\ 3 & 3 & 4 & : & 20 \\ 2 & 1 & 3 & : & 13 \end{bmatrix}$$

Apply $R_2 - 3R_1$ and $R_3 - 2R$

$$[A : B] = \begin{bmatrix} 1 & 1 & 1 & : & 6 \\ 0 & 0 & 1 & : & 2 \\ 0 & -1 & 1 & : & 1 \end{bmatrix}$$

Solve, the second point is zero, we have $R_2 \leftrightarrow R_3$.

$$[A : B] = \begin{bmatrix} 1 & 1 & 1 & : & 6 \\ 0 & -1 & 1 & : & 1 \\ 0 & 0 & 1 & : & 2 \end{bmatrix}$$

Gauss $$x + y + z = 6$$
$$-y + z = 1$$
$$z = 2$$

Solution is $x = 3$, $y = 1$, $z = 2$.

Example 9. Using Gauss' elimination method, solve the system:

$$3.15x - 1.96y + 3.85z = 12.95$$
$$2.13x + 5.12y - 2.89z = -8.61$$
$$5.92x + 3.05y + 2.15z = 6.88$$

Solution:
$$[A : B] = \left[\begin{array}{ccc|c} 3.15 & -1.96 & 3.85 & 12.95 \\ 2.13 & 5.12 & -2.89 & -8.61 \\ 5.92 & 3.05 & 2.15 & 6.88 \end{array}\right]$$

$$\frac{R_2 + \left(-\dfrac{2.13}{3.15}\right)R_1}{R_3 + \left(\dfrac{-5.92}{3.15}\right)R_1} \left[\begin{array}{ccc|c} \boxed{3.15} & -1.96 & 3.85 & 12.95 \\ 0 & \boxed{6.4453} & -5.4933 & -17.3666 \\ 0 & 6.7335 & -5.0855 & -17.4578 \end{array}\right]$$

$$\sim \left[\begin{array}{ccc|c} 3.15 & -1.96 & 3.85 & 12.95 \\ 0 & \boxed{6.4453} & -5.4933 & -17.3666 \\ 0 & 0 & 0.6534 & 0.6853 \end{array}\right]$$

$\therefore$
$$3.15x - 1.96y + 3.85z = 12.95$$
$$6.4453y - 5.4933z = -17.3666$$
$$0.6534z = 0.6853$$

$\therefore$
$$z = \frac{0.6853}{0.6534} = 1.0488$$

$$y = \frac{5.4933 \times 1.0488 - 17.3666}{6.4453}$$
$$= -1.8005$$

$$x = \frac{1.96 \times (-1.8005) - 3.85\,[1.0488] + 12.95}{3.15}$$
$$= 1.7089$$

$\therefore$ $x = 1.7089, \quad y = -1.8005, \ z = 1.0488$

Example 10. Solve the system by Gauss'-Jordan method:

$$5x_1 + x_2 + x_3 + x_4 = 4$$
$$x_1 + 7x_2 + x_3 + x_4 = 12$$
$$x_1 + x_2 + 6x_3 + x_4 = -5$$
$$x_1 + x_2 + x_3 + 4x_4 = -6$$

Solution: Interchange the first and the last equation, so that the coefficient of x_1 in the first equation is 1. Then we write

$$[A:B] = \left[\begin{array}{cccc|c} 1 & 1 & 1 & 4 & -6 \\ 1 & 7 & 1 & 1 & 12 \\ 1 & 1 & 6 & 1 & -5 \\ 5 & 1 & 1 & 1 & 4 \end{array}\right]$$

Apply

$$R_2 \to R_2 + R_1\ (-1)$$
$$R_3 \to R_3 + R_1\ (-1)$$
$$R_4 \to R_4 + R_1\ (-\pi)$$

$$\begin{bmatrix} 1 & 1 & 1 & 4 & -6 \\ 0 & \boxed{6} & 0 & -3 & 18 \\ 0 & 0 & 5 & -3 & 1 \\ 0 & -4 & -4 & -19 & 34 \end{bmatrix}$$

Apply $R_2 \to R_2\left(\frac{1}{6}\right)$ to make the point as 1 $\to \left[\begin{array}{cccc|c} 1 & 1 & 1 & 4 & -6 \\ 0 & \boxed{1} & 0 & -0.5 & 3 \\ 0 & 0 & 5 & -3 & 1 \\ 0 & -4 & -4 & -19 & 34 \end{array}\right]$

$$\frac{R_1 \to R_1 + (-1)\,R_2}{R_4 \to R_4 + (4)\,R_2} \left[\begin{array}{cccc|c} 1 & 0 & 1 & 4.5 & -9 \\ 0 & 1 & 0 & -0.5 & 3 \\ 0 & 0 & \boxed{5} & -3 & 1 \\ 0 & 0 & -4 & -21 & 46 \end{array}\right]$$

$$R_3 \to R_3\left(\frac{1}{5}\right) \left[\begin{array}{cccc|c} 1 & 0 & 1 & 4.5 & -9 \\ 0 & 1 & 0 & -0.5 & 3 \\ 0 & 0 & 1 & -0.6 & 0.2 \\ 0 & 0 & -4 & -21 & 46 \end{array}\right]$$

$$\frac{R_1 \to R_1 + R_3\,(-1)}{R_4 \to R_4 + R_3\,(4)} \left[\begin{array}{cccc|c} 1 & 0 & 0 & 5.1 & -9.2 \\ 0 & 1 & 0 & -0.5 & 3 \\ 0 & 0 & 1 & -0.6 & 0.2 \\ 0 & 0 & 0 & -23.4 & 46.8 \end{array}\right]$$

$$R_4 \to \left(\frac{1}{23.4}\right) R_4 \left[\begin{array}{cccc|c} 1 & 0 & 0 & 5.1 & -9.2 \\ 0 & 1 & 0 & -0.5 & 3 \\ 0 & 0 & 1 & -0.6 & 0.2 \\ 0 & 0 & 0 & -1 & 2 \end{array}\right]$$

$$R_3 \to R_3 + R_4 \left(-\frac{1}{2}\right)$$

$$R_2 \to R_2 + R_4 \left(-\frac{1}{2}\right)$$

$$R_1 \to R_1 + R_4\ (5.1)$$

$$\to \left[\begin{array}{cccc|c} 1 & 0 & 0 & 0 & 1 \\ 0 & 1 & 0 & 0 & 2 \\ 0 & 0 & 1 & 0 & -1 \\ 0 & 0 & 0 & -1 & 2 \end{array}\right]$$

$$x_1 = 1,\ x_2 = 2,\ x_3 = -1,\ x_4 = -2.$$

Inverse of a Matrix

The Gauss' elimination process may be employed to find the inverse of a non-singular matrix.

If X is the invese of A, then $AX = I$ where I is the unit matrix of order 3. Now we have to find the element of X.

Let $$A = \begin{bmatrix} a_{11} & a_{12} & a_{13} \\ a_{21} & a_{22} & a_{23} \\ a_{31} & a_{32} & a_{33} \end{bmatrix}$$

and $$X = \begin{bmatrix} x_{11} & x_{12} & x_{13} \\ x_{21} & x_{22} & x_{23} \\ x_{31} & x_{32} & x_{33} \end{bmatrix}$$

$\therefore$ $AX = I$ reduces to

$$\begin{bmatrix} a_{11} & a_{12} & a_{13} \\ a_{21} & a_{22} & a_{23} \\ a_{31} & a_{32} & a_{33} \end{bmatrix} \begin{bmatrix} x_{11} & x_{12} & x_{13} \\ x_{21} & x_{22} & x_{23} \\ x_{31} & x_{32} & x_{33} \end{bmatrix} = \begin{bmatrix} 1 & 0 & 0 \\ 0 & 1 & 0 \\ 0 & 0 & 1 \end{bmatrix} \quad (1)$$

This equation is equivalent to three equations given below.

$$\begin{bmatrix} a_{11} & a_{12} & a_{13} \\ a_{21} & a_{22} & a_{23} \\ a_{31} & a_{32} & a_{33} \end{bmatrix} \begin{bmatrix} x_{11} \\ x_{21} \\ x_{31} \end{bmatrix} = \begin{bmatrix} 1 \\ 0 \\ 0 \end{bmatrix} \tag{2}$$

$$\begin{bmatrix} a_{11} & a_{12} & a_{13} \\ a_{21} & a_{22} & a_{23} \\ a_{31} & a_{32} & a_{33} \end{bmatrix} \begin{bmatrix} x_{12} \\ x_{22} \\ x_{32} \end{bmatrix} = \begin{bmatrix} 0 \\ 1 \\ 0 \end{bmatrix} \tag{3}$$

$$\begin{bmatrix} a_{11} & a_{12} & a_{13} \\ a_{21} & a_{22} & a_{23} \\ a_{31} & a_{32} & a_{33} \end{bmatrix} \begin{bmatrix} x_{13} \\ x_{23} \\ x_{33} \end{bmatrix} = \begin{bmatrix} 0 \\ 0 \\ 1 \end{bmatrix} \tag{4}$$

From equations (2), (3), (4), we can solve for the vectors

$$\begin{bmatrix} x_{11} \\ x_{21} \\ x_{31} \end{bmatrix}, \begin{bmatrix} x_{12} \\ x_{22} \\ x_{33} \end{bmatrix} \text{ and } \begin{bmatrix} x_{13} \\ x_{23} \\ x_{33} \end{bmatrix}$$

by using Gauss' elimination method or even by Gauss'-Jordan method. The solution set of each system (2), (3), (4) will be the corresponding column of the inverse matrix X.

Example 11. Find, by Gauss' elimination method, the inverse of

$$A = \begin{bmatrix} 3 & -1 & 1 \\ -15 & +6 & -5 \\ 5 & -2 & 2 \end{bmatrix}$$

Solution:

Step 1. We write down the augmented system $(A : I)$
That is,

$$(A : I) = \left[\begin{array}{ccc|ccc} 3 & -1 & 1 & 1 & 0 & 0 \\ -15 & 6 & -5 & 0 & 1 & 0 \\ 5 & -2 & 2 & 0 & 0 & 1 \end{array}\right] \tag{1}$$

Step 2. $$\frac{R_2 + 5R_1}{R_3 + \left(\frac{-5}{3}\right)R_1} \left[\begin{array}{ccc|ccc} 3 & -1 & 1 & 1 & 0 & 0 \\ 0 & 1 & 0 & 5 & 1 & 0 \\ 0 & \frac{-1}{3} & \frac{1}{3} & -\frac{5}{3} & 0 & 1 \end{array}\right] \tag{2}$$

Note: When we reduce the element below a_{11} in A to zero, only the first column of I is changed. Write the second and third columns unchanged.

Step 3. (1) reduces to if

$$R_3 + \left(\frac{1}{3}\right) R_2 \qquad \left[\begin{array}{ccc|ccc} 3 & -1 & 1 & 1 & 0 & 0 \\ 0 & 1 & 0 & 5 & 1 & 0 \\ 0 & 0 & \frac{1}{3} & 0 & \frac{1}{3} & 1 \end{array}\right] \tag{3}$$

Step 4. Now the system is equivalent to the 3 systems

$$\left[\begin{array}{ccc|c} 3 & -1 & 1 & 1 \\ 0 & 1 & 0 & 5 \\ 0 & 0 & \frac{1}{3} & 0 \end{array}\right] \tag{4}$$

$$\left[\begin{array}{ccc|c} 3 & -1 & 1 & 0 \\ 0 & 1 & 0 & 1 \\ 0 & 0 & \frac{1}{3} & \frac{1}{3} \end{array}\right] \tag{5}$$

and

$$\left[\begin{array}{ccc|c} 3 & -1 & 1 & 0 \\ 0 & 1 & 0 & 0 \\ 0 & 0 & \frac{1}{3} & 1 \end{array}\right] \tag{6}$$

That is,

$$\left.\begin{aligned} 3x_{11} - x_{21} + x_{31} &= 1 \\ x_{21} &= 5 \\ \frac{1}{3} x_{31} &= 0 \end{aligned}\right\} \Rightarrow \begin{aligned} x_{11} &= 0 \\ x_{21} &= 5 \\ x_{11} &= 2 \end{aligned}$$

$$\left.\begin{aligned} 3x_{12} - x_{22} + x_{32} &= 0 \\ x_{22} &= 1 \\ \frac{1}{3} x_{32} &= \frac{1}{3} \end{aligned}\right\} \Rightarrow \begin{aligned} x_{32} &= 1 \\ x_{22} &= 1 \\ x_{12} &= 0 \end{aligned}$$

$$\left.\begin{aligned} 3x_{13} - x_{23} + x_{33} &= 0 \\ x_{23} &= 0 \\ \frac{1}{3} x_{33} &= 1 \end{aligned}\right\} \Rightarrow \begin{aligned} x_{33} &= 3 \\ x_{23} &= 0 \\ x_{13} &= -1 \end{aligned}$$

Hence,

$$A^{-1} = \begin{bmatrix} 2 & 0 & -1 \\ 5 & 1 & 0 \\ 0 & 1 & 3 \end{bmatrix}$$

Example 12. By Gauss' elimination, find the inverse of

$$A = \begin{bmatrix} 0 & 1 & 1 \\ 1 & 2 & 0 \\ 3 & -1 & -4 \end{bmatrix}$$

Solution: The augmented system $(A : I)$ is

$$(A : I) = \left[\begin{array}{ccc|ccc} 0 & 1 & 1 & 1 & 0 & 0 \\ 1 & 2 & 0 & 0 & 1 & 0 \\ 3 & -1 & -4 & 0 & 0 & 1 \end{array}\right] \quad (1)$$

Since the element $a_{11} = 0$, we will interchange the first and second row. The reduced system is

$$\left[\begin{array}{ccc|ccc} 1 & 2 & 0 & 0 & 1 & 0 \\ 0 & 1 & 1 & 1 & 0 & 0 \\ 3 & -1 & -4 & 0 & 0 & 1 \end{array}\right] \quad (2)$$

Apply
$R_3 \to R_3 + (-3)\,R_1$, we get

$$\left[\begin{array}{ccc|ccc} 1 & 2 & 0 & 0 & 1 & 0 \\ 0 & 1 & 1 & 1 & 0 & 0 \\ 0 & -7 & -4 & 0 & -3 & 1 \end{array}\right] \quad (3)$$

Apply

$$R_3 \to R_3 + 7R_2 \left[\begin{array}{ccc|ccc} 1 & 2 & 0 & 0 & 1 & 0 \\ 0 & 1 & 1 & 1 & 0 & 0 \\ 0 & 0 & 3 & 7 & -3 & 1 \end{array}\right]$$

Thus,

$$\left.\begin{aligned} x_{11} + 2x_{21} &= 0 \\ x_{21} + x_{31} &= 1 \\ 3x_{31} &= 7 \end{aligned}\right\} \Rightarrow \quad \begin{aligned} x_{31} &= \frac{7}{3} \\ x_{21} &= -\frac{4}{3} \\ x_{11} &= \frac{8}{3} \end{aligned}$$

$$\left.\begin{aligned} x_{12} + 2x_{22} &= 1 \\ x_{22} + x_{32} &= 0 \\ 3x_{32} &= -3 \end{aligned}\right\} \Rightarrow \begin{aligned} x_{32} &= -1 \\ x_{22} &= 1 \\ x_{12} &= -1 \end{aligned}$$

$$\left.\begin{aligned} x_{13} + 2x_{23} &= 0 \\ x_{23} + x_{33} &= 0 \\ 3x_{33} &= 0 \end{aligned}\right\} \Rightarrow \begin{aligned} x_{33} &= \frac{1}{3} \\ x_{23} &= -\frac{1}{3} \\ x_{13} &= \frac{2}{3} \end{aligned}$$

Hence,
$$A^{-1} = \begin{bmatrix} \frac{8}{3} & -1 & \frac{2}{3} \\ \frac{-4}{3} & 1 & -\frac{1}{3} \\ \frac{7}{3} & -1 & \frac{1}{3} \end{bmatrix}$$

GAUSS'-JORDAN METHOD

In Gauss' method, we convert the coefficient matrix to an upper triangular matrix. But in Gauss'-Jordan method, we reduce A to a diagonal matrix or even to a unit matrix. This avoids back substitution and the solution is got directly from the modified b' column. For large systems the number of operations (multiplications, additions) involved are 50% more in Gauss'-Jordan method than Gauss' method and hence, it is not economical to use Gauss'-Jordan method.

Just as Gauss' method, Gauss'-Jordan method can also be used to find the determinant of a square matrix or to invert a non-singular matrix.

The system $AX = b$ will reduce to the form

$$\left[\begin{array}{ccccc|c} a_{11} & 0 & 0 & 0 & 0 & b_1 \\ 0 & b_{22} & 0 & 0 & 0 & c_2 \\ . & . & . & . & . & d_3 \\ 0 & 0 & 0 & 0 & \alpha_{nn} & k_n \end{array}\right] \qquad (1)$$

From (1), $x_n = \dfrac{k_n}{\alpha_{nm}}, \ldots, x_2 = \dfrac{c_2}{b_{22}}, x_1 = \dfrac{b_1}{a_{11}}$

Note: By this method, the values of $x_1, x_2 \ldots x_n$ are got immediately without using the process of back substitution.

Example 13. Solve: $3x + 2y = 4$, $2x - 3y = 7$ by Gauss'-Jordan method.

Solution: $\left[\begin{array}{cc|c} 3 & 2 & 4 \\ 2 & -3 & 7 \end{array}\right]$

$$R_2 - \frac{2}{3}R_1 \rightarrow \left[\begin{array}{cc|c} 3 & 2 & 4 \\ 0 & -\frac{13}{3} & \frac{13}{3} \end{array}\right]$$

$$R_1 + \frac{6}{13}R_2 \rightarrow \left[\begin{array}{cc|c} 3 & 0 & 6 \\ 0 & -\frac{13}{3} & \frac{13}{3} \end{array}\right]$$

$$-\frac{3}{13}R_2 \rightarrow \left[\begin{array}{cc|c} 1 & 0 & 2 \\ 0 & 1 & -1 \end{array}\right]$$

Hence, $x = 2$, $y = -1$.

Example 14. Solve by Gauss'-Jordan method:

$$\begin{aligned} 5x - 2y + 3z &= 18 \\ x + 7y - 3z &= -22 \\ 2x - y + 6z &= 22 \end{aligned}$$

Solution: $\left[\begin{array}{ccc|c} 5 & -2 & 3 & 18 \\ 1 & 7 & -3 & -22 \\ 2 & -1 & 6 & 22 \end{array}\right]$

$$\frac{1}{5}R_2 \rightarrow \left[\begin{array}{ccc|c} 1 & -0.4 & 0.6 & 3.6 \\ 1 & 7 & -3 & -22 \\ 2 & -1 & 6 & 22 \end{array}\right]$$

$$\frac{R_2 - R_1}{R_3 - 2R_1} \rightarrow \left[\begin{array}{ccc|c} 1 & -0.4 & 0.6 & 3.6 \\ 1 & 7.4 & -3.6 & -25.6 \\ 0 & -1.2 & 4.8 & 14.8 \end{array}\right]$$

$$\frac{1}{7.4}R_2 \rightarrow \left[\begin{array}{ccc|c} 1 & -0.4 & 0.6 & 3.6 \\ 0 & 1 & -0.49 & -3.46 \\ 0 & -0.2 & 4.8 & 14.8 \end{array}\right]$$

$$\frac{R_1 + 0.4R_2}{R_3 + 0.2R_2} \rightarrow \left[\begin{array}{ccc|c} 1 & 0 & 0.41 & 2.22 \\ 0 & 1 & -0.49 & -3.46 \\ 0 & 0 & 4.70 & 14.11 \end{array}\right]$$

$$\frac{1}{4.7}R_3 \rightarrow \left[\begin{array}{ccc|c} 1 & 0 & 0.41 & 2.22 \\ 0 & 1 & -0.49 & -3.46 \\ 0 & 0 & 1 & 3 \end{array}\right]$$

$$\frac{R_1 - 0.41R_3}{R_2 + 0.49R_3} \rightarrow \left[\begin{array}{ccc|c} 1 & 0 & 0 & 1 \\ 0 & 1 & 0 & -2 \\ 0 & 0 & 1 & 3 \end{array}\right]$$

Hence, $x = 1, y = -2, z = 3.$

Example 15. Find $|A|$ and A^{-1} if

$$A = \begin{bmatrix} 1 & 2 & -1 \\ 4 & 1 & 0 \\ 2 & -1 & 3 \end{bmatrix}, \text{ using Gauss'-Jordan method.}$$

Solution:

$$|A| = \begin{bmatrix} 1 & 2 & -1 \\ 4 & 1 & 0 \\ 2 & -1 & 3 \end{bmatrix}$$

$$\frac{R_2 - 4R_1}{R_3 - 2R_1} \rightarrow \begin{bmatrix} 1 & 2 & -1 \\ 0 & -7 & 4 \\ 0 & -5 & 5 \end{bmatrix}$$

$$\frac{R_1 + \frac{2}{7}R_2}{R_3 - \frac{5}{7}} \rightarrow \begin{bmatrix} 1 & 0 & \frac{1}{7} \\ 0 & -7 & 4 \\ 0 & 0 & \frac{15}{7} \end{bmatrix}$$

$$\frac{R_1 - \frac{1}{15}R_3}{R_2 - \frac{28}{15}R_3} \rightarrow \begin{bmatrix} 1 & 0 & 0 \\ 0 & -7 & 0 \\ 0 & 0 & \frac{15}{7} \end{bmatrix}$$

$$= 1\ (7)\left(\frac{15}{7}\right) = -15.$$

To find the inverse:

Let $$\left[\begin{array}{ccc|ccc} 1 & 2 & -1 & 1 & 0 & 0 \\ 4 & 1 & 0 & 0 & 1 & 0 \\ 2 & -1 & 3 & 0 & 0 & 1 \end{array}\right]$$

$$= \left[\begin{array}{ccc|ccc} 1 & 2 & -1 & 1 & 0 & 0 \\ 0 & -7 & 4 & -4 & 1 & 0 \\ 0 & -5 & 5 & -2 & 0 & 1 \end{array}\right]$$

$$= \left[\begin{array}{ccc|ccc} 1 & 0 & \frac{1}{7} & -\frac{1}{7} & \frac{2}{7} & 0 \\ 0 & -7 & 4 & -4 & 1 & 0 \\ 0 & 0 & \frac{15}{7} & \frac{6}{7} & -\frac{5}{7} & 1 \end{array}\right]$$

$$= \left[\begin{array}{ccc|ccc} 1 & 0 & 0 & -\frac{1}{5} & \frac{1}{3} & -\frac{1}{15} \\ 0 & -7 & 0 & -\frac{28}{5} & \frac{7}{3} & -\frac{28}{15} \\ 0 & 0 & \frac{15}{7} & \frac{6}{7} & -\frac{5}{7} & 1 \end{array}\right]$$

$$= \left[\begin{array}{ccc|ccc} 1 & 0 & 0 & -\frac{1}{5} & \frac{1}{3} & -\frac{1}{15} \\ 0 & 1 & 0 & \frac{4}{5} & -\frac{1}{3} & \frac{4}{15} \\ 0 & 0 & 1 & \frac{2}{5} & -\frac{1}{3} & \frac{7}{15} \end{array}\right]$$

Hence,

$$A^{-1} = \begin{bmatrix} -\frac{1}{5} & \frac{1}{3} & -\frac{1}{15} \\ \frac{4}{5} & -\frac{1}{3} & \frac{4}{15} \\ \frac{2}{5} & -\frac{1}{3} & \frac{7}{15} \end{bmatrix}$$

EXERCISES

I. Solve by Gauss'-Jordan method.

(i) $13x + 2y = 40$

$12x - 3y = 70$ $\left(\text{Ans: } \left(\frac{260}{63}, -\frac{430}{63}\right)\right)$

(ii) $4x - y - z = -7,$

$x - 5y + z = -10,$ (Ans: −1, 2, 1)

$x + 2y + 6z = 9.$

(iii) $2x - y + 2z = 2$

$x + 10y - 3z = 5$ (Ans: 2, 0, –1)

$x - y - z = 3$

(iv) $2x - y + 3z = 8$

$-x + 2y + z = 4$ (Ans: 2, 2, 2)

$3x + y - 4z = 0$

(v) $-3x + 6y - 11z = 14$

$3x - 4y + 6z = -5$ (Ans: 3, 2, –1)

$4x - 8y + 13z = -17$

(vi) $10x - y - z = -33$

$-x + 10y - z = 22$ (Ans: –3, 2, 1)

$x + y - 10z = -11$

II. Find $|A|$ and A^{-1} by Gauss'-Jordan method if

(i) $A = \begin{bmatrix} 1 & -1 & 1 \\ 1 & -2 & 4 \\ 1 & 2 & 2 \end{bmatrix}$; (ii) $A = \begin{bmatrix} 2 & 1 & 1 \\ 1 & 0 & -1 \\ 2 & -1 & 2 \end{bmatrix}$

$$\left(\text{Ans: (i) } -15, \begin{bmatrix} -\frac{1}{5} & \frac{1}{3} & -\frac{1}{15} \\ \frac{4}{5} & -\frac{1}{3} & \frac{4}{15} \\ -\frac{2}{5} & -\frac{1}{3} & \frac{7}{15} \end{bmatrix}\right)$$

$$\left(\text{Ans: (ii) } -7, \frac{1}{7}\begin{bmatrix} 1 & 3 & 1 \\ 4 & -2 & -3 \\ 1 & -4 & 1 \end{bmatrix}\right)$$

CROUT'S METHOD

This method is also called LU decomposition or triangularisation. It is a direct method. But we will follow a different technique as suggested by Crout.

To solve $AX = B$ (1)

where $A = \begin{bmatrix} a_{11} & a_{12} & a_{13} \\ a_{21} & a_{22} & a_{23} \\ a_{31} & a_{32} & a_{33} \end{bmatrix}$, $X = \begin{bmatrix} x_1 \\ x_2 \\ x_3 \end{bmatrix}$ and $B = \begin{bmatrix} b_1 \\ b_2 \\ b_3 \end{bmatrix}$

We split the matrix A into two factors

$$A = LU \tag{2}$$

where
$$[L] = \begin{bmatrix} l_{11} & 0 & 0 \\ l_{21} & l_{22} & 0 \\ l_{31} & l_{32} & l_{33} \end{bmatrix} \text{ and } [U] = \begin{bmatrix} 1 & u_{12} & u_{13} \\ 0 & 1 & u_{23} \\ 0 & 0 & 1 \end{bmatrix}$$

Since
$$AX = B$$
$$LUX = B$$

∴
$$LY = B$$

where
$$UX = Y \tag{3}$$

$LU = A$ reduces to

$$\begin{bmatrix} l_{11} & 0 & 0 \\ l_{21} & l_{22} & 0 \\ l_{31} & l_{32} & l_{33} \end{bmatrix} \begin{bmatrix} 1 & u_{12} & u_{13} \\ 0 & 1 & u_{23} \\ 0 & 0 & 1 \end{bmatrix} = \begin{bmatrix} a_{11} & a_{12} & a_{13} \\ a_{21} & a_{22} & a_{23} \\ a_{31} & a_{32} & a_{33} \end{bmatrix} \tag{4}$$

$$\Rightarrow \begin{bmatrix} l_{11} & l_{11}u_{12} & l_{11}u_{13} \\ l_{21} & l_{21}u_{12} + l_{22} & l_{21}u_{13} + l_{22}u_{23} \\ l_{31} & l_{31}u_{12} + l_{32} & l_{31}u_{13} + l_{32}u_{23} + l_{33} \end{bmatrix} = \begin{bmatrix} a_{11} & a_{12} & a_{13} \\ a_{21} & a_{22} & a_{23} \\ a_{31} & a_{32} & a_{33} \end{bmatrix}$$

Equating coefficients and simplifying, we get

$$l_{11} = a_{11}; \quad l_{21} = a_{21}; \quad l_{31} = a_{31}$$

$$u_{12} = \frac{a_{12}}{a_{11}}; \quad u_{13} = \frac{a_{13}}{a_{11}}$$

$$l_{22} = a_{22} - l_{21}\, u_{12}; \quad l_{32} = a_{32} - l_{31}\, u_{12};$$

$$u_{23} = \frac{a_{23} - l_{21}u_{13}}{l_{22}}; \quad l_{33} = a_{33} - l_{31}\, u_{13} - l_{32}\, u_{23}$$

Now $[L]$ and $[U]$ are known.
Since $LY = B$, we get

$$\begin{bmatrix} l_{11} & 0 & 0 \\ l_{21} & l_{22} & 0 \\ l_{31} & l_{32} & l_{33} \end{bmatrix} \begin{bmatrix} y_1 \\ y_2 \\ y_3 \end{bmatrix} = \begin{bmatrix} b_1 \\ b_2 \\ b_3 \end{bmatrix}$$

Multiplying and equating coefficients, we obtain:

therefore

$$l_{11}\, y_1 = b_1 \quad \Rightarrow \quad y_1 = \frac{b_1}{a_{11}}$$

$$l_{21}\, y_1 + l_{22}\, y_2 = b_2 \quad \Rightarrow \quad y_2 = \frac{b_2 - l_{21}\, y_1}{l_{22}}$$

$$l_{31}\, y_1 + l_{32}\, y_2 + l_{33}\, y_3 = b_3 \quad \Rightarrow \quad y_3 = \frac{b_3 - l_{31}\, y_1 - l_{32}\, y_2}{l_{33}}$$

Knowing $[Y] = \begin{bmatrix} y_1 \\ y_2 \\ y_3 \end{bmatrix}$, $[L]$ and $[U]$

$[X]$ can be found out from $UX = Y$.

Note (i): We compute 12 unknowns:

$$\begin{Bmatrix} l_{11}, & l_{21,} & l_{22,} & l_{31,} & l_{32,} & l_{33} \\ u_{12,} & u_{13,} & u_{23,} & y_{1,} & y_{2,} & y_3 \end{Bmatrix} \text{ as a matrix, called}$$

$$\text{derived matrix: } DM = \begin{bmatrix} l_{11} & u_{12} & u_{13} & y_1 \\ l_{21} & l_{22} & u_{23} & y_2 \\ l_{31} & l_{32} & l_{33} & y_3 \end{bmatrix}$$

Note (ii): If L is lower triangular, then L^{-1} is also lower triangular. Also U is upper triangular, so is U^{-1}.

Since $\quad LL^{-1} = I$

and $\quad UU^{-1} = I$

Having known, L^{-1} and U^{-1}, we can find $A^{-1} = U^{-1}\, L^{-1}$
$(\because\ A^{-1} = (LU)^{-1} = U^{-1}\, L^{-1})$

Thus, if $A = LU$, then $A^{-1} = U^{-1}\, L^{-1}$ can be found. So, we can find inverse of a square matrix by Crout's method.

Example 16. Use Crout's method to solve the system:

$$x + y + z = 3$$
$$2x - y + 3z = 16$$
$$3x + y - z = -3$$

Solution:

$$\text{Here } [A:B] = \left[\begin{array}{ccc|c} 1 & 1 & 1 & 3 \\ 2 & -1 & 3 & 16 \\ 3 & 1 & -1 & -3 \end{array}\right]$$

$$\text{Let the derived matrix be: D.M.} = \left[\begin{array}{ccc|c} l_{11} & u_{12} & u_{13} & u_1 \\ l_{21} & l_{22} & u_{23} & y_2 \\ l_{31} & l_{32} & l_{33} & y_3 \end{array}\right]$$

$$\text{Elements of first column of D.M. are} = \begin{bmatrix} 1 & \cdot & \cdot & \cdot \\ 2 & \cdot & \cdot & \cdot \\ 3 & \cdot & \cdot & \cdot \end{bmatrix}$$

Elements of first row of D.M:

$$U_{12} = \frac{1}{1} = 1; \quad u_{13} = \frac{1}{1} = 1; \quad y_1 \frac{3}{1} = 3$$

$$\therefore \qquad \text{D.M.} = \begin{bmatrix} 1 & 1 & 1 & 3 \\ 2 & \cdot & \cdot & \cdot \\ 3 & \cdot & \cdot & \cdot \end{bmatrix}$$

Elements of second column:

$$l_{22} = a_{22} - u_{12}\, l_{21} = -1 - 2 = -3$$

$$l_{32} = a_{32} - u_{12}\, l_{31} = 1 - 1 \times 3 = -2$$

$$\therefore \qquad \text{D.M.} = \begin{bmatrix} 1 & 1 & 1 & 3 \\ 2 & -3 & \cdot & \cdot \\ 3 & -2 & \cdot & \cdot \end{bmatrix}$$

Elements of second row:

$$u_{23} = \frac{3 - 1(2)}{-3} = \frac{-1}{3}$$

$$y_2 = \frac{16 - 3 \times 2}{-3} = \frac{-10}{3}$$

$$\therefore \qquad \text{D.M.} = \begin{bmatrix} 1 & 1 & 1 & 3 \\ 2 & -3 & \frac{-1}{3} & \frac{-10}{3} \\ 3 & -2 & \cdot & \cdot \end{bmatrix}$$

Elements of third column:

$$l_{33} = 1 - 1\ (3) - (-1/3)\ (-2) = -\frac{14}{3}$$

$$y_3 = \frac{-3-(3)(3)-(-2)\left(\frac{-10}{3}\right)}{\frac{-14}{3}} = 4$$

$$\text{D.M.} = \begin{bmatrix} 1 & 1 & 1 & 3 \\ 2 & -3 & -1/3 & \frac{-10}{3} \\ 3 & -2 & -\frac{14}{3} & 4 \end{bmatrix}$$

The solution is obtained from $UX = Y$ i.e.,

$$\begin{bmatrix} 1 & 1 & 1 \\ 0 & 1 & -1/3 \\ 0 & 0 & 1 \end{bmatrix} \begin{bmatrix} x \\ y \\ z \end{bmatrix} = \begin{bmatrix} 3 \\ -10/3 \\ 4 \end{bmatrix}$$

$$x + y + z = 3$$

$$y - \frac{1}{3}\ z = \frac{-10}{3}$$

$$z = 4$$

By back substitution, $z = 4$, $y = -2$, $x = 1$.

Example 17. Solve the system of equation:

$$3x_1 - x_2 + 2x_3 = 12$$
$$x_1 + 2x_2 + 3x_3 = 11$$
$$2x_1 - 2x_2 - x_3 = 2$$

by Crout's reduction technique.

Solution: The matrix equation is

$$AX = B,$$

where

$$A = \begin{bmatrix} 3 & -1 & 2 \\ 1 & 2 & 3 \\ 2 & -2 & -1 \end{bmatrix}, \quad X = \begin{bmatrix} x_1 \\ x_2 \\ x_3 \end{bmatrix}, \quad B = \begin{bmatrix} 12 \\ 11 \\ 2 \end{bmatrix}$$

Suppose that $LU = A$, then

$$\begin{bmatrix} l_{11} & 0 & 0 \\ l_{21} & l_{22} & 0 \\ l_{31} & l_{32} & l_{33} \end{bmatrix} \begin{bmatrix} 1 & u_{12} & u_{13} \\ 0 & 1 & u_{23} \\ 0 & 0 & 1 \end{bmatrix} = \begin{bmatrix} 3 & -1 & 2 \\ 1 & 2 & 3 \\ 2 & -2 & -1 \end{bmatrix}$$

Multiplying the matrices on the l.h.s. and comparing the corresponding elements, we get

$$l_{11} = 3,\ l_{21} = 1,\ l_{31} = 2$$

$$l_{11}\, u_{12} = -1,$$

$$\therefore \quad u_{12} = \frac{-1}{l_{11}} = -\frac{1}{3};$$

$$l_{22} = 2 - l_{21}\, u_{12} = 2 - 1\left(-\frac{1}{3}\right) = \frac{7}{3};$$

$$l_{32} = -2 - l_{31}\, u_{12} = -2 - (2)\left(-\frac{1}{3}\right) = -\frac{4}{3};$$

$$l_{11}\, u_{13} = 2;$$

$$l_{21}\, u_{13} + l_{22}\, u_{23} = 3;$$

$$l_{31}\, u_{13} + l_{32}\, u_{23} + l_{33} = -1$$

$$\Rightarrow \quad u_{13} = \frac{2}{l_{11}} = \frac{2}{3};$$

$$u_{23} = \frac{3 - l_{21}\, u_{13}}{l_{22}} = \frac{3-(1)\left(\frac{2}{3}\right)}{\left(\frac{7}{3}\right)} = \frac{9-2}{\left(\frac{7}{3}\right)} = \frac{7\times 3}{7} = 3;$$

$$l_{33} = -1 - l_{31}\, u_{13} - l_{32}\, u_{23} = -1 - (2)\left(\frac{2}{3}\right) - \left(\frac{-4}{3}\right) \quad (1)$$

$$= -1 - \frac{4}{3} + 4 = 5 - \frac{4}{3} = \frac{15-4}{3} = \frac{11}{3}$$

Thus, we have

$$L = \begin{bmatrix} 3 & 0 & 0 \\ 1 & \frac{7}{3} & 0 \\ 2 & \frac{-4}{3} & \frac{11}{3} \end{bmatrix} \quad \text{and} \quad U = \begin{bmatrix} 1 & \frac{-1}{3} & \frac{2}{3} \\ 0 & 1 & 3 \\ 0 & 0 & 1 \end{bmatrix}$$

Now, consider $LY = B$, namely

$$\begin{bmatrix} 3 & 0 & 0 \\ 1 & \frac{7}{3} & 0 \\ 2 & \frac{-4}{3} & \frac{11}{3} \end{bmatrix} \begin{bmatrix} y_1 \\ y_2 \\ y_3 \end{bmatrix} = \begin{bmatrix} 12 \\ 11 \\ 2 \end{bmatrix}$$

$$3y_1 = 12$$

$$\Rightarrow \quad y_1 = 4,$$

$$y_1 + \frac{7}{3} y_2 = 11$$

$$\Rightarrow \quad y_2 = \frac{3(11-4)}{7} = 3$$

$$2y_1 - \frac{4}{3} y_2 + \frac{11}{3} y_3 = 2$$

$$8 - 4 + \frac{11}{3} y_3 = 2$$

$$y_3 = \frac{(2-4)^3}{11} = -\frac{6}{11}$$

Next, consider $UX = Y$, namely

$$\begin{bmatrix} 1 & -\frac{1}{3} & \frac{2}{3} \\ 0 & 1 & 3 \\ 0 & 0 & 1 \end{bmatrix} \begin{bmatrix} x_1 \\ x_2 \\ x_3 \end{bmatrix} = \begin{bmatrix} 4 \\ 3 \\ \frac{-6}{11} \end{bmatrix}$$

This yields

$$x_1 - \frac{1}{3} x_2 + \frac{2}{3} x_3 = 4, \; x_2 + 3x_3 = 3; \; x_3 = -\frac{6}{11}$$

by back substitution, we get

$$x_2 = 3 + \frac{3(6)}{11} = 3 + \frac{18}{11} = \frac{33+18}{11} = \frac{51}{11}$$

$$x_1 = 4 + \frac{51}{3\times 11} - \frac{2}{3}\left(\frac{-6}{11}\right)$$

$$= 4 + \frac{17}{11} + \frac{4}{11}$$

$$= \frac{44+17+4}{11} = \frac{65}{11}$$

$$\therefore \quad x_1 = \frac{65}{11}, \; x_2 = \frac{51}{11}, \; x_3 = -\frac{6}{11}$$

Example 18. Solve the system of equations:

$$\begin{aligned} 10x_1 + 7x_2 + 8x_3 + 7x_4 &= 32 \\ 7x_1 + 5x_2 + 6x_3 + 5x_4 &= 23 \\ 8x_1 + 6x_2 + 10x_3 + 9x_4 &= 3 \\ 7x_1 + 5x_2 + 9x_3 + 10x_4 &= 31 \end{aligned}$$

by Crout's reduction technique.

Solution: The given equations are written in the matrix form as:

$$\begin{bmatrix} 10 & 7 & 8 & 7 \\ 7 & 5 & 6 & 5 \\ 8 & 6 & 10 & 9 \\ 7 & 5 & 9 & 10 \end{bmatrix} \begin{bmatrix} x_1 \\ x_2 \\ x_3 \\ x_4 \end{bmatrix} = \begin{bmatrix} 32 \\ 23 \\ 3 \\ 31 \end{bmatrix}$$

The $[L]$ and $[M]$ are multiplied as shown below:

$$\begin{bmatrix} l_{11} & 0 & 0 & 0 \\ l_{21} & l_{22} & 0 & 0 \\ l_{31} & l_{32} & l_{33} & 0 \\ l_{41} & l_{42} & l_{43} & l_{44} \end{bmatrix} \begin{bmatrix} 1 & u_{12} & u_{13} & u_{14} \\ 0 & 1 & u_{23} & u_{24} \\ 0 & 0 & 1 & u_{34} \\ 0 & 0 & 0 & 1 \end{bmatrix} = \begin{bmatrix} 10 & 7 & 8 & 7 \\ 7 & 5 & 6 & 5 \\ 8 & 6 & 10 & 9 \\ 7 & 5 & 9 & 10 \end{bmatrix}$$

$$l_{11} = 10, \qquad l_{21} = 7,$$
$$l_{31} = 8, \qquad l_{41} = 7 \tag{1}$$
$$l_{11}\, u_{12} = 7, \qquad l_{21}\, u_{12} + l_{22} = 5,$$
$$l_{31}\, u_{12} + l_{32} = 6, \qquad l_{41}\, u_{12} + l_{42} = 5 \tag{2}$$
$$l_{11}\, u_{13} = 8, \qquad l_{21}\, u_{13} + l_{22}\, u_{23} = 6,$$
$$l_{31}\, u_{13} + l_{32}\, u_{23} + l_{33} = 10, \qquad l_{41}\, u_{13} + l_{42}\, u_{23} + l_{43} = 9 \tag{3}$$
$$l_{11}\, u_{14} = 7, \qquad l_{21}\, u_{14} + l_{22}\, u_{24} = 5,$$
$$l_{31}\, u_{14} + l_{32}\, u_{24} + l_{33}\, u_{34} = 9 \qquad l_{41}\, u_{14} + l_{42}\, u_{24} + l_{43}\, u_{34} + l_{44} = 10 \tag{4}$$

Solving the above equations, we obtain from (1), (2), (3) and (4):

$$l_{22} = \frac{1}{10}, \qquad l_{32} = \frac{2}{5}, \qquad l_{42} = \frac{1}{10},$$
$$l_{33} = 2, \qquad l_{43} = 3, \qquad l_{44} = \frac{27}{5} \tag{5}$$
$$u_{12} = \frac{7}{10}, \qquad u_{13} = \frac{4}{5}, \qquad u_{23} = 4,$$
$$u_{14} = \frac{7}{10}, \qquad u_{24} = 1, \qquad u_{34} = \frac{3}{2} \tag{6}$$

$$\underset{[L]}{\begin{bmatrix} 10 & 0 & 0 & 0 \\ 7 & 1/10 & 0 & 0 \\ 8 & 2/5 & 2 & 0 \\ 7 & 1/10 & 3 & 1/2 \end{bmatrix}} \underset{[U]}{\begin{bmatrix} 1 & 7/10 & 4/5 & 7/10 \\ 0 & 1 & 4 & 1 \\ 0 & 0 & 1 & 3/2 \\ 0 & 0 & 0 & 1 \end{bmatrix}} = \underset{[A]}{\begin{bmatrix} 10 & 7 & 8 & 7 \\ 7 & 5 & 6 & 5 \\ 8 & 6 & 10 & 9 \\ 7 & 5 & 9 & 10 \end{bmatrix}}$$

Now, consider the matrix equation $LY = B$, namely

$$\begin{bmatrix} 10 & 0 & 0 & 0 \\ 7 & 1/10 & 0 & 0 \\ 8 & 2/5 & 2 & 0 \\ 7 & 1/10 & 3 & 1/2 \end{bmatrix} \begin{bmatrix} y_1 \\ y_2 \\ y_3 \\ y_4 \end{bmatrix} = \begin{bmatrix} 32 \\ 23 \\ 33 \\ 31 \end{bmatrix}$$

which yields

$$10y_1 = 32; \quad \therefore \quad y_1 = \frac{16}{5}$$

$$7y_1 + \frac{1}{10} y_2 = 23; \quad \therefore \quad y_2 = 6$$

$$8y_1 + \frac{2}{5} y_2 + 2y_3 = 33; \quad \therefore \quad y_3 = 5/2$$

$$7y_1 + \frac{1}{10} y_2 + 3y_3 + \frac{1}{2} y_4 = 31, \quad \therefore \quad y_4 = 1$$

The solution is obtained from $UX = Y$, namely

$$\begin{bmatrix} 1 & 7/10 & 4/5 & 7/10 \\ 0 & 1 & 4 & 1 \\ 0 & 0 & 1 & 3/2 \\ 0 & 0 & 0 & 1 \end{bmatrix} \begin{bmatrix} x_1 \\ x_2 \\ x_3 \\ x_4 \end{bmatrix} = \begin{bmatrix} 16/5 \\ 6 \\ 5/2 \\ 1 \end{bmatrix}$$

$$\Rightarrow \quad x_1 + \frac{7}{10} x_2 + \frac{4}{5} x_3 + \frac{7}{10} x_4 = \frac{16}{5}$$

$$x_2 + 4x_3 + x_4 = 6$$

$$x_3 + \frac{3}{2} x_4 = 5/2, \; x_4 = 1$$

By back substitution, we obtain:

$$x_1 = x_2 = x_3 = x_4 = 1.$$

is the solution of the given system.

EXERCISES

Using Crout's method, solve:

1. $3x + 4y = 19$; $2x - 3y = 7$.

[Ans: (5, 1)]

2. $10x - y - z = -33$; $x - 10y + z = -22$; $x + y - 10z = -11$

[Ans: (–3, 2, 1)]

3. $3x - 6y + 11z = -14$; $3x - 4y + 6z = -5$; $4x - 8y + 13z = -17$

[Ans: (3, 2, –1)]

4. $x + 2y - z = -3$, $2x + 3y - z = -2$, $x - y + 2z = 9$

[Ans: (3, –2, 2)]

5. $4x - y - z = -7$; $x - 5y + z = -10$; $x + 2y + 6z = 9$

[Ans: (–1, 2, 1)]

6. $2.51x + 1.48y + 4.53z = 0.05$
   $1.48x + 0.93y - 1.30z = 1.03$
   $2.68x + 3.04y - 1.48z = -0.53$

[Ans: (1.453, –1.589, –0.275)]

7. Using Crout's method, find det A and A^{-1} if

$$\text{(i) } A = \begin{bmatrix} 2 & 1 & 1 \\ 1 & 0 & -1 \\ 2 & -1 & 2 \end{bmatrix}, \qquad \text{(ii) } A = \begin{bmatrix} 1 & 2 & -1 \\ 4 & 1 & 0 \\ 2 & -1 & 3 \end{bmatrix}$$

$$\left(\text{Ans: (i)} -7, \frac{1}{7}\begin{bmatrix} 1 & 3 & 1 \\ 4 & -2 & -3 \\ 1 & -4 & 1 \end{bmatrix}; \text{(ii)} -15, \frac{1}{15}\begin{bmatrix} -3 & 5 & -1 \\ 12 & -5 & 4 \\ 6 & -5 & 7 \end{bmatrix}\right)$$

PARTITION METHOD

Partition method is used to find the inverse of a matrix. Let the coefficient matrix A of the matrix equation

$$AX = B \tag{1}$$

where

$$A = \begin{bmatrix} a_{11} & a_{12} & \dots & a_{1n} \\ a_{21} & a_{22} & \dots & a_{2n} \\ \dots & \dots & \dots & \dots \\ a_{n1} & a_{2n} & \dots & a_{nn} \end{bmatrix} \tag{2}$$

be partitioned as

$$A = \left[\begin{array}{c|c} C & D \\ \hline E & F \end{array}\right] \tag{3}$$

where C is an $r \times r$ matrix, D is an $r \times s$ matrix, E is an $s \times r$ matrix and D is an $s \times s$ matrix where r and s are positive integers such that $r + s = n$.

Further, we also assume that A^{-1} is partitioned similarly as

$$A^{-1} = \left[\begin{array}{c|c} U & V \\ \hline Y & Z \end{array}\right] \tag{4}$$

where U, V, Y and Z are matrices of the same order as C, D, E, F respectively then

$$AA^{-1} = \left[\begin{array}{c|c} C & D \\ \hline E & F \end{array}\right] \left[\begin{array}{c|c} U & V \\ \hline Y & Z \end{array}\right] = \left[\begin{array}{c|c} I_1 & 0 \\ \hline 0 & I_2 \end{array}\right] \tag{5}$$

where I_1 and I_2 are unit matrices of orders r and s respectively.

From (5), we obtain:

$$CU + DY = I_1 \tag{6}$$
$$CV + DZ = 0 \tag{7}$$
$$EU + FY = 0 \tag{8}$$
$$EV + FZ = I_2 \tag{9}$$

From (7), we get

$$V = -C^{-1} DZ \tag{10}$$

which when substituted into (9) yields

$$(F - EC^{-1} D) Z = I_2$$

so that

$$Z = (F - EC^{-1} D)^{-1} \tag{11}$$

From (6), we get

$$U = C^{-1} (J, - DY) = (C^{-1} - C^{-1} DY) \tag{12}$$

which when substituted into (8) yields

$$E (C^{-1} - C^{-1} DY) + FY = 0$$

so that

$$Y = - (F - EC^{-1} D)^{-1} EC^{-1}$$
$$= - Z E C^{-1} \tag{13}$$

Substituting (10) to (13), we obtain A^{-1}.

Then the solution of the system (1) is

$$X = A^{-1} B \tag{14}$$

This method of solving the equation (1) is known as partition method. The method essentially consists in determining the invevse A^{-1} of the coefficient matrix A of the system (L).

Example 19. Using the partition method, solve the system of equations:

$$2x + 4y + 3z = 4$$
$$y + z = 1$$
$$2x + 2y - z = -2$$

Solution: Let A be the coefficient matrix of the given system. We first find A^{-1} by the partition method. Let us partition A as shown below:

$$A = \left[\begin{array}{c} C \mid D \\ \hline E \mid F \end{array}\right] = \left[\begin{array}{cc|c} 2 & 4 & 3 \\ 0 & 1 & 1 \\ \hline 2 & 2 & -1 \end{array}\right]$$

Then, we find

$$C^{-1} = \begin{bmatrix} 2 & 4 \\ 0 & 1 \end{bmatrix}^{-1} = \begin{bmatrix} \frac{1}{2} & -2 \\ 0 & 1 \end{bmatrix}$$

$$\therefore \qquad F - EC^{-1} D = [-1] - [2\ 2] \begin{bmatrix} \frac{1}{2} & -2 \\ 0 & 1 \end{bmatrix} \begin{bmatrix} 3 \\ 1 \end{bmatrix} = [-2]$$

We put

$$Z = (F - EC^{-1} D)^{-1} = \left[-\frac{1}{2}\right]$$

$$Y = -ZEC^{-1} = -\left[-\frac{1}{2}\right] [22] \begin{bmatrix} \frac{1}{2} & -2 \\ 0 & 1 \end{bmatrix}$$

$$= \left[\frac{1}{2} - 1\right]$$

$$V = -C^{-1} DZ = -\begin{bmatrix} \frac{1}{2} & -2 \\ 0 & 1 \end{bmatrix} \begin{bmatrix} 3 \\ 1 \end{bmatrix} \left[-\frac{1}{2}\right]$$

$$= \begin{bmatrix} -1/4 \\ 1/2 \end{bmatrix}$$

and

$$U = C^{-1} - C^{-1} DY$$

$$= \begin{bmatrix} \frac{1}{2} & -2 \\ 0 & 1 \end{bmatrix} - \begin{bmatrix} \frac{1}{2} & -2 \\ 0 & 1 \end{bmatrix} \begin{bmatrix} 3 \\ 1 \end{bmatrix} \left[\frac{1}{2} - 1\right]$$

$$= \begin{bmatrix} \frac{1}{2} & -2 \\ 0 & 1 \end{bmatrix} - \begin{bmatrix} -\frac{1}{4} & \frac{1}{2} \\ \frac{1}{2} & -1 \end{bmatrix} = \begin{bmatrix} \frac{3}{4} & -\frac{5}{2} \\ \frac{-1}{2} & 2 \end{bmatrix}$$

Then, we get

$$A^{-1} = \left[\begin{array}{c|c} U & V \\ \hline Y & Z \end{array}\right] = \begin{bmatrix} \frac{3}{4} & \frac{-5}{2} & \frac{-1}{4} \\ -\frac{1}{2} & 2 & \frac{1}{2} \\ \frac{1}{2} & -1 & -\frac{1}{2} \end{bmatrix}$$

Using this, A^{-1}, the matrix equation $X = A^{-1} B$ gives

$$\begin{bmatrix} x \\ y \\ z \end{bmatrix} = \begin{bmatrix} \frac{3}{4} & \frac{-5}{2} & \frac{-1}{4} \\ -\frac{1}{2} & 2 & \frac{1}{2} \\ \frac{1}{2} & -1 & -\frac{1}{2} \end{bmatrix} \begin{bmatrix} 4 \\ 1 \\ -2 \end{bmatrix}$$

$$= \begin{bmatrix} 1 \\ -1 \\ 2 \end{bmatrix}$$

so that $x = 1$, $y = -1$, $z = 2$ is the required solution.

Example 20. Determine the inverse of the matrix $\begin{bmatrix} 1 & 1 & 1 \\ 4 & 3 & -1 \\ 3 & 5 & 3 \end{bmatrix}$ by using the partition method. Hence, find the solution of the system of equations

$$\begin{aligned} x_1 + x_2 + x_3 &= 1 \\ 4x_1 + 3x_2 - x_3 &= 6 \\ 3x_1 + 5x_2 + 3x_3 &= 4 \end{aligned}$$

Solution: We first find A^{-1} by the partition method. Let us partition A as shown below:

$$A = \left[\begin{array}{c|c} C & D \\ \hline E & F \end{array}\right] = \left[\begin{array}{cc|c} 1 & 1 & 1 \\ 4 & 3 & -1 \\ \hline 3 & 5 & 3 \end{array}\right]$$

and $$C^{-1} = \begin{bmatrix} 1 & 1 \\ 4 & 3 \end{bmatrix}^{-1} = -\begin{bmatrix} 3 & -1 \\ -4 & 1 \end{bmatrix}$$

$\therefore$ $$F - EC^{-1} D = 3 + [35] \begin{bmatrix} 3 & -1 \\ -4 & 1 \end{bmatrix} \begin{bmatrix} 1 \\ -1 \end{bmatrix} = -10$$

We put

$$Z = (F - EC^{-1} D)^{-1} = \left[-\frac{1}{10}\right]$$

$$Y = - ZEC^{-1} = - \left[-\frac{1}{10}\right] [35] \begin{bmatrix} 3 & -1 \\ -4 & 1 \end{bmatrix}$$

$$= - \frac{1}{10} [-11 \;\; 2]$$

$$V = -C^{-1} DZ = - \begin{bmatrix} 3 & -1 \\ -4 & 1 \end{bmatrix} \begin{bmatrix} 1 \\ -1 \end{bmatrix} \left(-\frac{1}{10}\right)$$

$$= \frac{-1}{10} \begin{bmatrix} 4 \\ -5 \end{bmatrix}$$

and

$$U = C^{-1} - C^{-1} DY$$

$$= - \begin{bmatrix} 3 & -1 \\ -4 & 1 \end{bmatrix} + \begin{bmatrix} 3 & -1 \\ -4 & 1 \end{bmatrix} \begin{bmatrix} 1 \\ -1 \end{bmatrix} \left(\frac{-1}{10}\right) [-11 \;\; 2]$$

$$= \begin{bmatrix} 1.4 & 0.2 \\ -1.5 & 0 \end{bmatrix}$$

Then, we get

$$A^{-1} = \left[\frac{U \mid V}{Y \mid Z}\right] = \begin{bmatrix} 1.4 & 0.2 & -0.4 \\ -1.5 & 0 & 0.5 \\ 1.1 & -0.2 & -0.1 \end{bmatrix}$$

The solution of the given system is

$$\begin{bmatrix} x_1 \\ x_2 \\ x_3 \end{bmatrix} = \begin{bmatrix} 1.4 & 0.2 & -0.4 \\ -1.5 & 0 & 0.5 \\ 1.1 & -0.2 & -0.1 \end{bmatrix} \begin{bmatrix} 1 \\ 6 \\ 4 \end{bmatrix} = \begin{bmatrix} 1 \\ 0.5 \\ -0.5 \end{bmatrix}$$

so that

$$x = 1, y = 0.5, z = - 0.5$$

is the required solution.

Example 21. By the partition method solve the system

$$2x + y = 3$$
$$x + 2y + z = 2$$
$$y + 2z + u = 1$$
$$z + 2u = 3$$

Solution: Let the coefficient matrix be partitioned as shown below.

$$A = \left[\begin{array}{c|c} C & D \\ \hline E & F \end{array}\right] = \left[\begin{array}{cc|cc} 2 & 1 & 0 & 0 \\ 1 & 2 & 1 & 0 \\ \hline 0 & 1 & 2 & 1 \\ 0 & 0 & 1 & 2 \end{array}\right]$$

Then
$$C^{-1} = \begin{bmatrix} 2 & 1 \\ 1 & 2 \end{bmatrix}^{-1} = \frac{1}{3}\begin{bmatrix} 2 & -1 \\ -1 & 2 \end{bmatrix}$$

and
$$F - EC^{-1}D = \begin{bmatrix} 2 & 1 \\ 1 & 2 \end{bmatrix} - \frac{1}{3}\begin{bmatrix} 0 & 1 \\ 0 & 0 \end{bmatrix}\begin{bmatrix} 2 & -1 \\ -1 & 2 \end{bmatrix}\begin{bmatrix} 0 & 0 \\ 1 & 0 \end{bmatrix}$$

$$= \begin{bmatrix} 2 & 1 \\ 1 & 2 \end{bmatrix} - \begin{bmatrix} 2/3 & 0 \\ 0 & 0 \end{bmatrix} = \begin{bmatrix} 4/3 & 1 \\ 1 & 2 \end{bmatrix}$$

We take

$$Z = (F - EC^{-1}D)^{-1} = \begin{bmatrix} 4/3 & 1 \\ 1 & 2 \end{bmatrix}^{-1} = \frac{3}{5}\begin{bmatrix} 2 & -1 \\ -1 & 4/3 \end{bmatrix}$$

$$Y = -ZEC^{-1} = -\frac{1}{5}\begin{bmatrix} 2 & -1 \\ -1 & 4/3 \end{bmatrix}\begin{bmatrix} 0 & 1 \\ 0 & 0 \end{bmatrix}\begin{bmatrix} 2 & -1 \\ -1 & 2 \end{bmatrix}$$

$$= \begin{bmatrix} 2/5 & -4/5 \\ -1/5 & 2/5 \end{bmatrix}$$

$$V = -C^{-1}DZ = -\frac{1}{5}\begin{bmatrix} 2 & -1 \\ -1 & 2 \end{bmatrix}\begin{bmatrix} 0 & 0 \\ 1 & 0 \end{bmatrix}\begin{bmatrix} 2 & -1 \\ -1 & 4/3 \end{bmatrix}$$

$$= \begin{bmatrix} 2/5 & -1/5 \\ -4/5 & 2/5 \end{bmatrix}$$

EXERCISES

1. Find the inverse of the matrix by using the partition method:

(1) $\begin{bmatrix} 1 & 2 & 1 \\ 2 & 3 & -1 \\ 2 & -1 & 3 \end{bmatrix}$; $\left(\text{Ans: } \frac{1}{10}\begin{bmatrix} -8 & 7 & -5 \\ 8 & -1 & -3 \\ 8 & -5 & 1 \end{bmatrix}\right)$

(2) $\begin{bmatrix} 3 & 2 & 1 \\ 2 & 3 & 2 \\ 1 & 2 & 2 \end{bmatrix}$; $\left(\text{Ans: } \frac{1}{3}\left[\begin{array}{cc|c} 2 & -2 & 1 \\ -2 & 5 & -4 \\ \hline 1 & -4 & 5 \end{array}\right]\right)$

(3) $\begin{bmatrix} 2 & 1 & 0 & 0 \\ 1 & 2 & 1 & 0 \\ 0 & 1 & 2 & 1 \\ 0 & 0 & 1 & 2 \end{bmatrix}$; $\left(\text{Ans: } \frac{1}{5}\begin{bmatrix} 4 & -3 & 2 & -1 \\ -3 & 6 & -4 & 2 \\ 2 & -4 & 6 & -3 \\ -1 & 2 & -3 & 4 \end{bmatrix}\right)$

(4) $\begin{bmatrix} 1 & 2 & 3 & 1 \\ 1 & 3 & 3 & : \\ 2 & 4 & 3 & 3 \\ 1 & 1 & 1 & 1 \end{bmatrix}$; $\left(\text{Ans: } \begin{bmatrix} 1 & -2 & 1 & 0 \\ 1 & -2 & 2 & -3 \\ 0 & 1 & -1 & 1 \\ -2 & 3 & -2 & 3 \end{bmatrix}\right)$

(5) $\begin{bmatrix} 1 & 3 & 3 & 2 \\ 1 & 4 & 3 & 4 \\ 1 & 3 & 4 & 5 \\ 2 & 5 & 3 & 2 \end{bmatrix}$; $\left(\text{Ans: } \frac{1}{9}\begin{bmatrix} -15 & -14 & 12 & 13 \\ 3 & 7 & -6 & -2 \\ 9 & -6 & 0 & -3 \\ -6 & 1 & 3 & -1 \end{bmatrix}\right)$

2. Solve the following systems of equation by the partition method.

(1) $3x + y + 2z = 3$
$2x - 3y - z = -3$ (Ans: $x = 1, y = 2, z = -1$)
$x + 2y + z = 4$

(2) $3x + 2y + 4z = 7$
$2x + y + z = 4$ (Ans: $x = 9/4, y = -9/8, z = 5/8$)
$x + 3y + 5z = 2$

(3) $2x - y + z = -1$
$2y - z + u = 1$ (Ans: $x = -1, y = 0, z = 1, u = 2$)
$x + 2z - u = -1$
$x + y + 2u = 3$

(4) $x - 3y + z = -1$
$2x + y - u = -1$ (Ans: $x = y = z = u = 1$)
$3x - 2y - z - 2u = -2$
$4x - y + 3u = 6$

(5) $x - 2y + 3z + 4u = 4.5$
$3x - y + 2z + 5u = 9.5$ (Ans: $x = -(1/2); y = 2, z = -1, u = 3$)

$$2x + 4y - 5z + u = 15$$
$$4x + 2y - z + 3u = 12$$

JACOBI'S ITERATION METHOD

A system of 3 linear equations in 3 unknowns is given by

$$\begin{aligned} a_{11} x_1 + a_{12} x_2 + a_{13} x_3 &= b_1 \\ a_{21} x_1 + a_{22} x_2 + a_{23} x_3 &= b_2 \\ a_{31} x_1 + a_{32} x_2 + a_{33} x_3 &= b_3 \end{aligned} \quad (1)$$

where the diagonal elements are not zero and are large compared to other coefficients. Such a system is called a diagonally dominant system.

We write (1) as

$$\begin{aligned} x_1 &= \frac{1}{a_{11}} [b_1 - a_{12} x_2 - a_{13} x_3] \\ x_2 &= \frac{1}{a_{22}} [b_2 - a_{21} x_1 - a_{23} x_3] \\ x_3 &= \frac{1}{a_{33}} [b_3 - a_{31} x_1 - a_{32} x_2] \end{aligned} \quad (2)$$

Suppose $x_1^{(0)}$, $x_2^{(0)}$, $x_3^{(0)}$ are the initial approximate values of x_1, x_2, x_3 which satisfy equations (2). Substituting these values into the right side of equations (2), we obtain a system of first approximations of x_1, x_2, x_3 or first iterates given by

$$\begin{aligned} x_1^{(1)} &= \frac{1}{a_{11}} [b_1 - a_{12}x_2^{(0)} - a_{13}x_3^{(0)}] \\ x_2^{(1)} &= \frac{1}{a_{22}} [b_2 - a_{21}x_1^{(0)} - a_{23}x_3^{(0)}] \\ x_3^{(1)} &= \frac{1}{a_{33}} [b_3 - a_{31}x_1^{(0)} - a_{32}x_2^{(0)}] \end{aligned} \quad (3)$$

Substituting $x_1^{(1)}$, $x_2^{(1)}$, $x_3^{(1)}$ for x_1, x_2, x_3 in the right sides of 2.

Example 22. Solve by Jacobi iteration Method

$$14x_1 - 3x_2 = 8$$
$$x_1 + 5x_2 = 11$$

Solution: Let us solve for x_1, from the first, and for x_2 from the second equation, we get

$$x_1 = \frac{1}{14} (8 + 3x_2)$$

$$x_2 = \frac{1}{5} (11 - x_1)$$

Hence, the iteration is

$$x_1^{(n+1)} = \frac{1}{14}\,(8 + 3x_2^{(n)})$$

$$x_2^{(n+1)} = \frac{1}{5}\,(11 - x_1^{(k)})$$

with $x_1^{(0)} = x_2^{(0)} = 0$, we get

$$x_1^{(1)} = \frac{4}{7} = 0.57,$$

$x_2^{(1)} = 2.2$ and proceeding similarly.

The approximations are as shown below:

$$\begin{array}{llllll} x_1 & : & 0.57 & 1.04 & 1.02 & 1.00 \\ x_2 & : & 2.2 & 2.09 & 1.99 & 2.00 \end{array}$$

$\therefore$ Solution is $x_1 = 1, x_2 = 2$.

Example 23. Solve by Jacobi Iteration Method, correct to two decimal places

$$10x + y - z = 11.19$$
$$x + 10y + z = 28.08$$
$$-x + y + 10z = 35.61$$

Solution: Starting with $x = 0, y = 0, z = 0$, we get,

$$x^{(1)} = \frac{1}{10}\,(11.19 - y + z) = 1.119 \qquad \left\{\begin{matrix} \therefore \\ y=0 \\ z=0 \end{matrix}\right\}$$

$$y^{(1)} = \frac{1}{10}\,[28.08 - x - z] \qquad \left\{\begin{matrix} \therefore \\ x=0 \\ z=0 \end{matrix}\right\}$$

$$= \frac{28.08}{10} = 2.808$$

$$z^{(1)} = \frac{1}{10}\,[35.61 + x - y]$$

$$= \frac{1}{10}\,[35.61] = 3.561 \qquad \left\{\begin{matrix} \therefore \\ x=0 \\ y=0 \end{matrix}\right\}$$

These are the first approximations.

We proceed to find the second approximations as follows:

$$x^{(2)} = \frac{1}{10}\,[11.19 - y^{(1)} + z^{(1)}]$$

$$\equiv \frac{1}{10}\,[11.19 - 2.808 + 3.561]$$

$$\equiv \frac{1}{10}\,[11.943]$$

$$= 1.194$$

$$y^{(2)} \equiv \frac{1}{10}\,[28.08 - x^{(1)} - z^{(1)}]$$

$$\equiv \frac{1}{10}\,[28.08 - 1.119 - 3.561]$$

$$\equiv \frac{1}{10}\,[23.40] \equiv 2.340$$

$$z^{(2)} \equiv \frac{1}{10}\,[35.61 + x^{(1)} - y^{(1)}]$$

$$\equiv \frac{1}{10}\,[35.61 + 1.119 - 2.808]$$

$$\equiv \frac{1}{10}\,[33.929] \equiv 3.392$$

Similarly, the next approximations are given by

$$x^{(3)} = \frac{1}{10}\,[11.19 - y^{(2)} + z^{(2)}]$$

$$y^{(3)} = \frac{1}{10}\,[28.08 - x^{(2)} - z^{(2)}]$$

$$z^{(3)} \equiv \frac{1}{10}\,[35.61 + x^{(2)} - y^{(2)}]$$

or

$$x^{(3)} \equiv \frac{1}{10}\,[11.19 - 2.340 + 3.392]$$

$$\equiv \frac{1}{10}\,[12.242] = 1.224$$

$$y^{(3)} \equiv \frac{1}{10}\,[28.08 - 1.194 - 3.392]$$

$$\equiv 2.349$$

and

$$z^{(3)} \equiv \frac{1}{10}\,[35.61 + 1.194 - 2.340]$$

$$\equiv 3.446$$

and similarly the next approximations are

$$x^{(4)} \equiv \frac{1}{10}\,[11.19 - y^{(3)} + z^{(3)}]$$

$$\equiv \frac{1}{10}\,[11.19 - 2.349 + 3.446]$$

$$\equiv \frac{1}{10}\,[12.287] \equiv 1.229$$

$$y^{(4)} \equiv \frac{1}{10}\,[28.08 - x^{(3)} - z^{(3)}]$$

$$\equiv \frac{1}{10}\,[28.08 - 1.224 - 3.446]$$

$$\equiv \frac{1}{10}\,[23.410] \equiv 2.341$$

$$z^{(4)} \equiv \frac{1}{10}\,[35.61 + 1.224 - 2.349]$$

$$\equiv \frac{1}{10}\,[34.485] = 3.448$$

Hence,

$$x = 1.23$$

$$y = 2.34$$

and

$$z = 3.45$$

correct to two decimal places.

Example 24. Solve the following system by Jacobi's iteration method.

$$\begin{aligned} 27x + 6y - z &= 85 \\ 6x + 15y + 2z &= 72 \\ x + y + 54z &= 110 \end{aligned} \qquad \text{(i)}$$

Solution:

$$x^{(1)} = \frac{1}{27}\,[85 - 6y + z]$$

$$y^{(1)} = \frac{1}{15}\,[72 - 6x - 2z]$$

$$z^{(1)} \equiv \frac{1}{54}\,[110 - x - y]$$

Starting with $x = 0$, $y = 0$, $z = 0$,

We get

$$x^{(1)} \equiv \frac{85}{27} \equiv 3.15$$

$$y^{(1)} = \frac{72}{15} \equiv 4.8$$

$$z^{(1)} \equiv \frac{110}{54} \equiv 2.04$$

These are the first approximations.

Similarly, the next approximations are:

$$x^{(2)} = \frac{1}{27}[85 - 6y^{(1)} + z^{(1)}]$$

$$\equiv \frac{1}{27}[85 - 28.8 + 2.04] \equiv 2.16$$

$$y_{(2)} \equiv \frac{1}{15}[72 - 6x^{(1)} - 2z^{(1)}]$$

$$\equiv \frac{1}{15}[72 - 18.9 - 4.08]$$

$$\equiv 3.27$$

$$z^{(2)} \equiv \frac{1}{54}[110 - x^{(1)} - y^{(1)}]$$

$$\equiv \frac{1}{54}[110 - 3.15 - 4.8]$$

$$\equiv 1.89$$

Similarly, the next approximations are given by

$$x^{(3)} \equiv \frac{1}{27}[85 - 6y^{(2)} - z^{(2)}]$$

$$y^{(3)} \equiv \frac{1}{15}[72 - 6x^{(2)} - 2z^{(2)}]$$

$$z^{(3)} \equiv \frac{1}{54}[110 - x^{(2)} - y^{(2)}]$$

Continuing as above, further iterations can be obtained.

EXERCISES

1. Use Jacobi's iteration method to solve the following systems of equations:

(1) $20x + y - 2z = 17$
$3x + 20y - z = -18$
$2x - 3y + 20z = 25$

(2) $5x + 2y + z = 12$
$x + 4y + 2z = 15$
$x + 2y + 5z = 20$

(3) $20x + 2y + 6z = 28$
$x + 20y + 9z = -23$
$2x - 7y - 20z = -57$

(4) $6y - z + 27x = 85$
$6x + 15y + 2z = 72$
$x + y + 54z = 110$

(5) $83x + 11y - 4z = 95$
$7x + 52y + 13z = 104$
$3x + 8y + 29z = 71$

(6) $13x_1 + 5x_2 - 3x_3 + x_4 = 18$
$2x_1 + 12x_2 + x_3 - 4x_4 = 13$
$3x_1 - 4x_2 + 10x_3 + x_4 = 29$
$2x_1 + x_2 - 3x_3 + 9x_4 = 31$

2. Solve the equations $5x_1 - x_2 = 9, -x_1 + 5x_2 - x_3 = 4, x_2 - 5x_3 = 6$ by Jacobi's method with (1.8, 0.8, −1.2) as the initial approximation to the solution. Carry out 6 steps.
3. Solve the equations $5x - y + z = 10, x + 2y = 6, x + y + 5z = -1$ by Jacobi's method with (2, 3, 0) as the initial approximation to the solution.

Answers

1. (1) $x = 1, y = -1, z = 1$
(2) $x = 1.08, y = 1.95, z = 3.16$
(3) $x = 0.5149, y = -2.9452, z = 3.9323$
(4) $x = 2.426, y = 3.573, z = 1.93$
(5) $x = 1.06, y = 1.37, z = 1.96$
(6) $x_1 = 1, x_2 = 2, x_3 = 3, x_4 = 4$
2. $x_1 = 1.9999, x_2 = 0.9999, x_3 = -1.0001$
3. $x = 2.556, y = 1.722, z = -1.055$

ITERATIVE OR INDIRECT METHOD

GAUSS'-SEIDEL METHOD

This method is applicable only when each equation of the system contains one coefficient much larger than the others in that equation and the larger coefficients in different equations correspond to the different variables (unknowns).

Such system is known as diagonal system, can be arranged such that the largest coefficients lie on the diagonal.

For illustration, consider a 3 non-homogeneous linear equation in 3 variables such that

$$\left\{\begin{array}{l} a_{11}x_1 + a_{12}x_2 + a_{13}x_3 = b_1 \\ a_{21}x_1 + a_{22}x_2 + a_{23}x_3 = b_2 \\ a_{31}x_1 + a_{32}x_2 + a_{33}x_3 = b_3 \end{array}\right\} \tag{1}$$

Assuming that x_1, x_2, x_3 respectively have the largest coefficients in the 1st, 2nd, 3rd equations of (1) i.e. a_{11}, a_{22}, a_{33} are largest in (1).

We can write the system (1) as

$$x_1 = \frac{1}{a_{11}}[b_1 - a_{12}x_2 - a_{13}x_3]$$

$$x_2 = \frac{1}{a_{22}} [b_2 - a_{21}\, x_1 - a_{23}\, x_3]$$

$$x_3 = \frac{1}{a_{33}} [b_3 - a_{31}\, x_1 - a_{32}\, x_2]$$

Now, for the first iteration, starting from $x_2 = 0$, $x_3 = 0$, we have (trivial solution)

$$x_1^{(1)} = \frac{1}{a_{11}} [b_1]$$

$$x_2^{(1)} = \frac{1}{a_{22}} [b_2 - a_{21}x_1^{(1)} - 0] \qquad (3)$$

$$x_3^{(1)} = \frac{1}{a_{33}} [b_3 - a_{31}x_1^{(1)} - a_{32}x_2^{(1)}]$$

For the second iteration, we have

$$x_1^{(2)} = \frac{1}{a_{11}} [b_1 - a_{12}x_2^{(1)} - a_{13}x_3^{(1)}]$$

$$x_2^{(2)} = \frac{1}{a_{22}} [b_2 - a_{21}x_1^{(2)} - a_{23}x_3^{(1)}] \qquad (4)$$

$$x_3^{(2)} = \frac{1}{a_{33}} [b_3 - a_{31}x_1^{(2)} - a_{32}x_2^{(2)}]$$

Similarly, for the third iteration, we have

$$x_1^{(3)} = \frac{1}{a_{11}} [b_1 - a_{12}x_2^{(2)} - a_{13}x_3^{(2)}]$$

$$x_2^{(3)} = \frac{1}{a_{22}} [b_2 - a_{21}x_1^{(3)} - a_{23}x_3^{(2)}] \qquad (5)$$

$$x_3^{(3)} = \frac{1}{a_{33}} [b_3 - a_{31}x_1^{(3)} - a_{32}x_2^{(3)}]$$

and so on, so forth, until we get the roots within the desired degree of accuracy.

Note (1): Convergence criteria should be satisfied, i.e., in a set of 3 equations, the diagonal number should be greater than or equal to the sum of other two numbers

$$a_{11}x_1 + a_{12}x_2 + a_{13}x_3 = b_1 \qquad (1)$$
$$a_{21}x_1 + a_{22}x_2 + a_{23}x_3 = b_2 \qquad (2)$$
$$a_{31}x_1 + a_{32}x_2 + a_{33}x_3 = b_3 \qquad (3)$$

$$|a_{11}| \geq |a_{12}| + |a_{13}|$$
$$|a_{22}| \geq |a_{21}| + |a_{23}|$$
$$|a_{33}| \geq |a_{31}| + |a_{32}|$$

For all systems of equations, this method will not work (since convergence is not assumed). It converges only for special system of equations.

2. Iteration method is a self-connecting method. i.e, any error made in computation, is corrected in the subsequent iterations.
3. The iteration is stopped when the values of x, y, z start repeating with the required degree of accuracy.

Example 25. Solve the system of equations

$$8x_1 + 2x_2 + 3x_3 = 30 \quad (1)$$
$$x_1 - 9x_2 + 2x_3 = 1 \quad (2)$$
$$2x_1 + 3x_2 + 6x_3 = 31 \quad (3)$$

Solution:

$$|8| \geq |2| + |3|$$
$$|-9| \geq |1| + |2|$$
$$|6| \geq |2| + |3|$$

Convergence criteria is satisfied.

From (1), (2) and (3), we have

$$x_1 = \frac{1}{8}\,[30 - 2x_2 - 3x_3] \quad (4)$$

$$x_2 = \frac{1}{9}\,[x_1 + 2x_3 - 1] \quad (5)$$

$$x_3 = \frac{1}{6}\,[31 - 2x_1 - 3x_2] \quad (6)$$

The first approximate to $x_1^{(1)}$ can be obtained by assuming $x_2 = x_3 = 0$ in (4).

$$x_1^{(1)} = 3.75 = \frac{30}{8}$$

Now substituting for $x_1^{(1)}$ and $x_3 = 0$ in (5), we get

$$x_2^{(1)} = \frac{1}{9}\,[x_1^{(1)} + 2\,(0) - 1] = \frac{1}{9}\left(\frac{22}{8}\right)$$

$$x_2^{(1)} = \frac{22}{72}$$

Now using the values of $x_1^{(1)}$ and $x_2^{(1)}$ in (6)

$$x_3^{(1)} = \frac{1}{6}\left[31 - \frac{30}{\cancel{8}_{\,4}} \times \cancel{2} - \cancel{3} \cdot \frac{22}{\cancel{72}_{\,24}}\right]$$

$$= \frac{1}{6}\left[31 - \frac{30}{4} - \frac{11}{12}\right]$$

$$= \frac{1}{6}\left[\frac{372 - 90 - 11}{12}\right]$$

$$= \frac{1}{6} = \left[\frac{271}{12}\right] = \frac{271}{72}$$

Solve

$$27x + 6y - z = 85 \quad (1)$$
$$6x + 15y + 2z = 72 \quad (2)$$
$$x + y + 54z = 110 \quad (3)$$

by Gauss'-Seidel method.

$$|27| \geq |6| + |-1|$$
$$|15| \geq |6| + |2|$$
$$|54| \geq |1| + |1|$$

Convergene criteria is satisfied. From (1), (2) and (3) we have

$$x = \frac{1}{27}[85 - 6y + z] \quad (4)$$

$$y = \frac{1}{15}[72 - 6x - 2z] \quad (5)$$

$$z = \frac{1}{54}[110 - x - y] \quad (6)$$

The first approximate to $x^{(1)}$ can be obtained by assuming

$y = z = 0$ in (4)

$$x^{(1)} = \frac{1}{27}(85) = 3.15$$

Now substituting for $x^{(1)}$ and $z = 0$, in (5), we get

$$y^{(1)} = \frac{1}{15}[72 - 6x^{(1)} - 0]$$

$$y^{(1)} = \frac{1}{15}[72 - 6\,(3.15)] = 3.54.$$

Now using the values of $x^{(1)}$ and $y^{(1)}$ in (6)

$$z^{(1)} = \frac{1}{54}[110 - x^{(1)} - y^{(1)}], \text{ we get}$$

$$z^{(1)} = \frac{1}{54}[110 - 3.15 - 3.54] = 1.91$$

Thus,
$$x^{(1)} = 3.15$$
$$y^{(1)} = 3.54$$
$$z^{(1)} = 1.91$$

To find the second approximation, we get

$$x^{(2)} = \frac{1}{27}\,[85 - 6y^{(1)} + z^{(1)}]$$

$$x^{(2)} = \frac{1}{27}\,[85 - 6\,(3.54) + 1.91]$$
$$= 2.43$$

$$y^{(2)} = \frac{1}{15}\,[72 - 6x^{(2)} - 2z^{(1)}]$$

$$y^{(2)} = \frac{1}{15}\,[72 - 6\,(2.43) - 2\,(1.91)]$$
$$= 3.57$$

$$z^{(2)} = \frac{1}{54}\,[110 - x^{(2)} - y^{(2)}]$$

$$= \frac{1}{54}\,[110 - 2.43 - 3.57]$$
$$= 1.92$$

Thus,
$$x^{(2)} = 2.43$$
$$y^{(2)} = 3.57$$
$$z^{(2)} = 1.92$$

To find the third approximation, we get

$$x^{(3)} = \left[\frac{1}{27}\right]\,[85 - 6y^{(2)} + z^{(2)}]$$

$$= \frac{1}{27}\,[85 - 6\,(3.57) + 1.92]$$
$$= 2.426$$

$$y^{(3)} = \frac{1}{15}\,[72 - 6x^{(3)} - 2z^{(2)}]$$

$$= \frac{1}{15}\,[72 - 6\,(2.426) - 2\,(1.92)]$$
$$= 3.572$$

$$z^{(3)} = \frac{1}{54}\,[110 - x^{(3)} - y^{(3)}]$$

$$= \frac{1}{54}\,[110 - 2.426 - 3.572]$$
$$= 1.926$$

Thus, we get

$$x^{(3)} = 2.426$$
$$y^{(3)} = 3.572$$
$$z^{(3)} = 1.926$$

Since the values of x, y, z from the 2nd and 3rd approximations, we require

$$x = 2.426; \quad y = 3.572; \quad z = 1.926$$

Gauss'-Seidel

$$8x_1 + x_2 - x_3 = 8$$
$$2x_1 + x_2 + 9x_3 = 12$$
$$x_1 - 7x_2 + 3x_3 = -4$$

Solution:

$$|8| \geq |1| + |-1|$$
$$|1| \ngtr |2| + |9|$$
$$|3| \ngtr |7| + |1|$$

Convergent criteria is not satisfied.
Rearrange the equations as follows:

$$8x_1 + x_2 - x_3 = 8 \quad (1)$$
$$x_1 - 7x_2 + 3x_3 = -4 \quad (2)$$
$$2x_1 + x_2 + 9x_3 = 12 \quad (3)$$

$$|8| \geq |1| + |-1|$$
$$|-7| \geq |1| + |3|$$
$$|9| \geq |2| + |1|$$

Convergent criteria is satisfied.
Find the values of x_1, x_2 and x_3 from equations (1), (2) and (3) respectively

$$x_1 = \frac{1}{8}\,[8 - x_2 + x_3] \quad (a)$$

$$x_2 = \frac{1}{7}\,[4 + x_1 + 3x_3] \quad (b)$$

$$x_3 = \frac{1}{9}\,[12 - 2x_1 - x_2] \quad (c)$$

Iteration (1)
Put $x_2 = x_3 = 0$ in (a),

$$x_1^{(1)} = [1 - 0 - 0] = 1$$
$$x_2^{(1)} = 0.571 + 0.143\ (1) + 0 = 0.714$$
$$x_3^{(1)} = 1.333 - 0.222\ (1) - 0.111 = (0 - 4)$$
$$= 1.032.$$

Iteration (2)

Alogrithm for Gauss'-Seidel Iterative Method

(Apply to diagonally dominant equations)

Step No.	Step details	Remarks
1.	Ensure a_{ii}, $i = 1$ to n are numerically large coefficients	$a_{11}, a_{22} \ldots a_{nn}$ are diagonally dominant.
2.	Set $x_i^{(r+1)} = \dfrac{1}{a_{ii}} \left[b_1 - \sum_{j=1}^{i-1} a_{ij} x_j^{(r+1)} - \sum_{j=i+1}^{n} a_{ij} x_j^{(r)} \right]$	Express each of $x_1, x_2, \ldots x_n$ respectively in terms of the other.
3.	Choose initial approximation $X^{(0)} = x_i^{(0)}$ starting with zero	Use initial approximation $x_1 = 0, x_2 = 0 \ldots$
4.	Select permissiable error E or the degree of accuracy	Solution correct to the number of decimal places to be decided
5.	$\lvert x_i^{(r+1)} - x_i^{(r)} \rvert < E$ go to step 6 otherwise return to step 2	Desired accuracy achieved/ proceed to find $x_i^{(2)}$
6.	Read x_i, stop	Process completed

EXERCISES

Solve by Gauss'-Seidel method:

1. $10x + y = 12, \quad x + 10y = 21$ [Ans: (1, 2)]
2. $3x + y + z = 9.3$
   $2x + 5y - z = 10.5$ [Ans: (1.2, 2.3, 3.4)]
   $x - 2y + 10z = 30.6$
3. $6x - 3y + z = 11$
   $2x + y - 8z = -15$ [Ans: (1, –1, 2)]
   $x - 7y + z = 10$
4. $10x + y - z = 11.19$
   $x + 10y + z = 28.08$ [Ans: (1.23, 2.34, 3.45)]
   $-x + y + 10z = 35.61$

5. $\begin{bmatrix} 3 & 0 & & 11 \\ 0 & 2 & & 10 \\ 1 & 1 & 4 & 1 \\ 2 & 1 & 0 & 4 \end{bmatrix} \begin{bmatrix} x \\ y \\ z \\ t \end{bmatrix} = \begin{bmatrix} 3 \\ 2 \\ 4 \\ 4 \end{bmatrix}$; [Ans: (1.18, 1.32, 0.95, 0.87)]

6. $\begin{bmatrix} 10 & 2 & 0 & -1 \\ -1 & 20 & 2 & 0 \\ -1 & 0 & 10 & -1 \\ 0 & -1 & 2 & 20 \end{bmatrix} \begin{bmatrix} x \\ y \\ z \\ t \end{bmatrix} = \begin{bmatrix} 11.0 \\ 49.5 \\ 27.5 \\ 92.4 \end{bmatrix}$ [Ans: (1.1, 2.2, 3.3, 4.4)]

Chapter 4

Eigenvalues and Eigenvectors

INTRODUCTION

For a linear transformation T on a vector space V, an eigenvalue is a scalar λ for which there is a non-zero member v of V such that $T(v) = \lambda v$. The vector v is an eigenvector (or characteristic vector). For a matrix A, the eigenvalues are the roots of the characteristic equation of the matrix (they are also called characteristic roots or latent roots); the number λ being an eigenvalue means there is a non-zero vector $X = (x_1, x_2, \ldots, x_n)$ for which $AX = \lambda X$, where multiplication is matrix multiplication and X is considered to be a one-column matrix. In this chapter, we consider the concepts of eigenvalues, eigenvectors and numerical methods of finding them approximately.

PRELIMINARIES

Let A be a square matrix of order n with elements a_{ij}. We can find one-column matrix X and a constant λ such that $AX = \lambda X$ or $AX - \lambda IX = 0$

or $$[A - \lambda I]\, X = 0$$

where $$A = \begin{bmatrix} a_{11} & a_{12} & \cdots & a_{1n} \\ a_{21} & a_{22} & \cdots & a_{2n} \\ \cdots & \cdots & \cdots & \cdots \\ a_{n1} & a_{n2} & \cdots & a_{nn} \end{bmatrix} \tag{1}$$

If there exists one-column vector (column matrix)

$$X = \begin{bmatrix} x_1 \\ x_2 \\ \vdots \\ x_n \end{bmatrix} \tag{2}$$

such that

$$AX = \lambda X$$

or
$$[A - \lambda I]\ X = 0 \tag{3}$$

for some scalar λ (real or complex number), then X is called an eigenvector and λ is called the corresponding eigenvalue.

The matrix equation (3) represents n homogeneous linear equations:

$$\begin{aligned}
(a_{11} - \lambda)x_1 + a_{12}x_2 + \ldots + a_{1n}x_n &= 0 \\
a_{21}x_1 + (a_{22} - \lambda)x_2 + \ldots + a_{2n}x_n &= 0 \\
\ldots \quad \ldots \quad \ldots \quad \ldots \quad \ldots & \\
a_{n_1}x_1 + a_{n_2}x_2 + \ldots + (a_{nn} - x_n) &= 0
\end{aligned} \tag{4}$$

which will have a non-trivial solution if and only if the coefficients of determinant vanishes,

i.e.
$$\begin{vmatrix} (a_{11} - \lambda) & a_{12} & \ldots & a_{1n} \\ a_{21} & (a_{22} - \lambda) & \ldots & a_{2n} \\ \ldots & \ldots & \ldots & \ldots \\ a_{n1} & a_{n2} & \ldots & (a_{nn} - \lambda) \end{vmatrix} = 0$$

This is also equal to

$$|\lambda I - A| = 0 \tag{5}$$

Thus, λ is an eigenvalue of A if and only if it is a root of the equation (5).

Equation (5) is called the characteristic equation of A. The left-hand side of this equation, denoted briefly by $|A - \lambda I|$, is a polynomial of degree n; this polynomial is called the characteristic polynomial of A. The equation (5) is thus a polynomial equation of degree n and therefore has exactly n roots. Hence, it follows that A has exactly n eigenvalues. It can be proved that the eigenvalues of the transpose A^T of A are the same as those of A. By using the matrix equation (3), it can also be proved that if λ is an eigenvalue of a non-singular matrix A, then $\frac{1}{\lambda}$ is an eigenvalue of A^{-1}.

Definition: The spectral radius $\rho(B)$ of a matrix B is the largest of the moduli of its eigenvalues.

For $B = \begin{bmatrix} -1 & 2 \\ 4 & -3 \end{bmatrix}$, the eigenvalues are -5 and 1.

Hence, $\rho(B) = 5$.

and if
$$A = \begin{bmatrix} 1 & -1 & 2 \\ 1 & 0 & 0 \\ -1 & 1 & -1 \end{bmatrix}$$

For this matrix, the characteristic equation is

$$\begin{vmatrix} 1-\lambda & -1 & 2 \\ 1 & -\lambda & 0 \\ -1 & 1 & -1-\lambda \end{vmatrix} = 0$$

$$\Rightarrow \qquad (1-\lambda)(\lambda^2+1) = 0$$

$\Rightarrow \lambda = 1$ and $\pm\, i$ are the roots.

Thus, $\lambda_1 = 1$, $\lambda_2 = i$ and $\lambda_3 = -i$ are the eigenvalues of the matrix A.

Note: Here, the matrix A is purely real, two of its eigenvalues are complex numbers. Thus, a real square matrix can have complex eigenvalues.

Cayley-Hamilton Theorem

Arthur Cayley (1821–1895) algebraist, geometer, analyst, and William Rowan Hamilton (1805–1865) Great Irish algebraist, astronomer and physicist are credited for the invention of an important theorem known as CAYLEY-HAMILTON THEOREM which studies the determination of the powers of a square matrix and is used to find the inverse of non-singular matrices.

Statement: Every square matrix satisfies its own characteristic equation.

Proof:

Let $[a_{ij}]$ be a square matrix of order $n \times n$. Therefore, the characteristic equation of A is $|\lambda I - A| = 0$.

Let $|\lambda I - A| = \lambda^n + C_{n-1}\lambda^{n-1} + C_{n-2}\lambda^{n-2} + \ldots C_0$

where $C_{n-1}, C_{n-2}, \ldots C_0$ are constants.

Since the elements $(\lambda I - A)$ are at most of degree n in λ, the elements of *adj* $(\lambda I - A)$ are polynomials in λ of degree at most $(n-1)$. Therefore, *adj* $(\lambda I - A)$ can be expressed as a matrix polynomial of degree $(n-1)$ in λ.

Let $adj(\lambda I - A) = B_{n-1}\lambda^{n-1} + B_{n-2}\lambda^{n-2} + \ldots + B_1\lambda + B_0$

where $B_{n-1}, B_{n-2}, \ldots, B_1, B_0$ are matrices of order $n \times n$.

We know that

$$(\lambda I - A)\, adj(\lambda I - A) = |\lambda I - A|$$

$$\therefore \qquad (\lambda I - A)\,[B_{n-1}\lambda^{n-1} + B_{n-2}\lambda^{n-2} + \ldots + B_1\lambda + B_0]$$
$$= [\lambda^n + C_{n-1}\lambda_{n-1} + \ldots + C_1\lambda + C_0]$$

By equating the coefficients like powers of λ, we get

$$B_{n-1} = I$$

$$AB_{n-1} - B_{n-2} = C_{n-1} I$$
$$AB_{n-2} - B_{n-3} = C_{n-2} I$$
$$AB_0 = C_0 I$$

Pre-multiplying these equations respectively by $A^n, A^{n-1}, A^{n-2}, \ldots I$ and add, we get

$$0 = A^n + C_{n-1} A^{n-1} + C_{n-2} A^{n-2} + \ldots + C_0 I$$

i.e. A satisfies the equation

$$\lambda^n + C_{n-1} \lambda^{n-1} + C_{n-2} \lambda^{n-2} + \ldots + C_0 = 0$$

i.e. A satisfies the characteristic equation

$$|\lambda I - A| = 0$$

Hence, Cayley-Hamilton theorem is proved.

Example 1. Verify Cayley-Hamilton theorem, find A^{-1} if

$$A = \begin{bmatrix} 1 & 2 & -2 \\ 2 & 5 & -4 \\ 3 & 7 & -5 \end{bmatrix}$$

Solution: The characteristic equation is

$$\begin{vmatrix} 1-\lambda & 2 & 2 \\ 2 & 5-\lambda & -4 \\ 3 & 7 & -5-\lambda \end{vmatrix} = 0$$

$$\Rightarrow \quad \lambda^3 - \lambda^2 + 5\lambda - 1 = 0$$
$$A^3 - A^2 + 5A - I = 0$$
$$I = A^3 - A^2 + 5A$$
$$A^{-1} = A^2 - A + 5I$$

$$A^2 = \begin{bmatrix} 1 & 2 & -2 \\ 2 & 5 & -4 \\ 3 & 7 & -5 \end{bmatrix} \begin{bmatrix} 1 & 2 & -2 \\ 2 & 5 & -4 \\ 3 & 7 & -5 \end{bmatrix} = \begin{bmatrix} -1 & -2 & 0 \\ 0 & 1 & 4 \\ 2 & 6 & -9 \end{bmatrix}$$

$$A^{-1} = \begin{bmatrix} -1 & -2 & 0 \\ 0 & 1 & -4 \\ 2 & 6 & -9 \end{bmatrix} - \begin{bmatrix} 1 & 2 & -2 \\ 2 & 5 & -4 \\ 3 & 7 & -5 \end{bmatrix} + \begin{bmatrix} 5 & 0 & 0 \\ 0 & 5 & 0 \\ 0 & 0 & 5 \end{bmatrix}$$

$$A^{-1} = \begin{bmatrix} 3 & -4 & 2 \\ -2 & 1 & 0 \\ -1 & -1 & 1 \end{bmatrix}$$

Verification:

$$A^3 = A^2A = \begin{bmatrix} -5 & -12 & 10 \\ -10 & -23 & 16 \\ -13 & -29 & 17 \end{bmatrix}$$

$$A^3 - A^2 + 5A - I = \begin{bmatrix} -5 & -12 & 10 \\ -10 & -23 & 16 \\ -13 & -29 & 17 \end{bmatrix} - \begin{bmatrix} -1 & -2 & 0 \\ 0 & 1 & -4 \\ 2 & 6 & -9 \end{bmatrix}$$

$$+ \begin{bmatrix} 5 & 10 & -10 \\ 10 & 25 & -20 \\ 15 & 35 & -25 \end{bmatrix} - \begin{bmatrix} 1 & 0 & 0 \\ 0 & 1 & 0 \\ 0 & 0 & 1 \end{bmatrix}$$

$$= \begin{bmatrix} 0 & 0 & 0 \\ 0 & 0 & 0 \\ 0 & 0 & 0 \end{bmatrix} = 0$$

Hence, verified.

BOUNDS FOR EIGENVALUES

In this section, we discuss the bounds for eigenvalues λ of an arbitrary n rowed square matrix $A = (a_{ij})$ which are given by circular disks from Gerschgorin's circle theorem.

Gerschgorin's Circle Theorem (G.C.T)

Statement: Every eigenvalue λ of a square matrix $A = [a_{ij}]_{n\times n}$ satisfies the inequality:

$$|\lambda - a_{mm}| \le |a_{m1}| + |a_{m2}| + \ldots + |a_{m,m-1}| + |a_{m,m+1}| + \ldots \ldots + |a_{mn}|$$

$$(1 \le m \le n)$$

Proof: Consider the matrix equals

$$AX = \lambda X$$

which represents a system of l equations

$$\begin{aligned} a_{11}x_1 + a_{12}x_2 + \ldots + a_{1n}x_n &= \lambda x_1 \\ a_{21}x_1 + a_{22}x_2 + \ldots + a_{2n}x_n &= \lambda x_2 \\ a_{m1}x_1 + a_{m2}x_2 + \ldots + a_{mm}x_m + \ldots + a_{mn}x_n &= \lambda x_m \\ \ldots \ldots \ldots \ldots \ldots \ldots \ldots \ldots & \\ a_{n1}x_1 + a_{n2}x_2 + \ldots + a_{nn}x_n &= \lambda x_n \end{aligned}$$

$$\text{Let } |x_m| = \text{Max } |x_i|$$
$$\text{then } \left|\frac{x_i}{x_m}\right| \leq | \ (i = 1, 2, \ldots n) \tag{1}$$

The mth equation contains x_m, so divide mth equation by x_m.

We get,

$$a_{m_1}\left(\frac{x_1}{x_m}\right) + a_{m_2}\left(\frac{x_2}{x_m}\right) + \ldots + a_{mm} + \ldots + a_{mn}\left(\frac{x_n}{x_m}\right) = \lambda$$

or
$$\lambda - a_{mm} = a_{m1}\left(\frac{x_1}{x_m}\right) + a_{m2}\left(\frac{x_n}{x_m}\right) + \ldots + a_{mn}\left(\frac{x_n}{x_m}\right)$$

∴ on using inequality,

$$|a + b| \leq |a| + |b|$$

we get

$$|\lambda - a_{mm}| \leq |a_{m_1}|\left|\frac{x_1}{x_m}\right| + |a_{m_2}|\left|\frac{x_2}{x_m}\right| + \ldots + |a_{mn}|\left|\frac{x_n}{x_m}\right| \tag{3}$$

or

$$|\lambda - a_{mm}| \leq |a_{m_1}| + |a_{m_2}| + \ldots + |a_{mn}|$$

This completes the proof of the theorem.

If
$$|a_{m_1}| + |a_{m_2} + \ldots + |a_{m,m-1}| + |a_{m,m+1}| + \ldots + |a_{nn}| = s_m$$

then
$$|\lambda - a_{mm}| \leq S_m$$

Geometrical interpretation. The eigenvalue λ lies within or on the boundary of the circle centred at a_{mm} and radius (S_m).

The circle is called Gerschgorin's circle and S_m is called a *Gerschgorin bound.*

Sturm Sequence

Consider the matrix
$$B = \begin{bmatrix} b_1 & c_1 & & & & \\ c_1 & b_2 & c_2 & & & \\ & c_2 & b_3 & c_3 & & \\ & & & \ddots & & \\ & & & c_{n-1} & b_{n-1} & c_{n-1} \\ & & & & c_{n-1} & b_n \end{bmatrix}$$

Define: $f_n = |\lambda I - B| = \begin{bmatrix} \lambda - b_1 & -c_1 & & & \\ -c_1 & \lambda - b_2 & -c_2 & & \\ & & \ddots & & \\ & & -c_{n-2} & \lambda - b_{n-1} & -c_{n-1} \\ & & & -c_{n-1} & \lambda - b_n \end{bmatrix}$

Expanding by minors, the sequence $\{f_n\}$
satisfy $f_0 = 1, f_1 = \lambda - b_1$
and $\quad f_r = (\lambda - b_r)\, f_{r-1} - c_{r-1}^2 f_{r-2}\ (2 \le r \le n)$

Example 2. Find the bounds for eigenvalues of the matrix $A = \begin{bmatrix} 1 & 0 & -1 \\ 1 & 2 & 1 \\ 2 & 2 & 3 \end{bmatrix}$

Solution: The bounds for eigenvalues given by three Gerschgorin circular disks C_1, C_2, C_3.

C_1: Centre 1 and radius $|0| + |-1| = 1$
C_2: Centre 2 and radius $1 + 1 = 2$
C_3: Centre 3 and radius $2 + 2 = 4$

The actual eigenvalues 1, 2, 3 of A lie within these three circular disks C_1, C_2 and C_3.

Example 3. Find the bounds for the eigenvalues of the matrix

$$A = \begin{bmatrix} 1 & 2 & 3 \\ 0 & -1 & 2 \\ 0 & 0 & 3 \end{bmatrix}$$

by using Gerschgorin circle theorem.

Solution: Using Gerschgorin circle theorem to the given matrix A, we find that its eigenvalues satisfy the inequalities.

$$|\lambda - 1| \le |2| + |3| = 5 \tag{1}$$
$$|\lambda + 1| \le |0| + |2| = 2 \tag{2}$$
$$|\lambda - 3| \le |0| + |0| = 0 \tag{3}$$

Next, apply the theorem to

$$A^T = \begin{bmatrix} 1 & 0 & 0 \\ 2 & -1 & 0 \\ 3 & 2 & 3 \end{bmatrix}$$

we get
$$|\lambda - 1| \le |0| + |0| = 0 \quad (4)$$
$$|\lambda + 1| \le |2| + |0| = 2 \quad (5)$$
$$|\lambda - 3| \le |3| + |2| = 5 \quad (6)$$

we get $\lambda = 3$ and
$\lambda = 1$ are eigenvalues

It is seen that (2) and (5) are identical.

(4) is contained in (1)

(3) is contained in (6)

∴ the inequality
$$|\lambda - 1| \le 5$$
$$|\lambda - 3| \le 5$$

and $|\lambda + 1| \le 2$ give the bounds for all eigenvalues.

The intersection of the

Circles $|\lambda - 1| = 5$,
$|\lambda - 3| = 5$

and $|\lambda + 1| = 2$ gives the region in which all the eigenvalues lie.

Example 4. Find the bounds for the eigenvalues of the matrix.

$$A = \begin{bmatrix} 5 & 0 & 1 \\ 0 & -2 & 0 \\ 1 & 0 & 5 \end{bmatrix}$$

by using Gerschgorin circle theorem (G.C.T).

Solution: It is a symmetric matrix

∴ $$A = A'$$

Apply the G.C.T. to the elements a_{11}, a_{22}, a_{33}

We find the eigenvalues of A satisfy the inequality

$$|\lambda - 5| \le |0| + |1| = 1 \quad \text{(i)}$$
$$|\lambda + 2| \le |0| + |0| = 0 \quad \text{(ii)}$$
$$|\lambda - 5| \le |1| + |0| = 1 \quad \text{(iii)}$$

From (ii) we find $\lambda = -2$ is an eigenvalue.

Since (i) and (ii) are the same, the other two eigenvalues lie within or on the boundary of the circle $|\lambda - 5| = 1$.

Determine the bounds for the eigenvalues of the following matrices:

1. $A = \begin{bmatrix} 0 & 1 \\ 0 & 0 \end{bmatrix}$;

(Ans: $\lambda = 0$; circles C_1, C_2; centres 0, 0 and radii 1, 0 respectively)

2. $A = \begin{bmatrix} 3 & 4 \\ 4 & -3 \end{bmatrix}$;

(Ans: $\lambda = \pm 5$; centres: 3, –3; radii 4, 4 respectively)

3. $A = \begin{bmatrix} 6 & 0 & -3 \\ 0 & 6 & 3 \\ -3 & 3 & 2 \end{bmatrix}$;

(Ans: $\lambda = 6,\ 4 \pm \sqrt{22} = 6,\ 8.6904 - 0.6904$ circular disks with centres at 6, 6, 2 and radii $|-3|$, $|-3| + 3$ respectively)

4. $A = \begin{bmatrix} 26 & -2 & 2 \\ 2 & 21 & 4 \\ 4 & 2 & 28 \end{bmatrix}$;

(Ans: 30, 25, 20; circles: centres 26, 21, 28 radii $|-2| + 2$, $2 + 4$, $4 + 2$)

POWER METHOD

The power method developed by Von Mises is a very useful tool for finding the dominant eigenvalue.

Let A be a square matrix of order n and $\lambda_1, \lambda_2 \ldots, \lambda_n$ be its distinct eigenvalues and $e_1, e_2, \ldots, e_n$ be linearly independent eigenvectors. Any arbitrary vector $x^{(0)}$ can be expressed as

$$x^{(0)} = c_1 e_1 + c_2 e_2 + \ldots + c_n e_n \tag{1}$$

Then

$$A^k x^{(0)} = c_1 \lambda_1^k e_1 + c_2 \lambda_2^k e_2 + \ldots + c_n \lambda_n^k e_n$$

$$= \lambda_1^k \left[c_1 e_1 + c_2 \left(\frac{\lambda_2}{\lambda_1} \right)^k e_2 + \ldots \right] \tag{2}$$

Since $\left| \frac{\lambda_2}{\lambda_1} \right| < 1$, for large values of k, we get the approximation

$$A^k x^{(0)} = \lambda_1^k c_1 e_1 \tag{3}$$

Hence, $\lambda_1 = \lim\limits_{k \to \infty} \dfrac{[A^{k+1} x^{(0)}]_i}{[A^k x^{(0)}]_i}$, $i = 1, 2, \ldots$, where the index i denotes the ith component.

Neglecting the terms with $\lambda_3, \lambda_4, \ldots$, etc. in (1), we get $\dfrac{A^{k+1}\,x^{(0)}}{c_1\,\lambda_1^{\,k+1}} - e_1 = \dfrac{\lambda_2}{\lambda_1}$ $\left(\dfrac{A^k\,x^{(0)}}{c_1\,\lambda_1^k} - e_1\right)$ which shows that the convergence is linear with convergence ratio $\left|\dfrac{\lambda_2}{\lambda_1}\right|$. Thus, the convergence is faster if this ratio is small.

The iteration formula for power method can be stated as

$$Y^{(k+1)} = AX^{(k)},\ K = 0, 1, 2, \ldots$$
$$X^{(k+1)} = Y^{(k+1)}$$

divided by its numerically largest component.

Hence, $X^{(0)}$ is quite arbitrary. But generally we choose it as the vector with all entries ones.

Working Rule

1. Choose $X^{(0)} \neq 0$, an arbitrary real vector. Generally, we choose $x^{(0)}$ as:

   $$X^{(0)} = \begin{bmatrix} 1 \\ 1 \\ 1 \\ 1 \end{bmatrix} \text{ or any other value like } X^{(0)} = [10000]^T \text{ etc.,...}$$

   In other words, for the purpose of computation, we generally take $X^{(0)}$ with 1 as the first component, unless prescribed otherwise.
2. We determine $X^{(1)} = AX^{(0)}$ and put it in the form $X^{(1)} = \lambda^{(1)}e_1$. Then $\lambda^{(1)}$ is the first approximation to λ_1 and e_1 is the first approximation to the corresponding eigenvector. Then, we find $X^{(2)} = Ae_1$ and put it in the form $X^{(2)} = \lambda^{(2)}e_2$. Then $\lambda^{(2)}$ is the second approximation to λ_1 and e_2 is the second approximation to the corresponding eigenvector. Proceeding like this, we obtain successive approximation of λ_1. We stop the process when two successive approximations are almost equal. Thus, the final equation will be of the form $X^{(v)} = \lambda^{(v)}e_r$.

Then $\lambda^{(v)}$ is taken as the approximate value of λ_1 and e_r is taken as the corresponding eigenvector.

The iteration method of finding the dominant eigenvalue of a square matrix is called the Rayleigh's power method.

Note: At each stage of iteration, we take out the largest component from $Y^{(k)}$ to find e_k.

Numerically largest (or dominant) eigenvalue of *A*.

Let us assume the numerically largest eigenvalue λ_1 of the given matrix A. Then, we may write λ_1 as:

$$|\lambda_1| \geq |\lambda_2| \geq \ldots \geq |\lambda_n| \tag{4}$$

λ_1 is also called the dominant eigenvalue. Evidently, $\lambda_1 \neq 0$: otherwise, all eigenvalues become zero.

Note: λ_1 is dominant if

$$|\lambda_1| > |\lambda_2| \geq |\lambda_3| \geq \ldots |\lambda_n|.$$

Note (a) 1. To find the numerically smallest eigenvalue of A, obtain the numerically greatest eigenvalue of A^{-1} and then take its reciprocal.

2. The eigenvalue of $A - KI$ are $\lambda_1 - k$ where λ_1 are the eigenvalues of A.

3. Sum of the eigenvalues of a matrix is equal to the sum of the diagonal elements of the matrix (trace of the matrix).

Note (b) 1. To find the numerically smallest eigenvalue of A, obtain the dominant eigenvalue λ_1 of A and then find $B = A - \lambda_1$ and find the dominant eigenvalue of B. Then, the smallest eigenvalue of A is equal to the dominant eigenvalue of $B + \lambda_1$.

Example 5. Obtain by power method, the numerically largest eigenvalue of the matrix.

$$A = \begin{bmatrix} 2 & 3 \\ 5 & 4 \end{bmatrix}$$

Solution: Choose the initial vector

$$X^{(0)} = \begin{bmatrix} 1 \\ 1 \end{bmatrix}$$

Then

$$X^{(1)} = AX^{(0)} = \begin{bmatrix} 2 & 3 \\ 5 & 4 \end{bmatrix} \begin{bmatrix} 1 \\ 1 \end{bmatrix}$$

$$= \begin{bmatrix} 2\,(1) + 3\,(1) \\ 5\,(1) + 4\,(1) \end{bmatrix}$$

$$= \begin{bmatrix} 5 \\ 9 \end{bmatrix} = 9 \begin{bmatrix} 0.56 \\ 1 \end{bmatrix} = \lambda^{(1)} e_1$$

Now

$$X^{(2)} = AX^{(1)} = \begin{bmatrix} 2 & 3 \\ 5 & 4 \end{bmatrix} \begin{bmatrix} 0.56 \\ 1 \end{bmatrix} = \begin{bmatrix} 2\,(0.56) + 3\,(1) \\ 5\,(0.56) + 4\,(1) \end{bmatrix}$$

$$= \begin{bmatrix} 4.12 \\ 6.80 \end{bmatrix} = 6.8 \begin{bmatrix} 0.61 \\ 1 \end{bmatrix} = \lambda^{(2)} e_2$$

Then $X^{(3)} = AX^{(2)} = \begin{bmatrix} 2 & 3 \\ 5 & 4 \end{bmatrix} \begin{bmatrix} 0.61 \\ 1 \end{bmatrix}$

$$= \begin{bmatrix} 2\,(0.61) + 3\,(1) \\ 5\,(0.61) + 4(1) \end{bmatrix} = \begin{bmatrix} 4.32 \\ 7.05 \end{bmatrix} = 7.05 \begin{bmatrix} 0.6 \\ 1 \end{bmatrix} = \lambda^{(3)} e_3$$

$$X^{(4)} = AX^{(3)} = \begin{bmatrix} 2 & 3 \\ 5 & 4 \end{bmatrix} \begin{bmatrix} 0.6 \\ 1 \end{bmatrix} = \begin{bmatrix} 2\,(0.6) + 3\,(1) \\ 5\,(0.6) + 4\,(1) \end{bmatrix} = \begin{bmatrix} 4.2 \\ 7 \end{bmatrix} = 7 \begin{bmatrix} 0.6 \\ 1 \end{bmatrix} = \lambda^{(4)} e_4$$

where the largest (or dominant) eigenvalue λ_1 is taken as $\lambda_1 = 7$ and $e_3 = [0.6, 1]^T$ is the corresponding eigenvector.

Example 6. Using power method, finding the dominant eigenvalue and the corresponding eigenvector of the matrix

$$A = \begin{bmatrix} 1 & -3 & 2 \\ 4 & 4 & -1 \\ 6 & 3 & 5 \end{bmatrix}.$$

perform four iterations.

Solution: Choose the initial vector

$$X^{(0)} = \begin{bmatrix} 1 \\ 0 \\ 0 \end{bmatrix} = [100]^T$$

Then
$$X^{(1)} = AX^{(0)} = \begin{bmatrix} 1 & -3 & 2 \\ 4 & 4 & -1 \\ 6 & 3 & 5 \end{bmatrix} \begin{bmatrix} 1 \\ 0 \\ 0 \end{bmatrix} = \begin{bmatrix} 1 \\ 4 \\ 6 \end{bmatrix}$$

$$= 6 \begin{bmatrix} \frac{1}{6} \\ \frac{2}{3} \\ 1 \end{bmatrix} = \lambda^{(1)} e_1$$

$\Rightarrow$ 1st approximation to the required eigenvalue and the eigenvector is

$$\lambda^{(1)} = 6$$

and
$$e_1 = \begin{bmatrix} \frac{1}{6} & \frac{2}{3} & 1 \end{bmatrix}^T$$

Now

$$X^{(2)} = AX^{(1)} = \begin{bmatrix} 1 & -3 & 2 \\ 4 & 4 & -1 \\ 6 & 3 & 5 \end{bmatrix} \begin{bmatrix} \frac{1}{6} \\ \frac{2}{3} \\ 1 \end{bmatrix}$$

$$= \begin{bmatrix} 0.166 \\ 2.336 \\ 8.003 \end{bmatrix} = (8.003) \begin{bmatrix} 0.21 \\ 0.292 \\ 1 \end{bmatrix} = \lambda^{(2)} e_2$$

Then, $\lambda^{(2)} = 8.003$ and $e_2 = [0.21, 0.292, 1]^T$ are the second approximations to the required eigenvalue and the corresponding eigenvector.

$$X^{(3)} = AX^{(2)} = \begin{bmatrix} 1 & -3 & 2 \\ 4 & 4 & -1 \\ 6 & 3 & 5 \end{bmatrix} \begin{bmatrix} 0.21 \\ 0.292 \\ 1 \end{bmatrix}$$

$$= \begin{bmatrix} 1.145 \\ 0.252 \\ 6.002 \end{bmatrix} = (6.002) \begin{bmatrix} 0.191 \\ 0.042 \\ 1 \end{bmatrix} = \lambda^{(3)} e_3$$

$$X^{(4)} = AX^{(3)} = \begin{bmatrix} 1 & -3 & 2 \\ 4 & 4 & -1 \\ 6 & 3 & 5 \end{bmatrix} \begin{bmatrix} 0.191 \\ 0.042 \\ 1 \end{bmatrix}$$

$$= \begin{bmatrix} 2.065 \\ -0.068 \\ 6.272 \end{bmatrix} = (6.272) \begin{bmatrix} 0.329 \\ -0.011 \\ 1 \end{bmatrix} = \lambda^{(4)} e_4$$

Hence, after four iterations, we get $\lambda^{(4)} = 6.272$ as an approximate value of the required eigenvalue and $e_4 = [0.329, -0.011, 1]^T$ as the corresponding eigenvector.

Example 7. Find the dominant eigenvalue and the corresponding eigenvector of

$$A = \begin{bmatrix} 4 & 1 & -1 \\ 2 & 3 & -1 \\ -2 & 1 & 5 \end{bmatrix}$$

Solution: Choose the initial vector

$$X^{(0)} = \begin{bmatrix} 1 \\ 0 \\ 0 \end{bmatrix} = [100]^T$$

Then
$$X^{(1)} = AX^{(0)} = \begin{bmatrix} 4 & 1 & -1 \\ 2 & 3 & -1 \\ -2 & 1 & 5 \end{bmatrix} \begin{bmatrix} 1 \\ 0 \\ 0 \end{bmatrix} = \begin{bmatrix} 4 \\ 2 \\ -2 \end{bmatrix}$$

$$= 4 \begin{bmatrix} 1.0 \\ 0.5 \\ -0.5 \end{bmatrix} = \lambda^{(1)} e_1$$

Thus, $\lambda^{(1)} = 4$ and $e_1 = [1.0, 0.5, -0.5]^T$ are the first approximations to the required eigenvalue and the corresponding eigenvector.

Next, we have $X^{(2)} = AX^{(1)} = \begin{bmatrix} 4 & 1 & -1 \\ 2 & 3 & -1 \\ -2 & 1 & 5 \end{bmatrix} \begin{bmatrix} 1.0 \\ 0.5 \\ -0.5 \end{bmatrix} = \begin{bmatrix} 5 \\ 4 \\ -4 \end{bmatrix} = 5 \begin{bmatrix} 1.0 \\ 0.8 \\ -0.8 \end{bmatrix} = \lambda^{(2)} e_2$

Hence, $\lambda^{(2)} = 5$ and $e_2 = [1, 0.8, -0.8]^T$ are the second approximations to the required eigenvalue and the corresponding eigenvector.

Continuing like this, we obtain,

$$X^{(3)} = AX^{(2)} = (5.6) \begin{bmatrix} 1.000 \\ 0.93 \\ -0.93 \end{bmatrix} = \lambda^{(3)} e_3$$

$$X^{(4)} = AX^{(3)} = (5.86) \begin{bmatrix} 1.000 \\ 0.98 \\ -0.98 \end{bmatrix} = \lambda^{(4)} e_4$$

$$X^{(5)} = AX^{(4)} = (5.96) \begin{bmatrix} 1.000 \\ 0.99 \\ -0.99 \end{bmatrix} = \lambda^{(4)} e_5$$

$$X^{(6)} = AX^{(5)} = (5.98) \begin{bmatrix} 1 \\ 1 \\ -1 \end{bmatrix} = \lambda^{(6)} e_6$$

$$X^{(7)} = AX^{(6)} = 6\begin{bmatrix} 1 \\ 1 \\ -1 \end{bmatrix} = \lambda^{(7)} e_7$$

Thus, e_6 and e_7 are identical.

Therefore, we stop the iteration process and take $\lambda^{(7)} = 6$ as the largest eigenvalue and $e_7 = [1, 1, -1]^T$ as the corresponding eigenvector, for the given matrix.

Example 8. Find the dominant eigenvalue and the corresponding eigenvector of the matrix $A = \begin{bmatrix} 1 & 3 & -1 \\ 3 & 2 & 4 \\ -1 & 4 & 10 \end{bmatrix}$ starting with $\begin{bmatrix} 0 \\ 0 \\ 1 \end{bmatrix}$ as the initial vector, upto an accuracy of 0.01.

Solution: Let $X^{(0)} = \begin{bmatrix} 0 \\ 0 \\ 1 \end{bmatrix} = [0, 0, 1]^T$

$$\therefore \quad X^{(1)} = AX^{(0)} = \begin{bmatrix} 1 & 3 & -1 \\ 3 & 2 & 4 \\ -1 & 4 & 10 \end{bmatrix}\begin{bmatrix} 0 \\ 0 \\ 1 \end{bmatrix} = \begin{bmatrix} -1 \\ 4 \\ 10 \end{bmatrix} = 10\begin{bmatrix} -0.1 \\ 0.4 \\ 1 \end{bmatrix} = \lambda^{(1)} e_1$$

$$X^{(2)} = AX^{(1)} = \begin{bmatrix} 1 & 3 & -1 \\ 3 & 2 & 4 \\ -1 & 4 & 10 \end{bmatrix}\begin{bmatrix} -0.1 \\ 0.4 \\ 1 \end{bmatrix}$$

$$= \begin{bmatrix} 0.1 \\ 4.5 \\ 11.7 \end{bmatrix} = (11.7)\begin{bmatrix} 0.008547 \\ 0.384615 \\ 1 \end{bmatrix} = \lambda^{(2)} e_2$$

Proceeding like this, we obtain:

$$X^{(3)} = AX^{(2)} = (11.5299)\begin{bmatrix} 0.01408 \\ 0.41586 \\ 1 \end{bmatrix} = \lambda^{(3)} e_3$$

$$X^{(4)} = AX^{(3)} = (11.64936)\begin{bmatrix} 0.02246 \\ 0.41839 \\ 1 \end{bmatrix} = \lambda^{(4)} e_4$$

$$X^{(5)} = AX^{(4)} = (11.6511)\begin{bmatrix} 0.02383 \\ 0.42092 \\ 1 \end{bmatrix} = \lambda^{(5)} e_5$$

$$X^{(6)} = AX^{(5)} = (11.6599)\begin{bmatrix} 0.02458 \\ 0.42139 \\ 1 \end{bmatrix} = \lambda^{(6)} e_6$$

$$X^{(7)} = AX^{(6)} = (11.66098)\begin{bmatrix} 0.02476 \\ 0.42162 \\ 1 \end{bmatrix} = \lambda^{(7)} e_7$$

We find $\lambda^{(6)}$ and $\lambda^{(7)} \approx 11.66$.

Hence, we stop the iteration process and take $\lambda^{(7)} \approx 11.66$ as the dominant (numerically largest) eigenvalue of the given matrix and

$$X^{(7)} = [0.02476,\ 0.42162,\ 1]^T$$

as the corresponding eigenvector.

Example 9. Find the dominant eigenvalue and the corresponding eigenvector of $A = \begin{bmatrix} 2 & -1 & 0 \\ -1 & 2 & -1 \\ 0 & -1 & 2 \end{bmatrix}$ by power method.

Solution: Let $X^{(0)} = [111]^T$

Thus,

$$X^{(1)} = AX^{(0)} = \begin{bmatrix} 2 & -1 & 0 \\ -1 & 2 & -1 \\ 0 & -1 & 2 \end{bmatrix}\begin{bmatrix} 1 \\ 1 \\ 1 \end{bmatrix} = \begin{bmatrix} 1 \\ 0 \\ 1 \end{bmatrix} = \lambda^{(1)} e_1$$

$$X^{(2)} = AX^{(1)} = \begin{bmatrix} 2 & -1 & 0 \\ -1 & 2 & -1 \\ 0 & -1 & 2 \end{bmatrix}\begin{bmatrix} 1 \\ 0 \\ 1 \end{bmatrix} = \begin{bmatrix} 2 \\ -2 \\ 2 \end{bmatrix} = 2\begin{bmatrix} 1 \\ -1 \\ 1 \end{bmatrix} = \lambda^{(2)} e_2$$

$$X^{(3)} = AX^{(2)} = \begin{bmatrix} 2 & -1 & 0 \\ -1 & 2 & -1 \\ 0 & -1 & 2 \end{bmatrix}\begin{bmatrix} 1 \\ -1 \\ 1 \end{bmatrix} = \begin{bmatrix} 3 \\ -4 \\ 3 \end{bmatrix} = 4\begin{bmatrix} 0.75 \\ -1 \\ 0.75 \end{bmatrix} = \lambda^{(3)} e_3$$

Continuing like this, we find

$$X^{(4)} = AX^{(3)} = (3.5)\begin{bmatrix} 0.7143 \\ -1.0000 \\ 0.7143 \end{bmatrix} = \lambda^{(4)} e_4$$

$$X^{(5)} = AX^{(4)} = (3.4286)\begin{bmatrix} 0.7083 \\ -1.0000 \\ 0.7083 \end{bmatrix} = \lambda^{(5)} e_5$$

$$X^{(6)} = AX^{(5)} = (3.4166)\begin{bmatrix} 0.7073 \\ -1.0000 \\ 0.7073 \end{bmatrix} = \lambda^{(6)} e_6$$

$$X^{(7)} = AX^{(6)} = (3.4166)\begin{bmatrix} 0.7071 \\ -1.0000 \\ 0.7071 \end{bmatrix} = \lambda^{(7)} e_7$$

$$X^{(8)} = AX^{(7)} = (3.4142)\begin{bmatrix} 0.7071 \\ -1.0000 \\ 0.7071 \end{bmatrix} = \lambda^{(8)} e_8$$

Since $\lambda^{(7)} \equiv \lambda^{(8)}$ we take $\lambda^{(8)} = 3.4142$ as the approximate value of the dominant eigenvalue and

$$X^{(8)} = [0.7071, \; -\; 1.0000, \; 0.7071]^T$$

as the corresponding eigenvector.

Example 10. Using power method, find the dominant eigenvalue and the corresponding eigenvector of the matrix:

$$A = \begin{bmatrix} 2 & 1 & 1 & 0 \\ 1 & 1 & 0 & 1 \\ 1 & 0 & 1 & 1 \\ 0 & 1 & 1 & 2 \end{bmatrix}$$

Solution: Choosing the initial approximation to

$$X^{(0)} = [1111]^T = \begin{bmatrix} 1 \\ 1 \\ 1 \\ 1 \end{bmatrix}$$

Then $$X^{(1)} = AX^{(0)} = \begin{bmatrix} 2 & 1 & 1 & 0 \\ 1 & 1 & 0 & 1 \\ 1 & 0 & 1 & 1 \\ 0 & 1 & 1 & 2 \end{bmatrix} \begin{bmatrix} 1 \\ 1 \\ 1 \\ 1 \end{bmatrix} = \begin{bmatrix} 4 \\ 3 \\ 3 \\ 4 \end{bmatrix} = 4 \begin{bmatrix} 1 \\ 0.75 \\ 0.75 \\ 1 \end{bmatrix} = \lambda^{(1)} e_1$$

$$X^{(2)} = AX^{(1)} = \begin{bmatrix} 2 & 1 & 1 & 0 \\ 1 & 1 & 0 & 1 \\ 1 & 0 & 1 & 1 \\ 0 & 1 & 1 & 2 \end{bmatrix} \begin{bmatrix} 1 \\ 0.75 \\ 0.75 \\ 1 \end{bmatrix} = \begin{bmatrix} 3.5 \\ 2.75 \\ 2.75 \\ 3.5 \end{bmatrix}$$

$$= (3.5) \begin{bmatrix} 1 \\ 0.7857 \\ 0.7857 \\ 1 \end{bmatrix} = \lambda^{(2)} e_2$$

$$X^{(3)} = AX^{(1)} = (3.57) \begin{bmatrix} 1.00 \\ 0.78 \\ 0.78 \\ 1.00 \end{bmatrix} = \lambda^{(3)} e_3$$

$$X^{(4)} = AX^{(3)} = (3.56) \begin{bmatrix} 1.00 \\ 0.78 \\ 0.78 \\ 1.00 \end{bmatrix} = \lambda^{(4)} e_4$$

$$X^{(5)} = AX^{(4)} = (3.56) \begin{bmatrix} 1.00 \\ 0.78 \\ 0.78 \\ 1.00 \end{bmatrix} = \lambda^{(5)} e_5$$

We find $X^{(4)}$ and $X^{(5)}$ are identical. Hence, the dominant eigenvalue is taken as 3.56 and the corresponding eigenvector as $[1, 0.78, 0.78, 1]^T$. Find the dominant eigenvalue and the corresponding eigenvector of $A = \begin{bmatrix} 1 & 6 & 1 \\ 1 & 2 & 0 \\ 0 & 0 & 3 \end{bmatrix}$.

Find also the least latent root and hence the third eigenvalue also.

Let $X^{(0)} = \begin{bmatrix} 1 \\ 0 \\ 0 \end{bmatrix}$ be the initial eigenvector, we have

$$X^{(1)} = AX^{(0)} = \begin{bmatrix} 1 & 6 & 1 \\ 1 & 2 & 0 \\ 0 & 0 & 3 \end{bmatrix} \begin{bmatrix} 1 \\ 0 \\ 0 \end{bmatrix} = \begin{bmatrix} 1 \\ 1 \\ 0 \end{bmatrix} = 1 \begin{bmatrix} 1 \\ 1 \\ 0 \end{bmatrix} = 1.e_1$$

$$X^{(2)} = AX^{(1)} = \begin{bmatrix} 1 & 6 & 1 \\ 1 & 2 & 0 \\ 0 & 0 & 3 \end{bmatrix} \begin{bmatrix} 1 \\ 1 \\ 0 \end{bmatrix} = \begin{bmatrix} 7 \\ 3 \\ 0 \end{bmatrix} = 7 \begin{bmatrix} 1 \\ 0.4 \\ 0 \end{bmatrix} = 7.e_2$$

$$X^{(3)} = AX^{(2)} = \begin{bmatrix} 1 & 6 & 1 \\ 1 & 2 & 0 \\ 0 & 0 & 3 \end{bmatrix} \begin{bmatrix} 1 \\ 0.4 \\ 0 \end{bmatrix} = \begin{bmatrix} 3.4 \\ 1.8 \\ 0 \end{bmatrix} = 3.4 \begin{bmatrix} 1 \\ 0.53 \\ 0 \end{bmatrix} = 3.4e_3$$

$$X^{(4)} = AX^{(3)} = \begin{bmatrix} 1 & 6 & 1 \\ 1 & 2 & 0 \\ 0 & 0 & 3 \end{bmatrix} \begin{bmatrix} 1 \\ 0.53 \\ 0 \end{bmatrix} = \begin{bmatrix} 4.18 \\ 2.06 \\ 0 \end{bmatrix} = 4.18 \begin{bmatrix} 1 \\ 0.49 \\ 0 \end{bmatrix} = 4.18e_4$$

$$X^{(5)} = AX^{(4)} = \begin{bmatrix} 1 & 6 & 1 \\ 1 & 2 & 0 \\ 0 & 0 & 3 \end{bmatrix} \begin{bmatrix} 1 \\ 0.49 \\ 0 \end{bmatrix} = \begin{bmatrix} 3.94 \\ 1.98 \\ 0 \end{bmatrix} = 3.94 \begin{bmatrix} 1 \\ 0.5 \\ 0 \end{bmatrix} = 3.94e_5$$

$$X^{(6)} = AX^{(5)} = \begin{bmatrix} 1 & 6 & 1 \\ 1 & 2 & 0 \\ 0 & 0 & 3 \end{bmatrix} \begin{bmatrix} 1 \\ 0.5 \\ 0 \end{bmatrix} = \begin{bmatrix} 4 \\ 2 \\ 0 \end{bmatrix} = 4 \begin{bmatrix} 1 \\ 0.5 \\ 0 \end{bmatrix} = 4e_6$$

$$X^{(7)} = AX^{(6)} = \begin{bmatrix} 1 & 6 & 1 \\ 1 & 2 & 0 \\ 0 & 0 & 3 \end{bmatrix} \begin{bmatrix} 1 \\ 0.5 \\ 0 \end{bmatrix} = \begin{bmatrix} 3.9982 \\ 1.9994 \\ 0 \end{bmatrix} = 3.9982 \begin{bmatrix} 1 \\ 0.5 \\ 0 \end{bmatrix} = 3.9982\ e_7$$

$$X^{(8)} = AX^{(7)} = \begin{bmatrix} 1 & 6 & 1 \\ 1 & 2 & 0 \\ 0 & 0 & 3 \end{bmatrix} \begin{bmatrix} 1 \\ 0.5 \\ 0 \end{bmatrix} = \begin{bmatrix} 4 \\ 2 \\ 0 \end{bmatrix} = 4 \begin{bmatrix} 1 \\ 0.5 \\ 0 \end{bmatrix} 4e_8$$

$\therefore$ Dominant eigenvalue = 4; corresponding eigenvector is $[1, 0.5, 0]^T$

To find the smallest eigenvalue, let $B = A - 4I$, since $\lambda_1 = 4$

$$\therefore \qquad B = \begin{bmatrix} 1 & 6 & 1 \\ 1 & 2 & 0 \\ 0 & 0 & 3 \end{bmatrix} - \begin{bmatrix} 4 & 0 & 0 \\ 0 & 4 & 0 \\ 0 & 0 & 4 \end{bmatrix} = \begin{bmatrix} -3 & 6 & 1 \\ 1 & -2 & 2 \\ 0 & 0 & -2 \end{bmatrix}$$

We will find the dominant eigenvalue of B. Let $X^{(1)} = \begin{bmatrix} 1 \\ 0 \\ 0 \end{bmatrix}$ be the initial vector.

$$BX^{(1)} = \begin{bmatrix} -3 & 6 & 1 \\ 1 & -2 & 0 \\ 0 & 0 & -2 \end{bmatrix} \begin{bmatrix} 1 \\ 0 \\ 0 \end{bmatrix} = \begin{bmatrix} -3 \\ 1 \\ 0 \end{bmatrix} = -3 \begin{bmatrix} 1 \\ -0.3333 \\ 0 \end{bmatrix} = -3e_2$$

$$Be_2 = \begin{bmatrix} -3 & 6 & 1 \\ 1 & -2 & 0 \\ 0 & 0 & -2 \end{bmatrix} \begin{bmatrix} 1 \\ -0.3333 \\ 0 \end{bmatrix} = \begin{bmatrix} -5 \\ 1.6666 \\ 0 \end{bmatrix} = -5 \begin{bmatrix} 1 \\ -0.3333 \\ 0 \end{bmatrix} = -5e_3$$

$$\therefore \qquad Be_3 = \begin{bmatrix} -3 & 6 & 1 \\ 1 & -2 & 0 \\ 0 & 0 & -2 \end{bmatrix} \begin{bmatrix} 1 \\ -0.3333 \\ 0 \end{bmatrix} = \begin{bmatrix} -5 \\ 1.6666 \\ 0 \end{bmatrix} = -5 \begin{bmatrix} 1 \\ -0.3333 \\ 0 \end{bmatrix}$$

$\therefore$ Dominant eigenvalue of B is $= -5$.

Adding 4, smallest eigenvalue of $A = -5 + 4 = -1$

Sum of eigenvalues = Trace of $A = 1 + 2 + 3 = 6$

$$4 + (-1) + \lambda_3 = 6$$

$$\therefore \qquad \lambda_3 = 3$$

All the three eigenvalues $(4, 3, -1)$ are called the spectrum of A.

INVERSE POWER METHOD

Let us suppose that A is a non-singular matrix, then the eigenvalues of A^{-1} are the reciprocals of the eigenvalues of A. Hence, if $\lambda \neq 0$ is the numerically largest eigenvalue of A, then $\frac{1}{\lambda}$ is the numerically largest eigenvalue of A^{-1} and vice versa. This shows that the numerically least eigenvalue of A can be determined by employing the power method to A^{-1}. This method is called inverse power method.

Example 11. Find the smallest eigenvalue in magnitude of the matrix

$$A = \begin{bmatrix} 2 & -1 & 0 \\ -1 & 2 & -1 \\ 0 & -1 & 2 \end{bmatrix}$$

using four iterations of the inverse power method.

Solution: The inverse power method is defined by

$$Y^{(k+1)} = A^{-1}\ Y_k \quad k = 0, 1, 2,... \tag{1}$$

Since the given matrix A is non-singular, i.e. $|A| \neq 0$.

Hence, $$A^{-1} = \begin{bmatrix} 3/4 & 1/2 & 1/4 \\ 1/2 & 1 & 1/2 \\ 1/4 & 1/2 & 3/4 \end{bmatrix} = \frac{1}{4}\begin{bmatrix} 3 & 2 & 1 \\ 2 & 4 & 2 \\ 1 & 2 & 3 \end{bmatrix}$$

In order to find the numerically smallest eigenvalue of A, we employ the power method to A^{-1}.

Let us choose $X^{(0)} = [1\ 1\ 1]^T$, then we obtain:

$$X^{(1)} = A^{-1}\ X^{(0)} = \frac{1}{4}\begin{bmatrix} 3 & 2 & 1 \\ 2 & 4 & 2 \\ 1 & 2 & 3 \end{bmatrix}\begin{bmatrix} 1 \\ 1 \\ 1 \end{bmatrix} = \begin{bmatrix} 1.5 \\ 2 \\ 1.5 \end{bmatrix} = 2\begin{bmatrix} 0.75 \\ 1.00 \\ 0.75 \end{bmatrix} = \lambda^{(1)} e_1$$

$$X^{(2)} = A^{-1}X^{(1)} = \frac{1}{4}\begin{bmatrix} 3 & 2 & 1 \\ 2 & 4 & 2 \\ 1 & 2 & 3 \end{bmatrix}\begin{bmatrix} 0.75 \\ 1 \\ 0.75 \end{bmatrix} = \begin{bmatrix} 1.25 \\ 1.75 \\ 1.25 \end{bmatrix} = (1.75)\begin{bmatrix} 0.7143 \\ 1.0000 \\ 0.7143 \end{bmatrix} = \lambda^{(2)} e_2$$

$$X^{(3)} = A^{-1}\ X^{(2)} = (1.7143)\begin{bmatrix} 0.7083 \\ 1.0000 \\ 0.7083 \end{bmatrix} = \lambda^{(3)} e_3$$

$$X^{(4)} = A^{-1}\ X^{(3)} = (1.7083)\begin{bmatrix} 0.7073 \\ 1.0000 \\ 0.7073 \end{bmatrix} = \lambda^{(4)} e_4$$

$$X^{(5)} = A^{-1}\ X^{(4)} = (1.7073)\begin{bmatrix} 0.7071 \\ 1.0000 \\ 0.7071 \end{bmatrix} = \lambda^{(5)} e_5$$

$$X^{(6)} = A^{-1}\ X^{(5)} = (1.7071)\begin{bmatrix} 0.7071 \\ 1.0000 \\ 0.7071 \end{bmatrix} = \lambda^{(6)} e_6$$

We observe that e_5 and e_6 are identical. Therefore, we take $\lambda^{(6)} = 1.7071$ as the approximate value of the numerically largest eigenvalue of A^{-1}.

Hence, $\dfrac{1}{\lambda^{(6)}} = 0.5858$ is taken as the numerically least eigenvalue of A.

EXERCISES

I. Using power method, calculate the numerically largest (dominant) eigenvalues of the following matrices; and the corresponding eigenvectors.

1. $\begin{bmatrix} -4 & -5 \\ 1 & 2 \end{bmatrix}$, (Ans: -3; $[1, -0.2]^T$)

2. $\begin{bmatrix} 10 & -2 & 1 \\ -2 & 10 & -2 \\ 1 & -2 & 10 \end{bmatrix}$, (Ans: 9; $[1, 0, -1]^T$)

3. $\begin{bmatrix} 2 & -1 & 0 \\ -1 & 2 & -1 \\ 0 & -1 & 2 \end{bmatrix}$, (Ans: 3.414; $[-0.707, 1, 0.707]^T$)

4. $\begin{bmatrix} 10 & 2 & 1 \\ 2 & 10 & 1 \\ 2 & 1 & 10 \end{bmatrix}$, (Ans: 13; $[1, 1, 1]^T$)

5. $\begin{bmatrix} -15 & 4 & 3 \\ 10 & -12 & 6 \\ 20 & -4 & 2 \end{bmatrix}$, (Ans: -20; $[1, -0.5, -1]^T$)

6. $\begin{bmatrix} 1 & 2 & 3 \\ 0 & -4 & 2 \\ 0 & 0 & 7 \end{bmatrix}$, (Ans: 7; $[0.56, 0.18, 1]^T$)

7. $\begin{bmatrix} -1 & 1 & 2 \\ 0 & 1 & -1 \\ 4 & -2 & -9 \end{bmatrix}$, (Ans: 10.099; $[-0.229, 0.001, 1]^T$)

8. $\begin{bmatrix} 7 & 3 & -2 \\ 3 & 4 & -1 \\ -2 & -1 & 3 \end{bmatrix}$, (Ans: 9.6055; $[1.0, 0.6056, -0.3944]^T$)

9. $\begin{bmatrix} -15 & 4 & 3 \\ 10 & -12 & 6 \\ 20 & -4 & 2 \end{bmatrix}$, (Ans: 19.988; $[0.9998, -0.494, -1]^T$)

10. $\begin{bmatrix} 4 & 1 & 0 \\ 1 & 20 & 1 \\ 0 & 1 & 4 \end{bmatrix}$, (Ans: 20.124; $[0.06202, 1, 0.06202]^T$)

11. $\begin{bmatrix} 25 & 1 & 2 \\ 1 & 3 & 0 \\ 2 & 0 & -4 \end{bmatrix}$, (Ans: 25.1821; $[1, 0.0451, 0.0685]^T$)

II. 1. Find the dominant eigenvalue of the matrix

$$\begin{bmatrix} 25 & -41 & 10 & -6 \\ -4 & 68 & -17 & 10 \\ 10 & -17 & 5 & -3 \\ -6 & 10 & -3 & 2 \end{bmatrix}$$ (Ans: 79.1)

by the power method.

2. Find the numerically smallest eigenvalues of the matrices

3. $\begin{bmatrix} 1 & 6 & 1 \\ 0 & 2 & 0 \\ 0 & 0 & 3 \end{bmatrix}$

4. $\begin{bmatrix} 4 & 1 & -1 \\ 2 & 3 & -1 \\ -2 & 1 & 5 \end{bmatrix}$ (Ans: (3). 1. (4). 2)

III. By the power method, find all the eigenvalues of the matrices

5. $\begin{bmatrix} 1 & 6 & 1 \\ 1 & 2 & 0 \\ 0 & 0 & 3 \end{bmatrix}$ (Ans: (5). 4; 2.98; –0.98)

6. $\begin{bmatrix} 9 & 10 & 8 \\ 10 & 5 & -1 \\ 8 & -1 & 3 \end{bmatrix}$ (Ans: (6). 19.29; 7.08, –9.37)

JACOBI'S METHOD

Suppose A is a given real symmetric matrix. Its eigenvalues are real and there exists a real orthogonal matrix S such that $S^{-1}AS$ is a diagonal matrix whose elements are the eigenvalues of the matrix A. Jacobi's method consists in diagonalising the matrix A by applying a series of orthogonal transformations S_1, S_2, ..., S_{r-1}, S_r such that their product S satisfies the equation $B = S^{-1}AS$.

Procedure for computation of eigenvalues of the given real symmetric matrix A.

Step 1. The symmetric matrix A is transformed to a matrix B_1 by using the scheme

$$B_1 = S_1^{-1} A S_1 \tag{1}$$

where S_1 is an orthogonal matrix. Such a scheme is called a 'rotation'.

Step 2. The matrix B_1 is transformed to a matrix B_2 using the rotation

$$B_2 = S_2^{-1} B_1 S_2 \tag{2}$$

where S_2 is an orthogonal matrix.

Using (1) and (2), we get

$$B_2 = (S_2^{-1} S_1^{-1}) A(S_1 S_2) \tag{3}$$

Continuing the process like this, we obtain after r rotations,

$$\begin{aligned} B_r &= (S_r^{-1} S_{r-1}^{-1} \ldots S_1^{-1}) A(S_1 S_2 \ldots S_r) \\ &= (S_1 S_2 \ldots S_r)^{-1} A(S_1 S_2 \ldots S_r) \end{aligned} \tag{4}$$

If we let $S_1 S_2 \ldots S_r = S$ (5)

then (4) reduces to

$$B_r = S^{-1} AS \tag{6}$$

For large r, the matrix B_r becomes (almost) a diagonal matrix. The diagonal elements of this matrix are the eigenvalues A. Furthermore, the columns of S are the corresponding eigenvectors of A, and S is called the matrix of eigenvectors of A.

The orthogonal matrices S_1, S_2, ... S_r used in successive rotations are chosen as follows:

Let a_{ij} be the numerically largest non-diagonal elements of A. We choose $S_1 = [S_{ij}]$ such that

$$S_{ii} = \cos\theta,\ S_{ij} = -\sin\theta,\ S_{ji} = \sin\theta,\ S_{jj} = \cos\theta \tag{7}$$

and all other elements are the ones identical with those of the unit matrix I.

Using the rotation defined in (1), we find that the elements of D_1 on the (i, i), (i, j), (j, i) and (j, j) positions are respectively

$$\begin{aligned} b_{ii} &= a_{ii} \cos^2\theta + a_{ij} \sin 2\theta + a_{jj} \sin^2\theta \\ b_{ij} &= b_{ji} = (a_{jj} - a_{ii}) \sin\theta \cos\theta + a_{ij} \cos 2\theta \\ b_{jj} &= a_{ii} \sin^2\theta - a_{ij} \sin 2\theta + a_{jj} \cos^2\theta \end{aligned} \tag{8}$$

We choose 'θ' such that $b_{ij} = b_{ji} = 0$.

For this, we put

$$(a_{jj} - a_{ii}) \sin\theta \cos\theta + a_{ij} \cos 2\theta = 0$$

or

$$\frac{1}{2} (a_{jj} - a_{ii}) \sin(2\theta) + a_{ij} \cos(2\theta) = 0$$

$$\Rightarrow \qquad \boxed{\tan(2\theta) = \frac{2a_{ij}}{a_{ii} - a_{jj}}} \qquad (9)$$

Equation (9) gives four values of θ, we choose that value which lies between $-\pi/4$ and $\pi/4$.

For the values of θ found from (9), the off-diagonal elements in the matrix B_1 vanish.

Next, we identify the numerically largest off-diagonal element in the matrix B_1 and proceed as above to obtain the matrix B_2. The matrices B_3, B_4, B_5, ... are obtained similarly. We stop the process when we obtain a diagonal or an almost diagonal matrix B_r.

Example 12. Find all the eigenvalues and eigenvectors of the matrix

$$\begin{bmatrix} 1 & \sqrt{2} & 2 \\ \sqrt{2} & 3 & \sqrt{2} \\ 2 & \sqrt{2} & 1 \end{bmatrix}$$

by Jacobi's method.

Solution: The largest off-diagonal element is $a_{13} = a_{31} = 2$ and the other elements are $a_{11} = 1$ and $a_{33} = 1$.

$$\Rightarrow \qquad S_1 = \begin{bmatrix} \cos\theta & 0 & -\sin\theta \\ 0 & 1 & 0 \\ \sin\theta & 0 & \cos\theta \end{bmatrix}$$

where we have replaced

$$a_{11} = \cos\theta,\ a_{13} = -\sin\theta$$
$$a_{31} = \sin\theta,\ a_{33} = \cos\theta$$

and all other elements of S, are identical with those of unit matrix. Choose θ, t, we get

$$\therefore \qquad \tan 2\theta = \frac{2a_{13}}{a_{11} - a_{33}}$$

$$= \frac{2\times 2}{1-1} = \frac{4}{0} = \infty$$

$$\therefore \qquad 2\theta = \tan^{-1}(\infty) = \frac{\pi}{2}$$

$$\therefore \qquad \theta = \frac{\pi}{4}$$

$$\therefore \qquad S_1 = \begin{bmatrix} \frac{1}{\sqrt{2}} & 0 & \frac{-1}{\sqrt{2}} \\ 0 & 1 & 0 \\ \frac{1}{\sqrt{2}} & 0 & \frac{1}{\sqrt{2}} \end{bmatrix}$$

The 1st rotation given is

$$B_1 = S_1^{-1} A S_1$$

$$= \begin{bmatrix} \frac{1}{\sqrt{2}} & 0 & \frac{1}{\sqrt{2}} \\ 0 & 1 & 0 \\ \frac{-1}{\sqrt{2}} & 0 & \frac{1}{\sqrt{2}} \end{bmatrix} \begin{bmatrix} 1 & \sqrt{2} & 2 \\ \sqrt{2} & 3 & \sqrt{2} \\ 2 & \sqrt{2} & 1 \end{bmatrix} \begin{bmatrix} \frac{1}{\sqrt{2}} & 0 & \frac{-1}{\sqrt{2}} \\ 0 & 1 & 0 \\ \frac{1}{\sqrt{2}} & 0 & \frac{1}{\sqrt{2}} \end{bmatrix}$$

$$= \begin{bmatrix} 3 & 2 & 0 \\ 2 & 3 & 0 \\ 0 & 0 & -1 \end{bmatrix}$$

Again, the largest off-diagonal element in (B_1) is $a_{12} = a_{21} = 2$. The other elements are,

$$a_{11} = 3,\ a_{22} = 3$$

$$\theta = \frac{1}{2} \tan^{-1}\left(\frac{4}{0}\right) = \frac{\pi}{4}$$

and

$$S_2 = S_2^{-1} = S_2^T \begin{bmatrix} \frac{1}{\sqrt{2}} & \frac{-1}{\sqrt{2}} & 0 \\ \frac{1}{\sqrt{2}} & \frac{1}{\sqrt{2}} & 0 \\ 0 & 0 & 1 \end{bmatrix}$$

The 2nd rotation is

$$B_2 = S_2^{-1}\, B_1 S_2 = \begin{bmatrix} 5 & 0 & 0 \\ 0 & 1 & 0 \\ 0 & 0 & -1 \end{bmatrix}$$

we have the matrix of eigenvalues

$$S = S_1 S_2 = \begin{bmatrix} \frac{1}{\sqrt{2}} & 0 & \frac{-1}{\sqrt{2}} \\ 0 & 1 & 0 \\ \frac{1}{\sqrt{2}} & 0 & \frac{1}{\sqrt{2}} \end{bmatrix} \begin{bmatrix} \frac{1}{\sqrt{2}} & \frac{-1}{\sqrt{2}} & 0 \\ \frac{1}{\sqrt{2}} & \frac{1}{\sqrt{2}} & 0 \\ 0 & 0 & 1 \end{bmatrix}$$

$$= \begin{bmatrix} \frac{1}{2} & \frac{-1}{2} & \frac{-1}{\sqrt{2}} \\ \frac{1}{\sqrt{2}} & \frac{1}{\sqrt{2}} & 0 \\ \frac{1}{2} & \frac{-1}{2} & \frac{1}{\sqrt{2}} \end{bmatrix} \text{ column eigenvalue}$$

The eigenvalues are 5, 1, –1.

Example 13. $A = \begin{bmatrix} 1 & \sqrt{2} & 4 \\ \sqrt{2} & 3 & \sqrt{2} \\ 4 & \sqrt{2} & 1 \end{bmatrix}$

Solution: Note that the matrix A is a symmetric matrix.

1st Iteration

Consider the largest off-diagonal element $a_{13} = a_{31} = 4$ and $a_{11} = a_{33} = 1$

$$\Rightarrow \qquad S_1 = \begin{bmatrix} \cos\theta & 0 & -\sin\theta \\ 0 & 1 & 0 \\ \sin\theta & 0 & \cos\theta \end{bmatrix}$$

we know that

$$\tan 2\theta = \frac{2a_{13}}{a_{11} - a_{33}}$$

$$= \frac{8}{0} = \infty$$

$$\therefore \qquad \theta = \frac{1}{2}\tan^{-1}(\infty)$$

$$= \frac{1}{2}\left(\frac{\pi}{2}\right) = \left(\frac{\pi}{4}\right)$$

$$\therefore \qquad S_1 = \begin{bmatrix} \frac{1}{\sqrt{2}} & 0 & \frac{-1}{\sqrt{2}} \\ 0 & 1 & 0 \\ \frac{1}{\sqrt{2}} & 0 & \frac{1}{\sqrt{2}} \end{bmatrix} \text{ is orthogonal}$$

$$\therefore \qquad \boxed{S_1^{-1} = S^T}$$

$$\therefore \quad S_1^{-1} = \begin{bmatrix} \frac{1}{\sqrt{2}} & 0 & \frac{1}{\sqrt{2}} \\ 0 & 1 & 0 \\ \frac{-1}{\sqrt{2}} & 0 & \frac{1}{\sqrt{2}} \end{bmatrix}$$

$$\therefore \quad B_1 = S_1^{-1} A S_1$$

$$= \begin{bmatrix} \frac{1}{\sqrt{2}} & 0 & \frac{1}{\sqrt{2}} \\ 0 & 1 & 0 \\ \frac{-1}{\sqrt{2}} & 0 & \frac{1}{\sqrt{2}} \end{bmatrix} \begin{bmatrix} 1 & \sqrt{2} & 4 \\ \sqrt{2} & 3 & \sqrt{2} \\ 4 & \sqrt{2} & -1 \end{bmatrix} \begin{bmatrix} \frac{1}{\sqrt{2}} & 0 & \frac{-1}{\sqrt{2}} \\ 0 & 1 & 0 \\ \frac{1}{\sqrt{2}} & 0 & \frac{1}{\sqrt{2}} \end{bmatrix}$$

$$= \begin{bmatrix} 6 & 2 & 0 \\ 2 & 3 & 0 \\ 0 & 0 & -3 \end{bmatrix}$$

Here, the largest off-diagonal elements are B_1 is b_{12}

$$b_{12} = b_{21} = 2$$

and $\quad b_{11} = 6,\ b_{22} = 3.$

$$\therefore \quad S_2 = \begin{bmatrix} \cos\theta & -\sin\theta & 0 \\ \sin\theta & \cos\theta & 0 \\ 0 & 0 & 1 \end{bmatrix}$$

$\therefore$ Choose θ t

$$\tan 2\theta = \frac{2b_{12}}{b_{11} - b_{22}}$$

$$= \frac{2 \times 2}{6 - 3} = \frac{4}{3},$$

$$\therefore \quad 0 = \frac{1}{2} \tan^{-1}\left(\frac{4}{3}\right)$$

$$= 0.4636$$

$$\therefore \quad S_2 = \begin{bmatrix} 0.8944 & -0.4472 & 0 \\ 0.4472 & 0.8944 & 0 \\ 0 & 0 & 1 \end{bmatrix}$$

$$\therefore \quad S_2^{-1} = S_2^T$$

$$= \begin{bmatrix} 0.8944 & 0.4472 & 0 \\ -0.4472 & 0.8944 & 0 \\ 0 & 0 & 1 \end{bmatrix}$$

$$\therefore \quad B_2 = S_2^{-1}(B_1)S_2$$

$$= \begin{bmatrix} 0.8944 & 0.4472 & 0 \\ -0.4472 & 0.8944 & 0 \\ 0 & 0 & 1 \end{bmatrix} \begin{bmatrix} 6 & 2 & 0 \\ 2 & 3 & 0 \\ 0 & 0 & -3 \end{bmatrix} \begin{bmatrix} 0.8944 & -0.4420 \\ 0.4472 & 0 \end{bmatrix}$$

$$B_2 = \begin{bmatrix} 7 & 0 & 0 \\ 0 & 1.2 & 0 \\ 0 & 0 & -3 \end{bmatrix}$$

This is a diagonal matrix whose diagonal elements

$$\lambda_1 = 7,\ \lambda_2 = 1.2,\ \lambda_3 = -3$$

The eigenvectors are

$$S = S_1 S_2$$

$$= \begin{bmatrix} \frac{1}{\sqrt{2}} & 0 & \frac{-1}{\sqrt{2}} \\ 0 & 1 & 0 \\ \frac{1}{\sqrt{2}} & 0 & \frac{1}{\sqrt{2}} \end{bmatrix} \begin{bmatrix} 0.8944 & -0.4972 & 0 \\ 0.4472 & 0.8944 & 0 \\ 0 & 0 & 1 \end{bmatrix}$$

$$= \begin{bmatrix} 0.6324 & -0.3162 & -0.7071 \\ 0.4472 & 0.8944 & 0 \\ 0 & 0 & 0.7071 \end{bmatrix}$$

The 1st, 2nd, 3rd columns of S give the eigenvalues corresponding to $\lambda_1, \lambda_2, \lambda_3$.

Example 14. Find all the eigenvalues and eigenvectors of the matrix

$$A = \begin{bmatrix} 2 & 3 & 1 \\ 3 & 2 & 2 \\ 1 & 2 & 1 \end{bmatrix}$$

by the Jacobi's method.

Solution: Note that the matrix A is a symmetric matrix

1st Iteration

Consider the largest off-diagonal element

$$a_{12} = a_{21} = 3$$

and $$a_{11} = 2 = a_{22}$$

$$\Rightarrow \qquad S_1 = \begin{bmatrix} \cos\theta & -\sin\theta & 0 \\ \sin\theta & \cos\theta & 0 \\ 0 & 0 & 1 \end{bmatrix}$$

We know that

$$\tan 2\theta = \left[\frac{2\,a_{12}}{a_{11} - a_{22}}\right]$$

$$= \frac{6}{0} = \infty$$

$$\therefore \qquad \theta = \frac{1}{2}\ \tan^{-1}(\infty)$$

$$= \frac{1}{2}\left(\frac{\pi}{2}\right) = \frac{\pi}{4}$$

$$\therefore \qquad \left(\theta = \frac{\pi}{4}\right)$$

$$\therefore \qquad S_1 = \begin{bmatrix} \frac{1}{\sqrt{2}} & \frac{-1}{\sqrt{2}} & 0 \\ \frac{1}{\sqrt{2}} & \frac{1}{\sqrt{2}} & 0 \\ 0 & 0 & 1 \end{bmatrix}$$

$$\therefore \qquad S_1^{-1} = S_1^T$$

$$= \begin{bmatrix} \frac{1}{\sqrt{2}} & \frac{1}{\sqrt{2}} & 0 \\ -\frac{1}{\sqrt{2}} & \frac{1}{\sqrt{2}} & 0 \\ 0 & 0 & 1 \end{bmatrix}$$

$$\therefore \qquad B_1 = S_1^{-1} A S_1$$

$$= \begin{bmatrix} \frac{1}{\sqrt{2}} & \frac{1}{\sqrt{2}} & 0 \\ -\frac{1}{\sqrt{2}} & \frac{1}{\sqrt{2}} & 0 \\ 0 & 0 & 1 \end{bmatrix} \begin{bmatrix} 2 & 3 & 1 \\ 3 & 2 & 2 \\ 1 & 2 & 1 \end{bmatrix} \begin{bmatrix} \frac{1}{\sqrt{2}} & -\frac{1}{\sqrt{2}} & 0 \\ \frac{1}{\sqrt{2}} & \frac{1}{\sqrt{2}} & 0 \\ 0 & 0 & 1 \end{bmatrix}$$

$$= \begin{bmatrix} 5 & 0 & 2.1213 \\ 0 & -1 & 0.7071 \\ 2.1213 & 0.7071 & 1 \end{bmatrix}$$

Again, the largest off-diagonal element in B_1 is $b_{13} = b_{31} = 2.1213$, so that the corresponding orthogonal matrix S_2 is taken as

$$S_2 = \begin{bmatrix} \cos\theta & 0 & -\sin\theta \\ 0 & 1 & 0 \\ \sin\theta & 0 & \cos\theta \end{bmatrix}$$

Choose θ, t

$$\tan(2\theta) = \left[\frac{2(2.1213)}{5-1}\right]$$

$$\therefore \qquad S_2 = \begin{bmatrix} 0.9181 & 0 & -0.3962 \\ 0 & 1 & 0 \\ 0.3962 & 0 & 0.9181 \end{bmatrix}$$

$$\therefore \qquad S_2^{-1} = S_2^T = \begin{bmatrix} 0.9181 & 0 & 0.3962 \\ 0 & 1 & 0 \\ -0.3962 & 0 & 0.9181 \end{bmatrix}$$

$$\therefore \qquad B_2 = S_2^{-1}(B_1)S_2$$

$$= \begin{bmatrix} 0.9181 & 0 & 0.3962 \\ 0 & 1 & 0 \\ -0.3962 & 0 & 0.9181 \end{bmatrix} \begin{bmatrix} 5 & 0 & 2.1213 \\ 0 & -1 & 0.7071 \\ 2.1213 & 0.7071 & 1 \end{bmatrix}$$

$$\times \begin{bmatrix} 0.9181 & 0 & 0.3962 \\ 0 & 1 & 0 \\ -0.3962 & 0 & 0.9181 \end{bmatrix}$$

$$= \begin{bmatrix} 5.91555 & 0.2802 & 0 \\ 0.2802 & -1 & 0.6493 \\ 0 & 0.6493 & 0.0845 \end{bmatrix}$$

Now continuing the process 6 times, we may arrive at the diagonal matrix

$$B_6 = \begin{bmatrix} 5.9273 & 0 & 0 \\ 0 & -1.2993 & 0 \\ 0 & 0 & 0.3857 \end{bmatrix}$$

Hence, the eigenvalues of A are

$$\lambda_1 = 5.9273$$
$$\lambda_2 = -1.2993$$
$$\lambda_3 = 0.3857$$

The matrix of eigenvectors

$$S = S_1 S_2 S_3 S_4 S_5 S_6$$

$$= \begin{bmatrix} 0.6145 & -0.5504 & -0.5648 \\ 0.6814 & 0.7313 & 0.0287 \\ 0.03972 & -0.3710 & 0.8246 \end{bmatrix}$$

The first, the second and the third columns of S are the eigenvectors corresponding to λ_1, λ_2, λ_3.

Example 15. Find the eigenvalues and eigenvectors of

$$A = \begin{bmatrix} 2 & 1 & 0 \\ 1 & 4 & 1 \\ 0 & 1 & 4 \end{bmatrix}$$

by Jacobi's method.

Solution: Note that the matrix A is a symmetric matrix.

1st Iteration:

Consider the largest off-diagonal elements.

$$a_{12} = a_{21} = 1$$
$$a_{22} = a_{33} = 4$$

$$\Rightarrow \qquad S_1 = \begin{bmatrix} 1 & 0 & 0 \\ 0 & \cos\theta & -\sin\theta \\ 0 & \sin\theta & \cos\theta \end{bmatrix}$$

$$\tan 2\theta = \frac{2a_{12}}{a_{22} - a_{33}} = \frac{2}{0} = \infty$$

$$\therefore \qquad \theta = \frac{1}{2}\tan^{-1}(\infty)$$

$$= \frac{1}{2}\left(\frac{\pi}{2}\right) = \frac{\pi}{4}$$

$$\therefore \qquad S_1 = \begin{bmatrix} 1 & 0 & 0 \\ 0 & \frac{1}{\sqrt{2}} & -\frac{1}{\sqrt{2}} \\ 0 & \frac{1}{\sqrt{2}} & \frac{1}{\sqrt{2}} \end{bmatrix}$$

and
$$S_1^{-1} = S^T = \begin{bmatrix} 1 & 0 & 0 \\ 0 & \frac{1}{\sqrt{2}} & \frac{1}{\sqrt{2}} \\ 0 & -\frac{1}{\sqrt{2}} & \frac{1}{\sqrt{2}} \end{bmatrix}$$

$\Rightarrow$
$$B_1 = S_1^{-1} A S_1$$

$$= \begin{bmatrix} 1 & 0 & 0 \\ 0 & \frac{1}{\sqrt{2}} & \frac{1}{\sqrt{2}} \\ 0 & -\frac{1}{\sqrt{2}} & \frac{1}{\sqrt{2}} \end{bmatrix} \begin{bmatrix} 2 & 1 & 0 \\ 1 & 4 & 1 \\ 0 & 1 & 4 \end{bmatrix} \begin{bmatrix} 1 & 0 & 0 \\ 0 & \frac{1}{\sqrt{2}} & -\frac{1}{\sqrt{2}} \\ 0 & \frac{1}{\sqrt{2}} & \frac{1}{\sqrt{2}} \end{bmatrix}$$

$\Rightarrow$
$$B_1 = \begin{bmatrix} 2 & 0.7071 & -0.7071 \\ 0.7071 & 4.9999 & 0 \\ -0.7071 & 0 & 2.9999 \end{bmatrix}$$

Again the largest off-diagonal element is B_1 is

$$a_{12} = 0.7071 = a_{21}$$
$$a_{11} = 2,\ a_{22} = 2.9999$$

$$\theta = \frac{1}{2} \tan^{-1}\left(\frac{2 a_{12}}{a_{11} - a_{22}}\right)$$

$$= \frac{1}{2} \tan^{-1}\left(\frac{2 \times 0.7071}{2 - 2.9999}\right)$$

$$= -27^0 22^1$$

$\Rightarrow$
$$S_2 = \begin{bmatrix} \cos\theta & -\sin\theta & 0 \\ \sin\theta & \cos\theta & 0 \\ 0 & 0 & 1 \end{bmatrix}$$

$$= \begin{bmatrix} 0.8881 & 0.4596 & 0 \\ -0.4596 & 0.8881 & 0 \\ 0 & 0 & 1 \end{bmatrix}$$

$$S_2^{-1} = S_2^T = \begin{bmatrix} 0.8881 & -0.4596 & 0 \\ 0.4596 & 0.8881 & 0 \\ 0 & 0 & 1 \end{bmatrix}$$

$$B_2 = S_2^{-1}(B_1)S_2$$

$$= \begin{bmatrix} 2.0562 & -.8161 & -0.6279 \\ -0.8160 & 4.9431 & -0.3249 \\ -0.6279 & 0.3249 & 0 \end{bmatrix}$$

Now continuing and proving 12 times we may arrive at the diagonal matrix.

$$B_{12} = \begin{bmatrix} 1.52 & 0 & 0 \\ 0 & 5.17 & 0 \\ 0 & 0 & 3.31 \end{bmatrix}$$

Hence, the eigenvalues of matrix A are 1.52, 5.17, 3.31 and eigenvector are the column of

$$S = S_1 S_2 \ldots S_{12}$$

EXERCISES

Using Jacobi's method, calculate all the eigenvalues and the eigenvectors of the following matrices.

1. (i) $\begin{bmatrix} 3 & 2 & 2 \\ 2 & 5 & 2 \\ 2 & 2 & 3 \end{bmatrix}$, (ii) $\begin{bmatrix} 1 & -2 & 4 \\ -2 & 5 & -2 \\ 4 & -2 & 1 \end{bmatrix}$

(iii) $\begin{bmatrix} 1 & \sqrt{3} & 4 \\ \sqrt{3} & 5 & \sqrt{3} \\ 4 & \sqrt{3} & 1 \end{bmatrix}$, (iv) $\begin{bmatrix} 2 & \sqrt{2} & 4 \\ \sqrt{2} & 6 & \sqrt{2} \\ 4 & \sqrt{2} & 2 \end{bmatrix}$

2. $\begin{bmatrix} 8 & -6 & 2 \\ -6 & 7 & -4 \\ 2 & -4 & 3 \end{bmatrix}$ 3. $\begin{bmatrix} 3 & -1 & 1 \\ -1 & 5 & -1 \\ 1 & -1 & 3 \end{bmatrix}$

4. $\begin{bmatrix} 5 & 0 & 1 \\ 0 & -2 & 0 \\ 1 & 0 & 5 \end{bmatrix}$ 5. $\begin{bmatrix} 1 & \sqrt{4} & 4 \\ \sqrt{2} & 3 & \sqrt{2} \\ 4 & \sqrt{2} & 1 \end{bmatrix}$

6. $\begin{bmatrix} 2 & -1 & 0 \\ -1 & 2 & -1 \\ 0 & -1 & 2 \end{bmatrix}$ 7. $\begin{bmatrix} 4 & 2 & 2 \\ 2 & 5 & 1 \\ 2 & 1 & 6 \end{bmatrix}$

Answers:

(2) 0, 3, 15; $S = \begin{bmatrix} 1 & 2 & 2 \\ 2 & 1 & 2 \\ 2 & 2 & 1 \end{bmatrix}$

(3) 2, 3, 6; $S = \begin{bmatrix} 0.5773 & -0.4082 & -0.7071 \\ 0.5773 & 0.8165 & 0 \\ 0.5773 & -0.4082 & 0.7071 \end{bmatrix}$

(4) 4, –2, 6; $S = \begin{bmatrix} 1/\sqrt{2} & 0 & -1/\sqrt{2} \\ 0 & 1 & 0 \\ 1/\sqrt{2} & 0 & 1/\sqrt{2} \end{bmatrix}$

(5) 6.24, 1.76, –3; $S = \begin{bmatrix} 0.6015 & -0.3717 & -0.7071 \\ 0.5257 & 0.8506 & 0 \\ 0 & 0 & 0.7071 \end{bmatrix}$

(6) 0.5858, 2, 3.4142; $S = \begin{bmatrix} 0.5 & 0.7071 & 0.5 \\ 0.7071 & 0 & -0.7071 \\ 0.5 & -0.7071 & 0.5 \end{bmatrix}$

(7) 8.3874, 4.4867, 2.1260; $S = \begin{bmatrix} 0.8077 & 0.2170 & 1 \\ 0.7720 & 1 & -0.5673 \\ 1 & -0.9473 & -0.3698 \end{bmatrix}$

Given's Method for Symmetry Matrix

To find Eigenvalues and Eigenvector Procedure: Let A be symmetric matrices. We reduce the given symmetric matrix A to a tridiagonal form.

Step (1). Let $A = \begin{bmatrix} a_{11} & a_{12} & \dots & a_{1n} \\ a_{21} & \boxed{a_{22}} & \dots & a_{2n} \\ \dots & \dots & \dots & \dots \\ a_{n1} & a_{n2} & \dots & a_{nn} \end{bmatrix}$ (1)

be a given symmetric matrix. We replace

$$a_{22} = \cos\theta$$
$$a_{23} = -\sin\theta$$

$$a_{32} = \sin\theta$$
$$a_{33} = \cos\theta$$

and all other elements by the corresponding elements in the unit matrix I (of order n)

$$\therefore \text{ We write a matrix } S_1 \text{ as } S_1 = \begin{bmatrix} 1 & 0 & 0 & 0 & 0 \\ 0 & \cos\theta & -\sin\theta & \dots & 0 \\ 0 & \sin\theta & \cos\theta & \dots & 0 \\ 0 & 0 & 0 & 1 & 0 \\ 0 & 0 & 0 & 0 & 1 \end{bmatrix} \qquad (2)$$

Now choose θ, t

$$\tan\theta = \frac{a_{13}}{a_{12}};$$

$$\theta = \tan^{-1}\left(\frac{a_{13}}{a_{12}}\right)$$

$$\therefore \qquad \cos\theta = \frac{a_{12}}{\sqrt{a_{12}^2 + a_{13}^2}}; \ \sin\theta = \frac{a_{13}}{\sqrt{a_{12}^2 + a_{13}^2}}$$

Substitute the value of θ in S_1. Consider $S_1^{-1} = S_1^T$ (S_1 is orthogonal)

$[\therefore\ S_1 S_1^T = I]$

Step (2). Let us consider B_1 f

$$\boxed{B_1 = S_1^{-1} A S_1}$$

$$B_1 = \text{diag}\ (\lambda_1\ \lambda_2\ \lambda_n)$$

Replace the elements b_{22}, b_{24}, b_{42} and b_{44} by

$$b_{22} = \cos\theta$$
$$b_{24} = -\sin\theta$$
$$b_{42} = \sin\theta$$

and
$$b_{44} = \cos\theta$$

replace other elements of B_1 by the corresponding elements of the unit matrix I.

Step (3). Construction of Sturm Sequence

Finally, we produce the matrix.

$$B_1 = \begin{bmatrix} b_1 & c_1 & & & & \\ c_1 & b_2 & c_2 & & & \\ & c_2 & b_3 & c_3 & & \\ & & \ddots & & & \\ & & & c_{n-2} & b_{n-1} & c_{n-1} \\ & & & & c_{n-1} & b_n \end{bmatrix}$$

we define

$$f_n = |\lambda I - B_1|$$

$$= \begin{bmatrix} \lambda - b_1 & -c_1 & & & \\ -c_1 & \lambda - b_2 & -c_2 & & \\ & \ddots & & & \\ & & -c_{n-2} & \lambda - b_{n-1} & -c_{n-1} \\ & & & -c_{n-1} & \lambda - b_n \end{bmatrix}$$

Expanding we get the sequence $\{f_n\}$ which satisfies $f_0 = 1$, $f_1 = \lambda - b_1$

and $$\boxed{f_r = (\lambda - b_r) f_{r-1} - C_{r-1}^2 f_{r-2}}$$

$(2 \leq r \leq n)$

The eigenvalues B_1 are the same as those of the given matrix.

Example 16. Construct the Sturm sequence for the matrix by Given's method.

$$A = \begin{bmatrix} 3 & -2 & 0 \\ -2 & 3 & -2 \\ 0 & -2 & 3 \end{bmatrix}$$

Hence, verify that 3 is an eigenvalue.

Solution: The given matrix is symmetric and it is in a tridiagonal form. The characteristic equation of A is given by

$$f_n = |\lambda I - A|$$

$$= \begin{vmatrix} \lambda - 3 & -2 & 0 \\ -2 & \lambda - 3 & -2 \\ 0 & -2 & \lambda - 3 \end{vmatrix}$$

The Sturm sequence

$$f_r = (\lambda - b_r) f_{r-1} - C_{r-1}^2 f_{r-2} \quad 2 \leq r \leq n$$

$\{f_n\}$ is defined as

$$f_0(\lambda) = 1$$
$$f_1(\lambda) = \lambda - 3$$
$$f_2(\lambda) = (\lambda - 3)f_1(\lambda) - 4f_0(\lambda)$$
$$f_3(\lambda) = (\lambda - 3)f_2(\lambda) - 4f_1(\lambda)$$

Example 17. Transform the matrix

$$A = \begin{bmatrix} 1 & 2 & 4 \\ 2 & 1 & 2 \\ 4 & 2 & 1 \end{bmatrix}$$

to tridiagonal form by Given's method. Obtain the intervals of unit length, each containing one eigenvalue of A. Find the largest positive eigenvalue correct to two decimal places using the Newton-Raphson method.

Solution: Performing the orthogonal rotation with respect to $a_{22}, a_{23}, a_{32}, a_{33}$ submatrix, we get

$$\tan\theta = \frac{a_{13}}{a_{12}} = 2,\ \cos\theta = \frac{1}{\sqrt{5}},\ \sin\theta = \frac{2}{\sqrt{5}}.$$

$$B = \begin{bmatrix} 1 & 0 & 0 \\ 0 & 1/\sqrt{5} & 2/\sqrt{5} \\ 0 & -2/\sqrt{5} & 1/\sqrt{5} \end{bmatrix} \begin{bmatrix} 1 & 2 & 4 \\ 2 & 1 & 2 \\ 4 & 2 & 1 \end{bmatrix} \begin{bmatrix} 1 & 0 & 0 \\ 0 & 1/\sqrt{5} & -2/\sqrt{5} \\ 0 & 2/\sqrt{5} & 1/\sqrt{5} \end{bmatrix}$$

$$= \begin{bmatrix} 1 & 2\sqrt{5} & 0 \\ 2\sqrt{5} & 2.6 & -1.2 \\ 0 & -1.2 & -0.6 \end{bmatrix}$$

The Sturm sequence is obtained as

$$f_0 = 1,\ f_1 = \lambda - 1$$
$$f_2 = (\lambda - 2.6)(\lambda - 1) - 20 = \lambda^2 - 3.6\lambda - 17.4$$
$$f_3 = (\lambda + 0.6)(\lambda^2 - 3.6\lambda - 17.4) - 1.44(\lambda - 1)$$
$$= \lambda^3 - 3\lambda^2 - 21\lambda - 9.$$

We find $f_3(-3) = 0$, $f(-1) = 8 > 0$, $f(0) = -9 < 0$, $f(6) = -27 < 0$ and $f(7) = 40 > 0$. Therefore, one eigenvalue is -3 and the other eigenvalues lie in the intervals $(-1, 0)$ and $(6, 7)$.

Starting with $\lambda_0 = 6.5$ as initial approximation and using the Newton-Raphson method

$$\lambda_{k+1} = \lambda_k - \frac{f_3(\lambda_k)}{f_3'(\lambda_k)} = \lambda_k - \frac{\lambda_k^3 - 3\lambda_k^2 - 21\lambda_k - 9}{3\lambda_k^2 - 6\lambda_k - 21}, \ k = 0, 1, \ldots$$

we obtain $\lambda_1 = 6.464419, \ \lambda_2 = 6.464120.$

Therefore, the largest positive eigenvalue correct to two decimal places is 6.46.

Example 18. Find the eigenvalues and eigenvectors of the matrix

$$A = \begin{bmatrix} 1 & \frac{1}{2} & \frac{1}{3} \\ \frac{1}{2} & \frac{1}{3} & \frac{1}{4} \\ \frac{1}{3} & \frac{1}{4} & \frac{1}{5} \end{bmatrix}$$

by Given's method.

Solution: We reduce the given symmetric matrix to the tridiagonal form.

Step (1). Replacing the elements a_{22}, a_{23}, a_{32}, a_{33} by $\cos\theta$, $-\sin\theta$, $\sin\theta$ and $\cos\theta$ and the other elements by the corresponding elements of the unit matrix, I,

$$\therefore \text{ Let } S_1 = \begin{bmatrix} 1 & 0 & 0 \\ 0 & \cos\theta & -\sin\theta \\ 0 & \sin\theta & \cos\theta \end{bmatrix}$$

Choose θ, t

$$\tan\theta = \frac{a_{13}}{a_{12}} = \frac{\frac{1}{3}}{\frac{1}{2}} = \frac{2}{3}$$

$$\theta = \tan^{-1}\left(\frac{2}{3}\right)$$

$$= 33.6901$$

Then $$S_1 = \begin{bmatrix} 1 & 0 & 0 \\ 0 & .8321 & -0.5547 \\ 0 & 0.5547 & 0.8321 \end{bmatrix}$$

$\therefore$ $$S_1^{-1} = S_1^T$$

$$= \begin{bmatrix} 1 & 0 & 0 \\ 0 & 0.8321 & 0.5547 \\ 0 & -0.5547 & 0.8321 \end{bmatrix}$$

Hence, $B_1 = S_1^{-1} AS_1$

$$= \begin{bmatrix} 1 & 0 & 0 \\ 0 & 0.8321 & 0.5547 \\ 0 & -0.5547 & 0.8321 \end{bmatrix} \begin{bmatrix} 1 & \frac{1}{2} & \frac{1}{3} \\ \frac{1}{2} & \frac{1}{3} & \frac{1}{4} \\ \frac{1}{3} & \frac{1}{4} & \frac{1}{5} \end{bmatrix} \times \begin{bmatrix} 1 & 0 & 0 \\ 0 & 0.8321 & -0.5547 \\ 0 & 0.5547 & 0.8321 \end{bmatrix}$$

$$\tan\theta = 1 \text{ or } \theta = \frac{\pi}{4} \cos\theta = \frac{2}{\sqrt{2^2+2}} = \frac{1}{\sqrt{2}}$$

$$\therefore \quad \sin\theta = \frac{2}{\sqrt{2^2+2}} = \frac{2}{\sqrt{8}} = \frac{1}{2\sqrt{2}} = \frac{1}{\sqrt{2}}$$

$$S_1^{-1} = S_1^T$$

$$= \begin{bmatrix} 1 & 0 & 0 \\ 0 & \cos\theta & \sin\theta \\ 0 & -\sin\theta & \cos\theta \end{bmatrix}$$

$$= \begin{bmatrix} 1 & 0 & 0 \\ 0 & \frac{1}{\sqrt{2}} & -\frac{1}{\sqrt{2}} \\ 0 & \frac{1}{\sqrt{2}} & \frac{1}{\sqrt{2}} \end{bmatrix}$$

$$\therefore \quad B_1 = S_1^{-1} AS_1$$

$$= \begin{bmatrix} 1 & 0 & 0 \\ 0 & \frac{1}{\sqrt{2}} & \frac{1}{\sqrt{2}} \\ 0 & -\frac{1}{\sqrt{2}} & \frac{1}{\sqrt{2}} \end{bmatrix} \begin{bmatrix} 1 & 2 & 2 \\ 2 & 1 & 2 \\ 2 & 2 & 1 \end{bmatrix} \begin{bmatrix} 1 & 0 & 0 \\ 0 & \frac{1}{\sqrt{2}} & \frac{-1}{\sqrt{2}} \\ 0 & \frac{1}{\sqrt{2}} & \frac{1}{\sqrt{2}} \end{bmatrix}$$

$$= \begin{bmatrix} 0.0769 & 0.1667 & 0 \\ 0.1667 & 0.5231 & 0.0346 \\ 0 & 0.0346 & 0.0103 \end{bmatrix}$$

Clearly, this is a trigonal matrix.

The Sturm sequence for B_1 is given by

$$f_0(\lambda) = 1, \; f_1(\lambda) = 0.0769 - \lambda$$

$$f_2(\lambda) = (0.523 - \lambda) f_1(\lambda) - (0.1667)^2$$

$$f_3(\lambda) = (0.0103 - \lambda) f_2(\lambda) - (0.0346)^2 \qquad \text{(iv)}$$
$$f_2(\lambda) = (0.5237 - \lambda)(0.0769 - \lambda) - (6.1667)^2$$

to scan that the eigenvalues lie in the interval (0, 1) and are eigenvalue in (7, 8)

Example 19. Transform the matrix

$$A = \begin{bmatrix} 1 & 2 & 3 \\ 2 & 1 & -1 \\ 3 & -1 & 1 \end{bmatrix}$$

to tridiagonal forms by Given's method. Use exact arithmetic.

Solution: Replacing the elements $a_{22}, a_{23}, a_{32}, a_{33}$ by $\cos\theta$, $-\sin\theta$, $\sin\theta$, $\cos\theta$ and other elements by the corresponding elements of unit matrix.

$$\therefore \text{ Let } S_1 = \begin{bmatrix} 1 & 0 & 0 \\ 0 & \cos\theta & -\sin\theta \\ 0 & \sin\theta & \cos\theta \end{bmatrix}$$

Choose θ, t

$$\tan\theta = \frac{a_{13}}{a_{12}}$$
$$= \frac{3}{2}$$

$$\therefore \qquad \cos\theta = \frac{2}{\sqrt{13}}, \ \sin\theta = \frac{3}{\sqrt{13}}$$

Hence,

$$\therefore \qquad S_1 = \begin{bmatrix} 1 & 0 & 0 \\ 0 & \frac{2}{\sqrt{13}} & -\frac{3}{\sqrt{13}} \\ 0 & \frac{3}{\sqrt{13}} & \frac{2}{\sqrt{13}} \end{bmatrix}$$

$$S_1^{-1} = S_1^T$$

$$= \begin{bmatrix} 1 & 0 & 0 \\ 0 & \frac{2}{\sqrt{13}} & \frac{3}{\sqrt{13}} \\ 0 & \frac{-3}{\sqrt{13}} & \frac{2}{\sqrt{13}} \end{bmatrix}$$

$$\therefore \qquad B_1 = S_1^{-1} A S_1$$

$$= \begin{bmatrix} 1 & 0 & 0 \\ 0 & \dfrac{2}{\sqrt{13}} & \dfrac{3}{\sqrt{13}} \\ 0 & \dfrac{-3}{\sqrt{13}} & \dfrac{2}{\sqrt{13}} \end{bmatrix} \begin{bmatrix} 1 & 2 & 3 \\ 2 & 1 & -1 \\ 3 & -1 & 1 \end{bmatrix} \begin{bmatrix} 1 & 0 & 0 \\ 0 & \dfrac{2}{\sqrt{13}} & \dfrac{-3}{\sqrt{13}} \\ 0 & \dfrac{3}{\sqrt{13}} & \dfrac{2}{\sqrt{13}} \end{bmatrix}$$

$$= \begin{bmatrix} 1 & \sqrt{13} & 0 \\ \sqrt{13} & \dfrac{1}{13} & \dfrac{5}{13} \\ 0 & \dfrac{5}{13} & \dfrac{25}{13} \end{bmatrix}$$

is the required tridiagonal form.

Example 20. Transform the matrix

$$A = \begin{bmatrix} 1 & 2 & 2 \\ 2 & 1 & 2 \\ 2 & 2 & 1 \end{bmatrix}$$

to tridiagonal form by Given's method. Find the eigenvector corresponding to the largest eigenvalue from the eigenvector of the tridiagonal matrix by using Given's method.

Solution: We replace a_{22}, a_{23}, a_{32} and a_{33} by $\cos\theta$, $-\sin\theta$, $\sin\theta$, $\cos\theta$

$$\therefore \qquad S_1 = \begin{bmatrix} 1 & 0 & 0 \\ 0 & \cos\theta & -\sin\theta \\ 0 & +\sin\theta & \cos\theta \end{bmatrix}$$

Choose θ, t

$$\tan\theta = \frac{a_{13}}{a_{12}} = \frac{2}{2} = 1$$

$$B_1 = S_1^{-1} A S_1$$

$$= \begin{bmatrix} 1 & 0 & 0 \\ 0 & \dfrac{1}{\sqrt{2}} & \dfrac{1}{\sqrt{2}} \\ 0 & -\dfrac{1}{\sqrt{2}} & \dfrac{1}{\sqrt{2}} \end{bmatrix} \begin{bmatrix} 1 & 2 & 2 \\ 2 & 1 & 2 \\ 2 & 2 & 1 \end{bmatrix} \begin{bmatrix} 1 & 0 & 0 \\ 0 & \dfrac{1}{\sqrt{2}} & -\dfrac{1}{\sqrt{2}} \\ 0 & \dfrac{1}{\sqrt{2}} & \dfrac{1}{\sqrt{2}} \end{bmatrix}$$

$$= \begin{bmatrix} 1.1+0.2+0.2 & 1.2+0.1+0.2 & 2+0+0 \\ 0+\sqrt{2}+\sqrt{2} & 0+\frac{1}{\sqrt{2}}+\sqrt{2} & 0+\sqrt{2}+\frac{1}{\sqrt{2}} \\ 0-\sqrt{2}+\sqrt{2} & 0-\frac{1}{\sqrt{2}}+\sqrt{2} & 0-\sqrt{2}+\frac{1}{\sqrt{2}} \end{bmatrix} \begin{bmatrix} 1 & 0 & 0 \\ 0 & \frac{1}{\sqrt{2}} & -\frac{1}{\sqrt{2}} \\ 0 & \frac{1}{\sqrt{2}} & \frac{1}{\sqrt{2}} \end{bmatrix}$$

$$= \begin{bmatrix} 1 & 2 & 2 \\ 2\sqrt{2} & \frac{3}{\sqrt{2}} & \frac{3}{\sqrt{2}} \\ 0 & \frac{1}{\sqrt{2}} & -\frac{1}{\sqrt{2}} \end{bmatrix} \begin{bmatrix} 1 & 0 & 0 \\ 0 & \frac{1}{\sqrt{2}} & -\frac{1}{\sqrt{2}} \\ 0 & \frac{1}{\sqrt{2}} & \frac{1}{\sqrt{2}} \end{bmatrix}$$

$$= \begin{bmatrix} 1.1+2.0+2.0 & 1.0+\sqrt{2}+\sqrt{2} & 0-\sqrt{2}+\sqrt{2} \\ 2\sqrt{2}+0+0 & 0+\frac{3}{2}+\frac{3}{2} & 0+\frac{3}{2}+\frac{3}{2} \\ 0+0+0 & 0+\frac{1}{2}-\frac{1}{2} & 0-\frac{1}{2}-\frac{1}{2} \end{bmatrix}$$

$$= \begin{bmatrix} 1 & 2\sqrt{2} & 0 \\ 2\sqrt{2} & 3 & 0 \\ 0 & 0 & -1 \end{bmatrix}$$

which is the tridiagonal form.

The characteristic equation of B_1 is given by $f_n = |\lambda I - A|$

$$= \begin{bmatrix} \lambda-1 & -2\sqrt{2} & 0 \\ -2\sqrt{2} & \lambda-3 & 0 \\ 0 & 0 & \lambda+1 \end{bmatrix}$$

The Sturm sequence $\{f_n\}$ is defined as

$$f_0 = 1, f_1 = \lambda - b_1$$

and

$$f_r = [\lambda - b_r]\, f_{r-1} - c_{r-1}^2 f_{r-2} \; (2 \le r \le n)$$

$$f_n(\lambda) = (\alpha_n - \lambda)\, f_{n-1}(\lambda) - \beta_{n-1}^2\, f_{n-2}(\lambda)$$

$$f_0(\lambda) = 1$$

$$f_1(\lambda) = \lambda - 1$$

$$f_2(\lambda) = (\lambda - 3) f_1(\lambda) - [-2\sqrt{2}]^2 f_0$$

$$= \lambda^2 - 4\lambda - 5$$

$$f_3(\lambda) = (\lambda + 1) f_2 - (0)^2 f_1$$

$$= (\lambda + 1)(\lambda + 1)(\lambda - 5)$$

Since $f_3(-1) = 0$ and $f_3(5) = 0$

the eigenvalues of A are -1, -1 and 5.

The largest eigenvalues in magnitude is 5.
The eigenvector corresponding to $\lambda = 5$ of B_1 is

$$V_1 = \left[1\ \sqrt{2}\ 0\right]^T$$

Hence, the corresponding eigenvector of A is

$$V = S_1 V_1 = [111]^T$$

Example 21. Find the eigenvalues and eigenvectors of the matrix

$$A = \begin{bmatrix} 1 & \frac{1}{2} & \frac{1}{3} \\ \frac{1}{2} & \frac{1}{3} & \frac{1}{4} \\ \frac{1}{3} & \frac{1}{4} & \frac{1}{5} \end{bmatrix}$$

by Given's method.

Solution: We reduce the given symmetric matrix to the tridiagonal form. Replacing a_{22}, a_{23}, a_{32}, a_{33} by $\cos\theta$, $-\sin\theta$, $\cos\theta$ submatrix, we get

$$S_1 = \begin{bmatrix} 1 & 0 & 0 \\ 0 & \cos\theta & -\sin\theta \\ 0 & \sin\theta & \cos\theta \end{bmatrix}$$

Choose θ, t

$$\tan\theta = \frac{a_{13}}{a_{12}} = \frac{\frac{1}{3}}{\frac{1}{2}}$$

$$= \frac{2}{3}$$

$$\therefore \qquad \cos\theta = \frac{3}{\sqrt{4+9}} = \frac{3}{\sqrt{13}}$$

$$\sin\theta = \frac{2}{\sqrt{13}}$$

$$\therefore \qquad S_1 = \begin{bmatrix} 1 & 0 & 0 \\ 0 & \frac{3}{\sqrt{13}} & -\frac{2}{\sqrt{13}} \\ 0 & \frac{2}{\sqrt{13}} & \frac{3}{\sqrt{13}} \end{bmatrix}$$

$$\therefore \quad S_1^{-1} = S_1^T$$

$$= \begin{bmatrix} 1 & 0 & 0 \\ 0 & \frac{3}{\sqrt{13}} & \frac{2}{\sqrt{13}} \\ 0 & -\frac{2}{\sqrt{13}} & \frac{3}{\sqrt{13}} \end{bmatrix}$$

$$B_1 = S_1^{-1} AS_1$$

$$= \begin{bmatrix} 1 & 0 & 0 \\ 0 & \frac{3}{\sqrt{13}} & \frac{2}{\sqrt{13}} \\ 0 & -\frac{2}{\sqrt{13}} & \frac{3}{\sqrt{13}} \end{bmatrix} \begin{bmatrix} 1 & \frac{1}{2} & \frac{1}{3} \\ \frac{1}{2} & \frac{1}{3} & \frac{1}{4} \\ \frac{1}{3} & \frac{1}{4} & \frac{1}{5} \end{bmatrix} \begin{bmatrix} 1 & 0 & 0 \\ 0 & \frac{3}{\sqrt{13}} & \frac{-2}{\sqrt{13}} \\ 0 & \frac{2}{\sqrt{13}} & \frac{3}{\sqrt{13}} \end{bmatrix}$$

$$= \begin{bmatrix} 0.0769 & 0.1667 & 0 \\ 0.1667 & 0.5231 & 0.0346 \\ 0 & 0.0346 & 0.0103 \end{bmatrix}$$

Clearly, this is a tridiagonal matrix.

Now, the characteristic equation of B_1 is f_1 given by

$$f_n = |\lambda I - A| \text{ is characteristic equation}$$

$$= \begin{bmatrix} \lambda - 0.0769 & 0.1667 & 0 \\ 0.1667 & \lambda - 0.5231 & 0.0346 \\ 0 & 0.0346 & \lambda - 0.0103 \end{bmatrix}$$

The Sturm sequence $\{f_n\}$ is defined as

$$f_0 = 1$$

$$f_1 = \lambda - 0.0769$$

$$f_2 = (\lambda - 0.5231)\, f_1 - (0.1667)^2 f_0$$

$$= (\lambda - 0.5231)\, (\lambda - 0.0769) - (0.1667)^2$$

$$f_3 = (\lambda - 0.0103)\, f_2 - (0.0346)^2$$

$$= (\lambda - 0.0103)\, f_2 - (0.0346)^2$$

$$\therefore \quad f_3 = (\lambda - 0.0103)\, (\lambda - 0.5231)\, (\lambda - 0.0769) - (0.0346)^2$$

From these, we find that these eigenvalues in (0, 1) and one in (7, 8).

EXERCISES

1. Construct the Sturm sequences for the following tridiagonal matrices:

(1) $\begin{bmatrix} 2 & -1 & 0 \\ -1 & 2 & -1 \\ 0 & -1 & 2 \end{bmatrix}$ (2) $\begin{bmatrix} 3 & -2 & 0 \\ -2 & 3 & -2 \\ 0 & -2 & 3 \end{bmatrix}$

Hence, verify that (i) for the first matrix 2 is an eigenvalue and the other two eigenvalues lie in the intervals (0, 1) and (3, 4), and (ii) for the second matrix 3 is an eigenvalue and the other two eigenvalues lie in the intervals (0, 1) and (5, 6).

2. Transform the matrix $\begin{bmatrix} 1 & 2 & 2 \\ 2 & 1 & 2 \\ 2 & 2 & 1 \end{bmatrix}$ to the tridiagonal form by using Given's method. Hence, find its dominant eigenvalue and the corresponding eigenvector.

3. Apply Given's method to reduce the following matrices to tridiagonal forms and find the intervals in which the eigenvalues lie.

(1) $\begin{bmatrix} 1 & 1 & 1 \\ 1 & 2 & 3 \\ 1 & 3 & 6 \end{bmatrix}$ (2) $\begin{bmatrix} 1 & 2 & 3 \\ 2 & 1 & 2 \\ 3 & 2 & 1 \end{bmatrix}$

4. Using the Given's method, reduce the following matrices to the tridiagonal form.

(1) $\begin{bmatrix} 2 & 1 & 2 \\ 1 & 4 & 2 \\ 3 & 2 & 3 \end{bmatrix}$ (2) $\begin{bmatrix} 1 & 2 & 3 \\ 2 & 1 & -1 \\ 3 & -1 & 1 \end{bmatrix}$

(3) $\begin{bmatrix} 1 & 2 & 3 & 4 \\ 2 & 3 & 4 & 5 \\ 3 & 4 & 5 & 6 \\ 4 & 5 & 6 & 7 \end{bmatrix}$ (4) $\begin{bmatrix} 1 & \sqrt{2} & \sqrt{2} & 2 \\ \sqrt{2} & -\sqrt{2} & -1 & \sqrt{2} \\ \sqrt{2} & -1 & \sqrt{2} & \sqrt{2} \\ 1 & \sqrt{2} & \sqrt{2} & -3 \end{bmatrix}$

5. By employing the Given's method, reduce the matrix

$$A = \begin{bmatrix} 1 & 1 & 1 & 1 \\ 1 & 2 & 3 & 4 \\ 1 & 3 & 6 & 10 \\ 1 & 4 & 10 & 20 \end{bmatrix}$$

to the tridiagonal from.

$$B = \begin{bmatrix} 1 & 1.7321 & 0 & 0 \\ 1.7231 & 20.6687 & 10.2746 & 0 \\ 0 & 10.2746 & 7.1758 & 0.3648 \\ 0 & 0 & 0.3648 & 0.1580 \end{bmatrix}$$

Hence, verify that three of the eigenvalues lie in the intervals (0, 0.1), (0.1, 0.5) and (2, 3).

Answers:

1. (1) $f_0(\lambda) = 1,$ $\quad f_1(\lambda) = \lambda - 2,$

   $f_2(\lambda) = (\lambda - 2)\, f_1(\lambda) - f_0(\lambda)$ $\quad f_3(\lambda) = (\lambda - 2)\, f_2(\lambda) - f_1(\lambda)$

   (2) $f_0(\lambda) = 1,$ $\quad f_1(\lambda) = (3 - \lambda),$

   $f_2(\lambda) = (3 - \lambda)\, f_1(\lambda) - 4f_0(\lambda)$ $\quad f_3(\lambda) = (3 - \lambda)\, f_2(\lambda) - 4f_1(\lambda)$

   (3) $\begin{bmatrix} 1 & 7.25 & 3.5 & 3 \\ 4 & 14 & 8 & 6 \\ 0 & 2.5 & 1 & 1 \\ 0 & 0 & 0 & 0 \end{bmatrix}$ (4) $\begin{bmatrix} 1 & 2\sqrt{2} & 0 & 0 \\ 2\sqrt{2} & 0 & \sqrt{2} & 0 \\ 0 & \sqrt{2} & -0.5 & 2.5 \\ 0 & 0 & 2.5 & -2.5 \end{bmatrix}$

2. Eigenvalues and eigenvectors

3. $\begin{bmatrix} 1 & 2\sqrt{2} & 0 \\ 2\sqrt{2} & 3 & 0 \\ 0 & 0 & -1 \end{bmatrix}$; 5, $\begin{bmatrix} 1 \\ 1 \\ 1 \end{bmatrix}$

4. (1) (0, 1), (1, 2), (4, 5) (2) (0.1, 0.2), 1, (7.8, 7.9)

5. (1) $\begin{bmatrix} 2 & 3.16 & 0 \\ 3.16 & 4.3 & -1.9 \\ 0 & -1.9 & 3.9 \end{bmatrix}$ (2) $\begin{bmatrix} 1. & 3.60555 & 0 \\ 3.60555 & 0.07692 & 0.38462 \\ 0 & 0.38462 & 1.92308 \end{bmatrix}$

Chapter 5

Finite Differences

INTRODUCTION

In this chapter, we introduce the idea of finite differences and associated concepts, which have important applications in Numerical Applied Mathematics. The associated differences are called the forward, backward and central differences of a function $y = f(x)$. These differences are three standard examples of finite differences and are fundamental study of differences calculus.

FORWARD DIFFERENCES

Let $y_0, y_1, y_2, \ldots y_r, \ldots$ be the values of $y = f(x)$ corresponding to the values of $x_0, x_1, x_2, \ldots x_r, \ldots$ of x, respectively. The first forward difference of $y = f(x)$ are given by $y_1 - y_0, y_2 - y_1, \ldots$ and we denote them by $\Delta y_0, \Delta y_1, \ldots$, that is,

$$\begin{aligned} \Delta y_0 &= y_1 - y_0 \\ \Delta y_1 &= y_2 - y_1 \\ \Delta y_2 &= y_3 - y_2 \\ &\ldots \ldots \ldots \end{aligned}$$

In general,

$$\Delta y_r = y_{r+1} - y_r, \quad r = 0, 1, 2, \ldots \tag{1}$$

Here, the symbol Δ is called the forward difference operator.

The first forward differences of the first forward differences are called second forward differences, and are denoted by $\Delta^2 y_0, \Delta^2 y_1, \ldots$; that is, $\Delta^2 y_0 = \Delta y_1 - \Delta y_0$, $\Delta^2 y_1 = \Delta y_2 - \Delta y_1, \ldots$.

In general,

$$\Delta^2 y_r = \Delta y_{r+1} - \Delta y_r, \quad r = 0, 1, 2, \ldots \tag{2}$$

Similarly, the nth forward differences are defined by the formula

$$\Delta^n y_r = \Delta^{n-1} y_{r+1} - \Delta^{n-1} y_r, \; r = 0, 1, 2, \ldots \tag{3}$$

While using this formula for $n = 1$, we use the notation $\Delta^0 y_r = y_r$.

If $f(x)$ is a constant function, that is, if $f(x) = c$, a constant, then $y_0 = y_1 = y_2 = \ldots = c$, and we have $\Delta^n y_r = 0$ for $n = 1, 2, 3, \ldots$ and $r = 0, 1, 2, \ldots$.

The symbol Δ^n is referred to as the nth forward difference operator.

Forward Differences Table

Values of x x	Values of y $y = f(x_0)$	1st difference (Δy)	2nd difference $(\Delta^2 y)$	3rd difference $(\Delta^3 y)$
x_0	y_0	$y_1 - y_0 = \Delta y_0$	$y_2 - 2y_1 + y_0 = \Delta^2 y_0$	$y_3 - 3y_2 + 3y_1 - y_0 = \Delta^4 y_0$
$x_0 + h = x_1$	y_1	$y_2 - y_1$		
$x_0 + 2h = x_2$	y_2		$y_3 - 2y_2 + y_1$	
$x_0 + 3h = x_3$	y_3	$y_3 - y_2$		

In the above table, Δy_0, $\Delta^2 y_1$ ... are called the leading differences.

A Particular Case: If the values of x are equally spaced and h is the difference between two consecutive values of x, then for any initial value x_0 of x, we have

$$\begin{Bmatrix} x_1 = x_0 + h, \\ x_2 = x_0 + 2h, \\ \ldots \quad \ldots \\ \ldots \quad \ldots \\ x_r = x_0 + rh \end{Bmatrix}$$

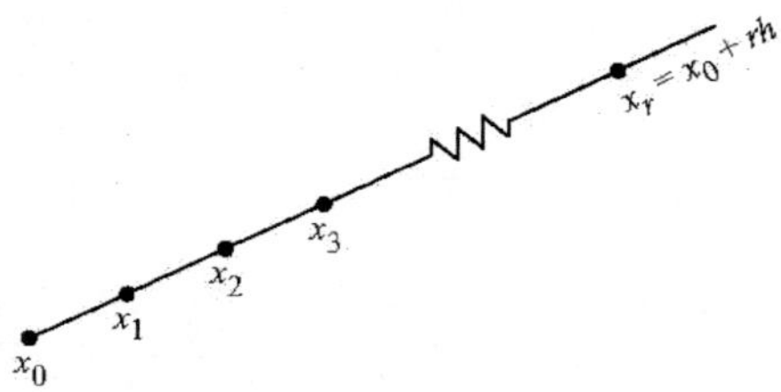

In this case,
$$y_0 = f(x_0),$$
$$y_1 = f(x_1) = f(x_0 + h),$$
$$y_2 = f(x_2) = f(x_0 + 2h) \ldots\ldots$$

In general,
$$y_r = f(x_r) = f(x_0 + rh),\ r = 0, 1, 2, \ldots \tag{4}$$

Then, (3) becomes
$$\Delta^n y_r = \Delta^n f(x_r) = \Delta^{n-1} f(x_r + h) - \Delta^{n-1} f(x_r)$$

This is often written as
$$\Delta^n y_n = \Delta^n f(x) = \Delta^{n-1} f(x + h) - \Delta^{n-1} f(x) \tag{5}$$

with the suffix r in x_r being understood.

The formula (5) gives the nth forward differences of $f(x)$ at a point $x = x_r = x_0 + rh$, for $r = 0, 1, 2, \ldots$. Thus, the formula for the first, the second, the third forward differences of $f(x)$ at x are, respectively,

$$\Delta y = \Delta f(x) = f(x + h) - f(x) \qquad \rightarrow \tag{6a}$$

$$\Delta^2 y = \Delta f(x + h) - \Delta f(x) \qquad \rightarrow \qquad (6b)$$
$$\Delta^3 y = \Delta^2 f(x + h) - \Delta^2 f(x) \qquad \rightarrow \qquad (6c)$$

A useful result: An important result which is needed in many computations is:

$$\Delta^n y_r = y_{n+r} - (^n c_1)\, y_{n+r-1} + (^n c_2)\, y_{n+r-2} - \ldots + (-1)^n\, y_r,$$
$$r = 0, 1, 2, \ldots \qquad (7)$$

Proof: We shall prove this result by the principle of mathematical induction on n.

Since $\Delta y_r = y_{r+1} - y_r$

for $n = 1$, the result is true.

Assume the result for $n = k$, we get

$$\Delta^k y_r = y_{k+r} - (^k c_1)\, y_{k+r-1} + (^k c_2)\, y_{k+r-2} + \ldots + (-1)^k y_r$$

Then
$$\begin{aligned}\Delta_{y_r}^{k+1} &= \Delta\, (\Delta^k y_r) \\ &= \Delta(y_{k+r}) - (^k c_1)\, \Delta(y_{k+r-1}) \\ &\quad + (^k c_2)\, \Delta(y_{k+r-2}) + \ldots + (-1)^k\, (\Delta y_r) \\ &= (y_{k+r+1} - y_{k+r}) - (^k c_1)\, (y_{k+r} - y_{k+r-1}) \\ &\quad + (^k c_2)\, (y_{k+r-1} - y_{k+r-2}) + \ldots + (-1)^k\, (y_{r+1} - y_r) \\ &= y_{k+r+1} - (1 + {^k c_1})\, y_{k+r} + (^k c_1 + {^k c_2})\, y_{k+r-1} + (-1)_{y_r}^{k+1}.\end{aligned}$$

This is the result (7) for $n = k + 1$ which is true for $n = k + 1$ as well. This shows that (7) is true for all positive integral values of n.

Again, we have,

$$y = f(x), \quad \Rightarrow \quad y_r = f(x_r)$$

(7) can be written as:

$$\Delta^n f(x_r) = f(x_{n+r}) - {^n c_1}\, f(x_{n+r-1}) + {^n c_2}\, f(x_{n+r-2}) - \ldots (-1)^n\, f(x_r)$$

$$= \sum_{k=0}^{n} (-1)^k\, (^n c_k)\, f(x_{n+r-k}) \qquad (8)$$

If the values of x are equally spaced with h as step-length, then

$$f(x_{n+r-k}) = f(x_r + \overline{n-k}\, h)$$

and hence 8 becomes

$$\Delta^n f(x_r) = \sum_{k=0}^{n} (-1)^k\, (^n c_k)\, f(x_r + \overline{n-k}\, h)$$

Replacing x_r by x, we get

$$\Delta^n f(x) = \sum_{k=0}^{n} (-1)^k\, (^n c_k)\, f(f + \overline{n-k}\, h)$$

Properties of the operator Δ

P_1: (Linearity property): For any two constants a and b and for any two functions $f(x)$ and $g(x)$, we have

$$\Delta[af(x) + bg(x)] = a\Delta f(x) + b\Delta g(x) \tag{1}$$

Proof:
$$\begin{aligned}\Delta\,[af(x) + bg(x)] &= [af(x+h) + bg(x+h)] - [af(x) + bg(x)]\\ &= a[f(x+h) - f(x)] + b[g(x+h) - g(x)]\\ &= a\Delta f(x) + b\Delta g(x)\end{aligned}$$

P_2: (Index law): If m and n are positive integers, then

$$\Delta^m \Delta^n f(x) = \Delta^{m+n} f(x) \tag{2}$$

Proof: By using the definition of the Δ operator, we have

$$\begin{aligned}\Delta^m \Delta^n f(x) = \Delta^m\ \{\Delta^n f(x)\} &= (\Delta.\Delta\ \ldots\ mx)\ \Delta^n f(x)\\ &= (\Delta.\Delta\ \ldots\ (m-1)\text{ times})\ \Delta^{n+1} f(x)\\ &= (\Delta.\Delta\ \ldots\ (m-2)\text{ times})\ \Delta^{n+2} f(x)\\ &\quad \ldots\ \ldots\\ &\quad \ldots\ \ldots\\ &= \Delta\{\Delta^{n+(m-1)} f(x)\} = \Delta^{n+m} f(x)\\ &= \Delta^{m+n} f(x)\end{aligned}$$

P_3: (Differences of a polynomial): If $f(x) = a_0 x^n + a_1 x^{n-1} + \ldots + a_{n-1} x + a_n$ is an nth degree polynomial and the values of x are equally spaced, then $\Delta^n f(x)$ is a constant, that is, $\Delta^n f(x) = n!\ a_0 h^n =$ constant (3)

where $\Delta x = h$

Proof: Consider
$$\begin{aligned}\Delta f(x) &= f(x+h) - f(x)\\ &= [a_0(x+h)^n + a_1(x+h)^{n-1} + \ldots + a_n]\\ &\quad - [a_0 x^n + a_1 x^{n-1} + \ldots + a_n]\\ &= a_0[(x+h)^n - x^n] + a_1[(x+h)^{n-1} - x^{n-1}] + \ldots\\ &\quad \ldots + a_{n-1}\,[x + h - x]\end{aligned}$$

Expanding by binomial theorem

$$(x+h)^n - x^n = x^n + nhx^{n-1} + {}^n c_2\, h^2\, x^{n-2} + \ldots - x^n$$

Similar terms in the other brackets. Thus,

$$g(x) = \Delta f(x) = b_0 x^{n-1} + b_1 x^{n-2} + \ldots + b_{n-1}$$

which is a polynomial of $(n-1)$th degree.

Here $\quad b_0 = nha_0$

Now $\quad \Delta^2 f(x) = \Delta(\Delta f(x)) = g(x+h) - g(x)$

Expanding by binomial theorem, observe that $\Delta^2 f(x)$ is a polynomial of degree $n-2$:

$$\Delta^2 f(x) = c_0 x^{n-2} + c_1 x^{n-3} + \ldots + c_{n-2}$$

where $$c_0 = (n-1)hb_0 = n(n-1)h^2 a_0.$$

Continuing in this manner successively, we get

$$\Delta^n f(x) = n!\, a_0 h^n = \text{constant.} \tag{4}$$

Thus, the nth order differences of a polynomial of nth degree are constant.

Note 1. Since $\Delta^s f(x) = 0, \quad \text{for } s > n$ (5)

i.e. $(n + 1)$th order differences of a polynomial of nth degree are zero.

2. The converse of the above result is also true. That is, if $\Delta^n f(x)$ tabulated at equally spaced intervals is a constant, then the function $f(x)$ is a polynomial of degree n.

3. The result $\Delta^n f(x) = n!\ a_0 h^n$

$$= a_0 n(n-1)(n-2) \ldots 2.1.h^n$$

$$= a_0 h^n\ (n!)$$

is used in several computations. This result, when written explicitly, reads

$$\Delta^n (a_0 x^n + a_1 x^{n-1} + \ldots + a_{n-1} x + a_n) = a_0 h^n\, n! \tag{6}$$

Example 1. Construct the forward differences table and the horizontal differences table for the data below:

x	:	0	1	2	3	4
$f(x)$	:	10	15	20	28	34

Solution: The forward differences table is given below:

x	$f(x)$	$\Delta f(x)$	$\Delta^2 f(x)$	$\Delta^3 f(x)$	$\Delta^4 f(x)$
0	10				
		5			
1	15		0		
		5		3	
2	20		3		–8
		8		–3	
3	28		–2		
		6			
4	34				

The horizontal backward differences table is given below:

x	$f(x)$	$\Delta f(x)$	$\Delta^2 f(x)$	$\Delta^3 f(x)$	$\Delta^4 f(x)$
0	10				
		5			
1	15		0		
		5		3	
2	20		3		–8
		8		–5	
3	28		–2		
		6			
4	34				

Example 2. Form the forward differences table write down the values of $\Delta f(10)$, $\Delta^2 f(10)$, $\Delta^3 f(15)$ and $\Delta^4 f(15)$ from the data

x	10	15	20	25	30	35
y	19.97	21.51	22.47	23.52	24.65	25.89

Solution:

x	y	Δy	$\Delta^2 y$	$\Delta^3 y$	$\Delta^4 y$	$\Delta^5 y$
10	19.97					
		1.54				
15	21.51		−0.38			
		1.16		0.27		
20	22.47		−0.11		−0.08	
		1.05		0.19		−0.08
25	23.52		0.08		−0.16	
		1.13		0.03		
30	24.65		0.11			
		1.24				
35	25.89					

x are equally spaced with step-length $h = 5$ and $x_0 = 10, x_1 = 15, \ldots x_5 = 35$ and

$$y_0 = f(x_0) = 19.97,\ y_1 = f(x_1) = 21.51,\ \ldots\ y_5 = f(x_5) = 25.89$$

We find from the table, that

$$\Delta f(10) = \Delta y_0 = 1.54;\ \Delta^2 f(10) = \Delta^2 y_0 = -0.38$$
$$\Delta^3 f(15) = \Delta^3 y_1 = 0.19;\ \Delta^4 f(15) = \Delta^4 y_1 = -0.16$$

Example 3. Construct a table of forward differences for the function $x^3 + 5x - 7$ for $x = -1, 0, 1, 2, 3, 4, 5$. Show that the third differences are constants.

Solution: Let $f(x) = x^3 + 5x - 7$,

$$f(-1) = -13,\ f(0) = -7,\ f(s) = -1,\ f(2) = 11,$$
$$f(3) = 35,\ f(4) = 77,\ f(5) = 143.$$

Forward differences table is

x	$f(x)$	$\Delta f(x)$	$\Delta^2 f(x)$	$\Delta^3 f(x)$
−1	−13			
		6		
0	−7		0	
		6		6
1	−1		6	
		12		6
2	11		12	
		24		6
3	35		18	
		42		6
4	77		24	
		66		
5	143			

We find that all the third differences are the same (namely 6) and are a constant.

Example 4. Prove that

(i) $y_2 = y_0 + 2\Delta y_0 + \Delta^2 y_0$
(ii) $y_3 = y_0 + 3\Delta y_0 + 3\Delta^2 y_0 + \Delta^3 y_0$

Solution:

(i) By definition of Δy_0, Δy_1, $\Delta^2 y_0$, we get

$$y_1 = y_0 + \Delta y_0 \tag{1}$$
$$y_2 = y_1 + \Delta y_1 \tag{2}$$
$$\Delta y_1 = \Delta y_0 + \Delta^2 y_0 \tag{3}$$

Using (1) and (3) in (2), we obtain

$$y_2 = y_0 + 2\Delta y_0 + \Delta^2 y_0 \tag{4}$$

(ii) By definition of

$$\Delta y_2 = y_3 - y_2,\ \Delta^2 y_1 = \Delta y_2 - \Delta y_1,\ \Delta^3 y_0 = \Delta^2 y_1 - \Delta^2 y_0$$

therefore,

$$y_3 = y_2 + \Delta y_2 \tag{5}$$
$$\Delta y_2 = \Delta y_1 + \Delta^2 y_1 \tag{6}$$
$$\Delta^2 y_1 = \Delta^3 y_0 + \Delta^2 y_0 \tag{7}$$

Using (4), (6) and (7) in (5), we obtain

$$y_3 = y_0 + 3\Delta y_0 + 3\Delta^2 y_0 + \Delta^3 y_0$$

Example 5. Show that

(a) $\Delta(\tan^{-1} x) = \tan^{-1}\left[\dfrac{h}{1 + hx + x^2}\right]$

(b) $\Delta \log(1 + 4x) = \log\left[1 + \dfrac{4h}{1 + 4x}\right]$

Solution:

(a) $\Delta(\tan^{-1} x) = \tan^{-1}(x + h) - \tan^{-1} x$

$$= \tan^{-1}\left[\frac{(x + h) - x}{1 + x(x + h)}\right]$$

$$= \tan^{-1}\left[\frac{h}{1 + hx + x^2}\right]$$

(b) $\Delta \log (4x + 1) = \log (4x + 4h + 1) - \log (4x + 1)$

$$= \log \left[\frac{4x + 4h + 1}{4x + 1} \right]$$

$$= \log \left[1 + \frac{4h}{4x + 1} \right]$$

Example 6. Evaluate

(i) $\Delta^3(1 - x)(1 - 2x)(1 - 3x)$ if $h = 1$
(ii) $\Delta^{10}(1 - x)(1 - 2x)(1 - 3x) \ldots (1 - 10x)$ taking $h = 1$
(iii) $\Delta^{10}[(1 - x)(1 - 2x^2)(1 - 3x^3)(1 - 4x^4)]$ if $h = 2$
(iv) $\Delta^{10}[(1 - ax)(1 - bx^2)(1 - cx^3)(1 - dx^4)]$

Solution:

(i) $\Delta^3(1 - x)(1 - 2x)(1 - 3x) = \Delta^3[-6x^3 + \text{terms of lesser power}]$
$= (-6)\ 3! + 0 = -36$

(ii) $\Delta^{10}(1 - x)(1 - 2x)\ldots(1 - 10x) = \Delta^{10}[10!\,x^{10} + \text{terms involving lesser degree}]$
$= 10!\ 10! + 0 = (10!)^2$

(iii) $\Delta^{10}[(1 - x)(1 - 2x^2)(1 - 3x^3)(1 - 4x^4)]$
$= \Delta^{10}[24x^{10} + \text{terms of lesser degree}]$
$= 24\ (10!)\ 2^{10} + 0$, since $\Delta^n (a_0 x^n) = a_0 n!\, h^n$
$= 24\ (10!)\ 2^{10}$

(iv) $\Delta^{10}[(1 - ax)(1 - bx^2)(1 - cx^3)(1 - dx^4)]$
$= (abcd)2^{10}(10!)$

Example 7. If $f(x)$ and $g(x)$ are any two functions of x, prove that $\Delta[f(x)\ g(x)] = f(x)\ \Delta g(x) + g(x + h)\ \Delta f(x)$. Hence, evaluate $\Delta(x \sin x)$.

Solution: $\Delta[f(x)\ g(x)] = f(x + h)\ g(x + h) - f(x)\ g(x)$
$= f(x)\ [g(x + h) - g(x)] + g(x + h)\ [f(x + h) - f(x)]$
$= f(x)\ \Delta g(x) + g(x + h)\ \Delta f(x)$

Now, choosing $f(x) = x$ and $g(x) = \sin x$, we get

$\Delta(x \sin x) = x\Delta(\sin x) + (x + h)\ (\Delta x)$
$= x[\sin (x + h) - \sin x] + (x + h)\ [(x + h) - x]$
$= x[\sin(x + h) - \sin x] + h(x + h)$
$= x[\sin(x + h) - \sin (x + h)] + h^2$

Example 8. Evaluate

(i) $\Delta^n (a^{bx+c})$
(ii) $\Delta^n [\cos (ax + b)]$

Solution:

(i)

$$\Delta(a^{bx+c}) = a^{b(x+h)+c} - a^{bx+c}$$
$$= a^{bx+c}\,(a^{bh} - 1)$$
$$\Delta^2(a^{bx+c}) = a^{bx+c}(a^{bh} - 1)^2$$
$$\Delta^n(a^{bx+c}) = a^{bx+c}\,(a^{bh} - 1)^n$$

(ii) $\Delta[\cos(ax + b)]$

$$= \cos(ax + ah + b) - \cos(ax + b)$$
$$= -2\sin(ax + b + \frac{ah}{2})\sin\frac{ah}{2}$$
$$= 2\sin\frac{ah}{2}\cos\left(\frac{\pi}{2} + ax + b + \frac{ah}{2}\right)$$
$$= 2\sin\frac{ah}{2}\cos\left(ax + b + \frac{\pi + ab}{2}\right)$$

$$\Delta^2[\cos(ax + b)] = \left(2\sin\frac{ah}{2}\right)^2 \cos\left(ax + b + \frac{2(\pi + ah)}{2}\right)$$

Proceeding like this

$$\Delta^n\cos(ax + b) = \left(2\sin\frac{ah}{2}\right)^n \cos\left[ax + b + \frac{n(\pi + ah)}{2}\right]$$

FACTORIAL NOTATION

The factorial polynomial of degree n, with h as the step-length in x, denoted by $x^{(n)}$ or $[x]^n$ and is given by

$$x^{(n)} = [x]^n = x(x - h)(x - 2h) \ldots (x - \overline{n-1}\,h) \tag{1}$$

Thus,

$[x] = x$, $[x]^2 = x(x - h)$, $[x]^3 = x(x - h)(x - 2h)$, and so on. In the particular case, where $h = 1$, we have

$$x^{(n)} = [x]^n = x(x - 1)(x - 2) \ldots (x - n + 1) \tag{2}$$

We now prove the following formulas:

$$\Delta^r\,[x]^n = n(n - 1)(n - 2) \ldots (n - r + 1)h^r\,[x]^{n-r},\ r < n \tag{3}$$
$$\Delta^n\,[x]^n = h^n\,(n!) \tag{4}$$
$$\Delta^r\,[x]^n = 0, \quad r > n \tag{5}$$

Proof: We have

$$\Delta[x]^n = [x + h]^n - [x]^n$$
$$= (x + h)(x + h - 1)(x + h - 2) \ldots (x - h - \overline{n-1}\,h)$$
$$-x(x - h)(x - 2h) \ldots (x - \overline{n-1}\,h)$$

$$= x(x-h)(x-2h)\ldots(x-\overline{n-2}\,h)\;\{(x+h)-(x-\overline{n-1}\,h)\}$$
$$= nh\,[x]^{n-1}$$

Therefore,

$$\begin{aligned}\Delta^2\,[x]^n &= \Delta\{\Delta[x]^n\}\\ &= \Delta\{nh[x]^{n-1}\}\\ &= nh\Delta[x]^{n-1}\\ &= nh\{(n-1)h\,[x]^{n-2}\}\\ &= n(n-1)h^2\,[x]^{n-2}\end{aligned}$$

Proceeding like this, we get

$$\begin{aligned}\Delta^3[x]^n &= n(n-1)(n-2)h^3[x]^{n-3}\\ &\ldots\quad\ldots\quad\ldots\\ &\ldots\quad\ldots\quad\ldots\\ \Delta^r[x]^n &= n(n-1)(n-2)\ldots(n-r+1)h^r\,[x]^{n-r},\; r<n\end{aligned}$$

This is the formula (3). For $r = n - 1$, this formula becomes

$$\Delta^{n-1}\,[x]^n = n(n-1)(n-2)\ldots 2h^{n-1}x$$

Therefore, $\Delta^n[x]^h = n(n-1)(n-2)\ldots 2.1.\,h^{n-1}\,(\Delta x)$

$$= n(n-1)(n-2)\ldots 2.1.\,h^{n-1}\,(x+h-x)$$

$= h^n\,(n!)$. This is the formula (4).

As a direct consequence of formula (4), we have

$$\Delta^{n+1}\,[x]^n = 0.$$

Consequently, $\Delta^{n+2}[x]^n = 0.$

$$\Delta^{n+3}[x]^n = 0, \text{ and so on.}$$

Thus, $\Delta^r[x]^n = 0$, for $r > n$.

which is formula (5)

Note: (i) $\Delta^r[x]^n$ for $r < n$, $r = n$, $r > n$ are

analogous to $\dfrac{d^r}{dx^r}(x^n)$.

(ii) Using the factorial notation,

$$x^{(n)} = [x^n] = x(x-h)(x-2h)\ldots(x-\overline{n-1}\,h)$$

we can express a polynomial of degree n in x as a combination of $[x]$, $[x]^2$, … $[x]^n$, called a factorial polynomial, and thereby compute the differences of $f(x)$.

Example 9. Express $f(x) = x^3 - 2x^2 + x - 1$ as a factorial polynomial. Hence, compute $\Delta f(x)$, $\Delta^2 f(x)$ and $\Delta^3 f(x)$.

Solution: Let $f(x) = x^3 - 2x^2 + x - 1;$ (1)

$= x(x-1)(x-2) + Ax(x-1) + Bx + D;$ (2)

where A, B and D are constants to be determined.

Putting $x = 0$ in (2), we get

$$-1 = D$$

Putting $x = 1$ in (2), we get

$$-1 = B + D, \text{ so that } B = 0$$

Putting $x = 2$, (2) yields

$$1 = 2A + 2B + D$$

$$\Rightarrow \quad A = 1$$

Hence, (2) becomes

$$f(x) = x(x-1)(x-2) + x(x-1) - 1$$
$$= [x]^3 + [x]^2 - 1 \qquad (3)$$

r.h.s. of (3) is the factorial polynomial respectively $f(x)$, with $h = 1$. This yields on using formulas:

$$\Delta^r [x]^n = n(n-1)(n-2) \ldots (n-r+1) h^r [x]^{n-r}, \; r < n,$$
$$\Delta^n [x]^n = h^n (n!) \text{ and } \Delta^r [x]^n = 0, \; r > n$$

$$\Delta f(x) = \Delta[x]^3 + \Delta[x]^2 - \Delta(1) = 3[x]^2 + 2[x]$$
$$= 3x(x-1) + 2x = 3x^2 - x$$
$$\Delta^2 f(x) = 3\Delta[x]^2 + 2\Delta[x] = 3.2\,[x] + 2 = 6x + 2$$
$$\Delta^3 f(x) = 3! = 6.$$

Example 10. Prove that the polynomial $f(x) = ax^2 + bx + c$ is expressible as a factorial polynomial with single term, provided $b^2 - a^2 = 4ac$.

Solution: Let $h = 1$ be the step-length, and by data $f(x)$ is expressible as a polynomial with single term, if

$$ax^2 + bx + c = k[x + r]^2 = k(x + r)(x + r - 1) \qquad (1)$$

for some k and some r.

Comparing the coefficients on both sides of (1), we get

$$a = k, \; b = k(2r - 1), \; c = kr(r - 1) \qquad (2)$$

The first two of these yield $r = \dfrac{a+b}{2a}$, and the last gives $c = \dfrac{b^2 - a^2}{4a}$, or $b^2 - a^2 = 4ac$.

Thus, the representation (1) holds, provided $b^2 - a^2 = 4ac$.

EXERCISES

Express the following polynomials in the factorial rotations:

(1) $9x^2 + 11x + 5$ (2) $3x^2 - 5x + 2$

(3) $x^3 + 3x^2 + 5x + 12$ (4) $2x^3 - 3x^2 + 3x - 10$

(5) $x^4 - 12x^3 + 24x^2 - 30x + 9$ (6) $3x^4 - 4x^3 + 6x^2 + 2x + 1$

Answers:

(1) $3[x]^3 + 10[x]^2 + 5[x] + D$ (2) $3[x]^2 - 2[x] + 2$

(3) $\frac{1}{4}[x]^4 + 2[x]^3 + \frac{9}{2}[x]^2 + 12[x] + k$, k is a constant

(4) $2[x]^3 + 3[x]^2 + 2[x] - 10$

(5) $[x]^4 - 6[x]^3 + 13[x]^2 + [x] + 9$

(6) $3[x]^4 + 14[x]^3 + 15[x]^2 + 7[x] + 1$

The inverse operator Δ^{-1}

If $\Delta f(x) = g(x)$, then, $\Delta^{-1} g(x) = f(x)$,

Clearly $\Delta\{\Delta^{-1}\phi(x)\} = \phi(x)$ (1)

i.e. if $\Delta f(x)$ is $g(x)$, then $f(x)$ is the inverse forward difference of $g(x)$ and the symbol Δ^{-1}, often denoted by $1/\Delta$, is called the inverse forward difference operator.

Linearity of Δ^{-1}

that is,

$$\Delta^{-1}[a\phi(x) + b\Psi(x)] = a\,\Delta^{-1}\phi(x) + b\Delta^{-1}\Psi(x)$$

where a and b are constants.

Hence, if $\phi(x)$ is constant, then $\Delta^{-1}\phi(x) = 0$.

A Helpful Result: We shall state and prove the following result.

$$\Delta^{-1}\phi(x_r) = \sum_{k=1}^{r-1}\phi(x_k) + \text{constant.}$$

Proof: Let us assume $\Delta^{-1}\phi(x_r) = Z_r$, then

$$\phi(x_r) = \Delta Z_r = Z_{r+1} - Z_r$$

Hence,

$$\phi(x_{r-1}) = Z_r - Z_{r-1}$$

$$\phi(x_{r-2}) = Z_{r-1} - Z_{r-2}$$

$$\ldots \quad \ldots \quad \ldots$$

$$\ldots \quad \ldots \quad \ldots$$

$$\phi(x_1) = Z_2 - Z_1$$

Adding these, we get

$$\phi(x_{r-1}) + \phi(x_{r-2}) + \ldots + \phi(x_1) = Z_r - Z_1$$

or

$$Z_r = Z_1 + \sum_{k=1}^{r-1}\phi(x_k)$$

i.e.

$$\Delta^{-1}\phi(x_r) = \sum_{k=1}^{r-1}\phi(x_k) + \text{constant} \quad (1)$$

This completes the proof.

Equation (1) can also be put in the form

$$\phi(x_r) = \Delta \sum_{k=1}^{r-1} \phi(x_k) \tag{2}$$

Example 11. Prove that $\Delta^{-1} [x]^n = \dfrac{1}{(n+1)h}[x]^{n+1}$ + constant.

Hence, evaluate $\sum_{k=1}^{n} k^4$.

Solution: We know that

$$\Delta[x]^{n+1} = (n + 1)h\ [x]^n \tag{1}$$

$$\Rightarrow \qquad \Delta^{-1} [x]^n = \frac{1}{h(n+1)}[x]^{n+1} + \text{constant.}$$

Next, let $f(x_k) = k^4$

by using $\Delta^{-1} \phi(x_r) = \sum_{k=1}^{r-1} \phi(x_k)$ + constant, we get

$$\sum_{k=1}^{n} k^4 = \sum_{k=1}^{n} f(x_k) = \Delta^{-1} f(x_{n+1}) + C$$

$$= \Delta^{-1}(n + 1)^4 + C \tag{2}$$

where C is a constant.

Again, $(n + 1)^4$ is a polynomial of degree 4 in n, we can express it as a factorial polynomial as:

$$(n + 1)^4 = [n]^4 + 10[n]^3 + 25[n]^2 + 15[n] + 1$$

Hence, (2) gives

$$\sum_{k=1}^{n} k^4 = \Delta^{-1}[n]^4 + 10\Delta^{-1}[n]^3 + 25\ \Delta^{-1}[n]^2 + 15\ \Delta^{-1}[n] + \Delta^{-1}1 + C$$

$$= \frac{1}{5}[n]^5 + \frac{10}{4}[n]^4 + \frac{25}{3}[n]^3 + \frac{15}{2}[n]^2 + [n] + C \tag{3}$$

[on using (1) with $h = 1$]

for $n = 1$, (3) yields $1 = 1 + C$, so that $A = 0$.

$$\text{Hence, } \sum_{k=1}^{n} k^4 = \frac{1}{5}[n]^5 + \frac{10}{4}[n]^4 + \frac{25}{3}[n]^3 + \frac{15}{2}[n]^2 + [n]$$

$$= \frac{n}{30}\ (6n^4 + 15n^3 + 10n^2 - 1) \tag{4}$$

after a minor simplification.

Note: If $h = 1$, equation (1) gives

$$\int x^n\,dx = \frac{x^{n+1}}{n+1} + \text{constant which is analogous to the result (1)}$$

Example 12. Obtain the function whose first difference is $2x^3 + 3x^2 - 5x + 4$.

Solution: Let $f(x) = 2x^3 + 3x^2 - 5x + 4$

$$= 2[x]^3 + 9[x]^2 + 4 \quad (1)$$

(on representation the given function is a factorial polynomial)

If $g(x)$ be the polynomial which is to be obtained,

then $f(x) = \Delta g(x) \Rightarrow g(x) = \Delta^{-1} f(x)$,

and (1) gives $g(x) = 2\Delta^{-1}\,[x]^3 + 9\,\Delta^{-1}\,[x]^2 + \Delta^{-1}\,(4)$

Now, using the result $\Delta^{-1}\,[x]^n = \dfrac{1}{h(n+1)}\,[x]^{n+1}$ + constant, we get (on choosing $h = 1$), $g(x) = \dfrac{1}{2}[x]^4 + 3[x]^3 + 4[x] + C$

'C' is a constant.

EXERCISES

1. Obtain the function whose first differences are given as:

   (1) $9x^2 + 11x + 5$ (Ans: $3x^3 + x^2 + x + c$)

   (2) $x^3 + 3x^2 + 3x$ (Ans: $\dfrac{1}{4}[x]^4 + 2[x]^3 + (7/2)[x]^2 + c$)

   (3) $x^3 + 3x^2 + 5x + 12$ (Ans: $\dfrac{1}{4}[x]^4 + 2[x]^3 + \dfrac{9}{2}[x]^2 + 12x + c$)

   (4) $x(x - 1)$ (Ans: $\dfrac{1}{3}[x]^3 + c$)

2. Evaluate

   (a) $\sum\limits_{k=1}^{n} k^3$; (b) $\sum\limits_{k=1}^{n} (k^3 - 3k^2 + 1)$; (c) $\sum\limits_{k=1}^{n} k^2 2^k$

   (Ans: (a) $\dfrac{1}{4}n^2(n + 1)^2$, (b) $\dfrac{1}{4}n(n^3 - 2n^2 - 5n + 2)$, (c) $2^{n+1}(n^2 - 2n + 3) - 6$)

3. Prove the following:

   (i) $\Delta^{-1}(e^{ax}) = \dfrac{e^{ax}}{e^{ah} - 1}$

(ii) $\Delta^{-1}(xe^{ax}) = \dfrac{(x-h)e^{a(x+h)} - e^{ax}}{(e^{ah}-1)^2}$

(iii) $\Delta^{-1}\,(xa^x) = \dfrac{xa^x}{a^h - 1} - \dfrac{ha^{x+h}}{(a^h-1)^2}$

(iv) $\Delta^{-1}(\sin ax) = -\dfrac{\cos a(x-h/2)}{2\sin h/2}$

BACKWARD DIFFERENCES

Backward difference operator (∇) is defined as

$$\nabla f(x) = f(x) - f(x-h)$$

By definition $\nabla y_1 = y_1 - y_0$

$\nabla y_2 = y_2 - y_1$, etc ... $\nabla y_r = y_r - y_{r-1}$, ... $r = 1, 2, 3$ ... (1)

Second order differences:

Hence,
$$\begin{aligned}\nabla^2 f(x) &= \nabla[f(x) - f(x-h)] \\ &= \nabla f(x) - \nabla f(x-h) \\ &= f(x) - f(x-h) - [f(x-h) - f(x-2h)] \\ &= f(x) - 2f(x-h) + f(x-2h)\end{aligned}$$

or
$$\begin{aligned}\nabla^2 y_2 &= \nabla y_2 - \nabla y_1 \\ \nabla^2 y_3 &= \nabla y_3 - \nabla y_2, \ldots\end{aligned}$$

In general, $\nabla^2 y_r = \Delta y_r - \Delta y_{r-1}$, $r = 2, 3, \ldots$ (2)

Similarly, the nth backward differences are defined by the formula

$$\nabla^n y_r = \nabla^{n-1}_{y_r} - \nabla^{n-1}_{y_{r-1}}, \; r = n, n+1, \ldots \quad (3)$$

Note 1. (a) For using (3), for $n = 1$, we employ the notation $\Delta^0 y_r = y_r$.

(b) If $y = f(x)$ is a constant function, then $y = c$, a constant, $\forall x$, and we get $\Delta^n y_r = 0$ for all n.

The symbol ∇^n is referred to as the nth backward difference operator.

Backward differences table

x	y	∇y	$\nabla^2 y$	$\nabla^3 y$
x_0	y_0			
		∇y_1		
x_1	y_1		$\nabla^2 y_2$	
		∇y_2		$\nabla^3 y_3$
x_2	y_2		$\nabla^2 y_3$	
		∇y_3		
x_3	y_3			

Note 2. If the values of x are equally spaced with step-length h, then for any x at x_0, we have $x_1 = x_0 + h$; $x_2 = x_0 + 2h$; … $x_r = x_0 + rh$, … and the corresponding values of y are $y_0 = f(x_0)$, $y_1 = f(x_1) = f(x_0 + h)$, $y_2 = f(x_2) = f(x_0 + 2h)$, etc. … and equation (3) reads

$$\nabla^n f(x_r) = \nabla^{n-1} f(x_r) - \nabla^{n-1} f(x_{r-1}) \;\ldots \tag{4}$$

which can be rewritten as

$$\nabla^n f(x) = \nabla^{n-1} f(x) - \nabla^{n-1} f(x - h) \;\ldots \tag{5}$$

with r in x_r being suppressed (understood). This gives the nth backward differences of $f(x)$ at a point $x = x_r = x_0 + rh$, for $r = n, n + 1, \ldots$.

Thus, the formulas for the first, the second, the third backward differences of $f(x)$ at x are, respectively,

$$\nabla f(x) = f(x) - f(x - h) \tag{6a}$$

$$\nabla^2 f(x) = \nabla f(x) - \nabla f(x - h) \tag{6b}$$

$$\nabla^3 f(x) = \nabla^2 f(x) - \nabla^2 f(x - h) \tag{6c}$$

Note 1. The formula (5) on forward differences, that is, $\Delta^n f(x) = \Delta^{n-1} f(x + h) - \Delta^{n-1} f(x)$, the formula on backward differences (5), that is,

$$\nabla^n f(x) = \nabla^{n-1} f(x) - \nabla^{n-1} f(x - h)$$

and the formulas $6(a - c)$ are valid only when the values of x are equally spaced with h as the step-length.

Note 2. In order to compute backward differences, it is convenient to take a value of x that appears in the middle or in the later half of the backward differences table as x_0. Then we use the rotation:

$$x_{-r} = x_{0-rh} \tag{7}$$

$$y_{-r} = f(x - r) = f(x_0 - rh) \tag{8}$$

so that

$$\nabla^n y_{-r} = \nabla^{n-1} y_{-r} - \nabla^{n-1} y_{-r-1} \tag{10}$$

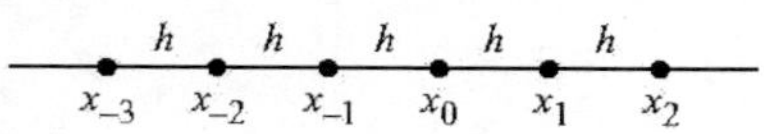

Illustration table:

x	y	∇y	$\nabla^2 y$	$\nabla^3 y$	$\nabla^4 y$	$\nabla^5 y$
x_{-3}	y_{-3}					
		∇y_{-2}				
x_{-2}	y_{-2}		$\nabla^2 y_{-1}$			
		∇y_{-1}		$\nabla^3 y_0$		
x_{-1}	y_{-1}		$\nabla^2 y_0$		$\nabla^4 y_1$	
		∇y_0		$\nabla^3 y_1$		$\nabla^5 y_2$
x_0	y_0		$\nabla^2 y_1$		$\nabla^4 y_2$	
		∇y_1		$\nabla^3 y_2$		
x_1	y_1		$\nabla^2 y_2$			
		∇y_2				
x_2	y_2					

Example 13. Construct a diagonal backward differences table for $y = f(x) = \log x$ given that:

x	10	20	30	40	50
y	1.0000	1.3010	1.4771	1.6021	1.6990

Also find the values of $\nabla^3 f(40)$ and $\nabla^4 f(50)$.

Solution: The differences table is

x	$y = f(x)$	$\nabla f(x)$	$\nabla^2 f(x)$	$\nabla^3 f(x)$	$\nabla^4 f(x)$
10	1.0000				
	$=f(10)$	0.3010			
		$=\nabla f(20)$	−0.1249		
20	1.3010		$=\nabla^2 f(30)$	0.0738	
	$=f(20)$	0.1761		$=\nabla^3 f(40)$	−0.0508
		$=\nabla f(30)$	−0.5011		$=\nabla^4 f(50)$
30	1.4771		$=\nabla^2 f(40)$	0.0230	
	$=f(30)$	0.1250		$=\nabla^3 f(50)$	
		$=\nabla f(40)$			
40	1.6021		−0.0281		
	$=f(40)$	0.0969	$=\nabla^2 f(50)$		
50	1.6990	$=\nabla f(50)$			
	$=f(50)$				

From the above table, we find that

$$\nabla^4 f(50) = -0.0508$$

and

$$\nabla^3 f(40) = 0.0738$$

Example 14. Prove that

(a) $y_{n-2} = y_n - 2\nabla y_n + \nabla^2 y_n$

(b) $y_{n-3} = y_n - 3\nabla y_n + 3\nabla^2 y_n - \nabla^3 y_n$

Solution: By definition, we have

$$\nabla y_n = y_n - y_{n-1},$$

$$\nabla y_{n-1} = y_{n-1} - y_{n-2}$$

and

$$\nabla^2 y_n = \nabla y_n - \nabla y_{n-1},$$

so that

$$y_{n-1} = y_n - \nabla y_n \qquad (1)$$

$$y_{n-2} = y_{n-1} - \nabla y_{n-1} \qquad (2)$$

$$\nabla y_{n-1} = \nabla y_n - \nabla^2 y_n \qquad (3)$$

Substituting from (1) and (3) in (2), we get

$$y_{n-2} = (y_n - \nabla y_n) - (\nabla y_n - \nabla^2 y_n)$$
$$= y_n - 2\nabla y_n + \nabla^2 y_n \qquad (4)$$

Next, we have (b)

$$\nabla y_{n-2} = y_{n-2} - y_{n-3}$$

or

$$y_{n-3} = y_{n-2} - \nabla y_{n-2}$$

Using (4), this becomes

$$y_{n-3} = (y_n - 2\nabla y_n + \nabla^2 y_n) - (\nabla y_n - 2\nabla^2 y_n + \nabla^3 y_n)$$
$$= y_n - 3\nabla y_n + 3\nabla^2 y_n - \nabla^3 y_n$$

Example 15. Prove that

$$(\Delta - \nabla)\, f(x) = \Delta\nabla f(x)$$

Solution: $\Delta\nabla f(x) = \Delta\{\nabla f(x)\} = \Delta\{f(x) - f(x-h)\}$

$$= \Delta f(x) - \Delta f(x-h)$$
$$= \Delta f(x) - \{f(x) - f(x-h)\}$$
$$= \Delta f(x) - \nabla f(x)$$
$$= (\Delta - \nabla)\, f(x)$$

Note: $\Delta\nabla$ and $\Delta - \nabla$ are equal operators

$\therefore$

$$\Delta\nabla = \Delta - \nabla$$

Example 16. Prove that

$$\Delta[f(x-1)\, \Delta g(x-1)] = \nabla[f(x)\, \Delta g(x)]$$

Solution: We know that

$$\boxed{\Delta[f(x)\, g(x)] = f(x)\, \Delta g(x) + g(x+h)\, \Delta f(x)} \qquad (1)$$

We have

$$\Delta[f(x-1)\, \Delta g(x-1)] = \Delta[f(x-1)\, \{g(x) - g(x-1)\}]$$

[taking $h = 1$ in (1)]

$$= \Delta[f(x-1)\, g(x)] - \Delta[f(x-1)\, g(x-1)]$$
$$= [g(x)\, \Delta f(x-1) + f(x)\, \Delta g(x)] - [g(x-1)\, \Delta f(x-1) + f(x)\, \Delta g(x-1)]$$

(using (1))

$$= g(x)\, [f(x) - f(x-1)] + f(x)\, [g(x+1) - g(x)]$$
$$-g(x-1)\, [f(x) - f(x-1)] - f(x)\, [g(x) - g(x-1)]$$
$$= f(x)\, g(x+1) - f(x-1)\, g(x) - f(x)\, g(x) + f(x-1)\, g(x-1)$$
$$= [f(x)\, g(x+1) - f(x-1)\, g(x)] - [f(x)\, g(x) + f(x-1)\, g(x-1)]$$
$$= \nabla[f(x)\, g(x+1)] - \nabla[f(x)\, g(x)]$$
$$= \nabla[f(x)\, \{g(x+1) - g(x)\}]$$
$$= \nabla[f(x)\, \Delta g(x)]$$

EXERCISES

1. Construct a diagonal backward differences table for $f(x) = \sin x$, given that

x	15°	20°	25°	30°
$f(x)$	0.2588	0.3420	0.4226	0.5000

Find the values of sin 10° from the table so constructed, it is being assumed that third differences are equal.

(Ans: sin 10° = 0.1736)

2. Construct the differences table for the data:

x	40	50	60	70	80	90
$f(x)$	184	204	226	250	276	304

Also, write down the values of $\nabla f(80)$, $\nabla^2 f(70)$ and $\nabla^5 f(90)$.

(Ans: 26; 2; 0)

3. Construct the differences table taking $x = 0, 1, 2$ and 3, given that $f(x) = x^3 - x + 1$, find $\nabla f(2)$ and $\nabla^2 f(3)$.

(Ans: 6; 12)

4. Prove the following:
   (a) ∇ is a linear operator
   (b) $\nabla [f(x - 1)\ \Delta g(x - 1)] = \Delta[f(x - 1)\ \Delta g(x)]$
5. Prove the following:
   (i) $y_{n-4} = y_n - 4\nabla y_n + 6\nabla^2 y_n - 4\nabla^3 y_n + \nabla^4 y_n$
   (ii) $y_{n-5} = y_n - 5\nabla y_n + 10\nabla^2 y_n - 10\nabla^3 y_n + 5\nabla^4 y_n - \nabla^5 y_n$
   (iii) $y_{-n} = \sum_{k=1}^{n} (-1)^k\ {}^n c_k\ \nabla^k y_0$
   (iv) $y_{n-r} = y_n - r\ (\nabla y_n) + \dfrac{r(r-1)}{2!}\ (\nabla^2 y_n) + \ldots + (-1)^r\ \nabla^r y_n$
6. Evaluate:
   (i) $\nabla(x^2 + 2x)$ (Ans: (1) $2h\ (x + 1) - h^2$)
   (ii) $\nabla^2(2x^2 + 3x - 5)$ (Ans: (ii) $4h^2$)

CENTRAL DIFFERENCES

The first central difference operator (δ) is defined by $\delta y_{1/2}$, $\delta y_{3/2}$, $\delta y_{5/2}$, ... as follows:

$$\delta y_{1/2} = y_1 - y_0; \qquad \delta y_{3/2} = y_2 - y_1,$$
$$\delta y_{5/2} = y_3 - y_2, \ldots \qquad \delta y_{r-1/2}, = y_r - y_{r-1} \tag{1}$$

Corresponding expressions $\Delta y_r = y_{r+1} - y_r$, $r = 0, 1, 2, \ldots$ with the expressions with $\Delta y_r = y_r - y_{r-1}$, $r = 1, 2, 3, \ldots$ we find that

$$\delta y_{1/2} = \Delta y_0 = \nabla y_1;$$
$$\delta y_{3/2} = \Delta y_1 = \nabla y_2;$$
$$\delta y_{5/2} = \Delta y_2 = \nabla y_3$$

and so on.

In general, $\delta y_{n+1/2} = \Delta y_n = \Delta y_{n+1}$, $n = 0, 1, 2, \ldots$ (2)

The first central differences of the first central differences are called the second central differences and are denoted by $\delta^2 y_1, \delta^2 y_2, \delta^2 y_3, \ldots$

Thus,
$$\delta^2 y_1 = \delta y_{3/2} - \delta y_{1/2}$$
$$\delta^2 y_2 = \delta y_{5/2} - \delta y_{3/2}$$
$$\ldots \quad \ldots \quad \ldots$$
$$\delta^2 y_n = \delta y_{n+1/2} - \delta y_{n-1/2} \tag{3}$$

Similarly, we can define the central differences of higher order.

In general, nth central differences are given by

$$\delta^n_{y_{r-1/2}} = \delta^{n-1}_{y_r} - \delta^{n-1}_{y_{r-1}}, \quad r = 1, 2, \ldots \tag{4}$$
(for odd n)

$$\delta^n_{y_r} = \delta^{n-1}_{y_{r+1/2}} - \delta^{n-1}_{y_{r-1/2}}, \quad r = 1, 2, \ldots \tag{5}$$
(for even n)

Note (1). For $n = 1$, in (4), we write $\delta^0_{y_r} = y_r$. and if y is a constant function, i.e. y = constant = c, then $\delta^n_{y_r} = 0$, for all $n \geq 1$.
δ^n is referred to as the nth central difference operator.

Note (2). If the values of x are *equally spaced* with $h > 0$, then $y_r = f(x_r) = f(x_0 + rh)$ and expression in (1) can be rewritten as:

$$\delta f\left(x_0 + \frac{h}{2}\right) = f(x_1) - f(x_0)$$

$$\delta f\left(x_1 + \frac{h}{2}\right) = f(x_2) - f(x_1)$$

$$\delta f\left(x_2 + \frac{h}{2}\right) = f(x_3) - f(x_2), \text{ and so on.}$$

In general, $\delta f\left(x_r + \frac{h}{2}\right) = f(x_{r+1}) - f(x_r) = f(x_r + h) - f(x_r)$

or $\delta f\left(x + \frac{h}{2}\right) = f(x + h) - f(x)$, (on dropping the siffines)

Replacing x by $x - (h/2)$, this can be rewritten as

$$\delta f(x) = f\left(x + \frac{h}{2}\right) - f\left(x - \frac{h}{2}\right)$$

Similarly, $$\delta^2 f(x+h) = \delta f\left(x+\frac{3h}{2}\right) - \delta f\left(x+\frac{h}{2}\right)$$

or $$\delta^2 f(x) = \delta f\left(x+\frac{h}{2}\right) - \delta f\left(x-\frac{h}{2}\right)$$

In general,

$$\delta^n f(x) = \delta^{n-1} f\left(x+\frac{h}{2}\right) - \delta^{n-1} f\left(x-\frac{h}{2}\right)$$

Note (3). It is convenient to take a value of x that appears in the middle of the differences table as x_0. Then, we write $y_{-r/2}$ for $f(x_0 - rh_{/2})$, $r = 1, 2, \ldots$ and calculate the central difference using

$$\delta^n_{y_{-r/2}} = \delta^{n-1}_{y_{(r+1)/2}} - \delta^{n-1}_{y_{(r\ 1)/2}}$$

Central differences table

x	y	δy	$\delta^2 y$	$\delta^3 y$	$\delta^4 y$
x_0	y_0				
		$\delta y_{1/2}$			
x_1	y_1		$\delta^2 y_1$		
		$\delta y_{3/2}$		$\delta^3 y_{3/2}$	
x_2	y_2		$\delta^2 y_2$		$\delta^4 y_2$
		$\delta y_{5/2}$		$\delta^3 y_{5/2}$	
x_3	y_3		$\delta^2 y_3$		
		$\delta y_{7/2}$			
x_4	y_4				

Example 17. Prove the following:

(a) $\delta^2 = \Delta\nabla = \Delta\nabla$

(b) $\delta^2 y_5 = y_6 - 2y_5 + y_4$

Solution: (a) By definitions,

i.e. $$\boxed{\delta^2 y_n = \delta y_{n+1/2} - \delta y_{n-1/2}} \quad (1)$$

and $$\boxed{\delta y_{r-1/2} = y_r - y_{r-1}} \quad (2)$$

We have,
$$\begin{aligned}
\nabla\Delta f(x) &= \nabla(f(x+h) - f(x)) \\
&= \nabla f(x+h) - \nabla f(x) \\
&= [f(x+h) - f(x)] - [f(x) - f(x-h)] \\
&= \delta f(x+h/2) - \delta f(x-h/2) \\
&= \delta^2 f(x)
\end{aligned}$$

$$\Rightarrow \quad \nabla\Delta = \delta^2 \quad (3)$$

Again
$$\begin{aligned}
\Delta\nabla f(x) &= \Delta[f(x) - f(x-h)] \\
&= \Delta f(x) - \Delta f(x-h)
\end{aligned}$$

$$= [f(x+h) - f(x)] - [f(x) - f(x-h)]$$
$$= \delta f(x + h/2) - \delta f(x - h/2)$$
$$= \delta^2 f(x)$$

$$\Rightarrow \qquad \nabla\Delta = \delta^2 \qquad (4)$$

From (3) and (4), we get

$$\boxed{\Delta\nabla = \nabla\Delta = \delta^2}$$

Again, from the formulas (1) and (2), we have

Solution (b) $\delta^2 y_5 = \delta y_{11/2} - \delta y_{9/2}$ (5)

$$\delta y_{11/2} = y_6 - y_5 \qquad (6)$$
$$\delta y_{9/2} = y_5 - y_4 \qquad (7)$$

Using (6) and (7) in (5), we get

$$\delta^2 y_5 = \delta y_{11/2} - \delta y_{9/2}$$
$$= (y_6 - y_5) - (y_5 - y_4)$$
$$\boxed{\delta^2 y_5 = y_6 - 2y_5 + y_4}$$

Example 18. Construct the central differences table and write the values of $\delta y_{-3/2}$, $\delta^2 y_0$ and $\delta^3 y_{1/2}$, by taking $x_0 = 0$ given that $f(-2) = 12, f(-1) = 16, f(0) = 15, f(1) = 18, f(2) = 20$.

Solution: Central differences table.

x	$y = f(x)$	δy	$\delta^2 y$	$\delta^3 y$	$\delta^4 y$
−2	12				
		4			
−1	16		−5		
		−1		9	
0	15		4		−14
		3		−5	
1	18		−1		
		2			
2	20				

Here $x_0 = 0$; $h = 1$

Hence, we have

$$y_{-r} = f(x_0 - rh) = f(-r)$$

From the table, we find

$$\delta y_{-3/2} = \delta f\left(-\frac{3}{2}\right) = 4$$
$$\delta^2 y_0 = 4, \; \delta^3 y_{1/2} = -5$$

Example 19. Prove the identity

$$\delta^2 f(x) = (\Delta - \nabla) f(x) = \Delta\nabla f(x)$$

Solution: By definitions of δ, Δ and ∇, we get

$$\delta^2 f(x) = \delta\{\delta f(x)\} = \delta\left\{f\left(x+\frac{h}{2}\right) - f\left(x-\frac{h}{2}\right)\right\}$$

$$= \delta f\left(x+\frac{h}{2}\right) - \delta f\left(x-\frac{h}{2}\right)$$

$$= \{f(x+h) - f(x)\} - \{f(x) - f(x-h)\}$$

$$= \Delta f(x) - \nabla f(x)$$

$$= (\Delta - \nabla) f(x)$$

Note: The operators δ^2, $\Delta - \nabla$ and $\Delta\nabla$ are equal. Hence, $\delta^2 = \Delta - \nabla = \Delta\nabla$

EXERCISES

1. Construct the central differences table for the following data:

x :	2.5	3.0	3.5	4.0	4.5
$f(x)$:	24.145	22.043	20.225	18.644	17.644

Find δy_0, $\delta^2 y_{-1}$, $\delta^3 y_{-1}$ $\delta^4 y_{-2}$ by taking $x_0 = 3.5$.

(Ans: −1.581, 0.284, 0.344, 0.391)

2. Construct the central differences table given that, $f(10) = 512, f(20) = 439$, $f(30) = 346, f(40) = 243$. Write down the values of $\delta f(15)$, $\delta^2 f(30)$, $\delta^3 f$ (25).

(Ans: −73, −10, 10)

3. Prove the following:
   (i) $\delta^2 y_0 = y_1 - 2y_0 + y_{-1}$
   (ii) $\delta^3 y_{1/2} = y_2 - 3y_1 + 3y_0 - y_{-1}$
   (iii) $\delta\{\nabla f(x)\} = \nabla\{\delta f(x)\}$
   (iv) δ is a linear operator.

4. Evaluate the following:
   (a) $\delta(x^2 + 2x)$, (Ans: $2h(x + 1)$
   (b) $\delta^2(2x^2 + 35 - 5)$, (Ans: $4h^2$)
   (c) $(\nabla^2 - 3\nabla\delta + 2\delta^2)(x^2 + 2x)$, (Ans: 0)

AVERAGING OPERATOR

The symbol μ is called the averaging operator and is defined as

$$\mu f(x) = \frac{1}{2}\left[f\left(x+\frac{1}{2}h\right) + f\left(x-\frac{1}{2}h\right)\right] \quad (1)$$

or

$$\mu f(x) = \frac{1}{2}[E^{1/2} f(x) + E^{-1/2} f(x)]$$

$(\because E^n f(x) = f(x + nh))$

$$= \frac{1}{2}\,[E^{1/2} + E^{-1/2}]\,f(x)$$

Therefore, $$\mu = \frac{1}{2}\,[E^{1/2} + E^{-1/2}]$$

If $f(x) = c$, a constant, we have

$$f\left(x+\frac{1}{2}h\right) = f\left(x-\frac{1}{2}h\right) = c;$$

$\therefore\ \mu f(x) = c =$ constant.

Example 20. Prove the following identities:

(a) $\mu^2 f(x) = f(x) + \frac{1}{4}\,\delta^2 f(x);\ (\mu^2 = 1 + \frac{1}{4}\delta^2)$

(b) $\mu\,\{\delta f(x)\} = \frac{1}{2}\,\{\Delta f(x) + \nabla f(x)\}$

Solution: By definition, we have

$$\mu^2 f(x) = \mu\,\{\mu + (x)\}$$

$$= \mu\left[\frac{1}{2}\left\{f\left(x+\frac{1}{2}h\right)+f\left(x-\frac{1}{2}h\right)\right\}\right]$$

$$= \frac{1}{2}\,\left\{\mu f\left(x+\frac{1}{2}h\right)+\mu f\left(x-\frac{1}{2}h\right)\right\}$$

$$= \frac{1}{2}\,\left[\frac{1}{2}\{f(x+h)+f(x)\}+\frac{1}{2}\{f(x)+f(x-h)\}\right]$$

$$= \frac{1}{4}\,[f(x+h) + 2f(x) + f(x-h)] \tag{1}$$

$$\delta^2 f(x) = \delta\{\delta f(x)\}$$

$$= \delta\left\{f\left(x+\frac{1}{2}h\right)-f\left(x-\frac{1}{2}h\right)\right\}$$

$$= \{f(x+h) - f(x)\} - \{f(x) - f(x-h)\}$$

$$= f(x+h) - 2f(x) + f(x-h) \tag{2}$$

From (1) and (2), we get

$$4\mu^2 f(x) - \delta^2 f(x) = 4f(x)$$

or $$\boxed{\mu^2 f(x) = f(x) + \frac{1}{4}\,\delta^2 f(x)}$$

or $$\boxed{\mu^2 = 1 + \frac{\delta^2}{4}}$$

(b) $\delta f(x) = f\left(x+\frac{1}{2}h\right) - f\left(x-\frac{1}{2}h\right)$

$$\therefore \quad \mu\{\delta f(x)\} = \mu\left\{f\left(x+\frac{1}{2}h\right) - f\left(x-\frac{1}{2}h\right)\right\}$$

$$= \mu\left\{f\left(x+\frac{1}{2}h\right) - \mu\left\{f\left(x-\frac{h}{2}\right)\right\}\right\}$$

$$= \frac{1}{2}\{f(x+h) + f(x)\} - \frac{1}{2}\{f(x) + f(x-h)\}$$

$$= \frac{1}{2}[\{f(x+h) - f(x)\} + [\{f(x) - f(x-h)\}]$$

$$= \frac{1}{2}[\Delta f(x) + \nabla f(x)]$$

or
$$\boxed{\mu\delta = \frac{1}{2}(\Delta + \nabla)}$$

Example 21. Prove that

$$\delta[f(x)\ g(x)] = \delta f(x)\ \mu g(x) + \delta g(x)\ \mu f(x) \qquad (1)$$

Hence, compute $\delta(x^3 + x^2 - 2x - 2)$.

Solution: By definition, we have

$$\delta f(x) = f\left(x+\frac{1}{2}h\right) - f\left(x-\frac{1}{2}h\right)$$

$$\mu g(x) = \frac{1}{2}\left[g\left(x+\frac{1}{2}h\right) + g\left(x-\frac{1}{2}h\right)\right]$$

$$\therefore \quad \delta f(x)\ \mu g(x) = \frac{1}{2} f\left(x+\frac{1}{2}h\right)\left\{g\left(x+\frac{1}{2}h\right) + g\left(x-\frac{1}{2}h\right)\right\}$$

$$-\frac{1}{2} f\left(x-\frac{1}{2}h\right)\left\{g\left(x+\frac{1}{2}h\right) + g\left(x-\frac{1}{2}h\right)\right\} \qquad (2)$$

Similarly, $\delta g(x)\ \mu f(x) = \frac{1}{2} g\left(x+\frac{1}{2}h\right)\left\{f\left(x+\frac{1}{2}h\right) + f\left(x-\frac{1}{2}h\right)\right\}$

$$-\frac{1}{2} g\left(x-\frac{1}{2}h\right)\left\{f\left(x+\frac{1}{2}h\right) + f\left(x-\frac{1}{2}h\right)\right\} \qquad (3)$$

Adding (2) and (3), we get

$$\delta f(x)\ \mu g(x) + \delta g(x)\ \mu f(x) = f\left(x+\frac{1}{2}h\right) g\left(x+\frac{1}{2}h\right) - f\left(x-\frac{1}{2}h\right) g\left(x-\frac{1}{2}h\right)$$

$$= \delta[f(x)\ g(x)]$$

Next, we have, $x^3 + x^2 - 2x - 2 = (x + 1)(x^2 - 2)$

Let $f(x) = x + 1$, $g(x) = x^2 - 2$

then $\delta f(x) = \delta(x + 1) = \left(x + \frac{h}{2} + 1\right) - \left(x - \frac{h}{2} + 1\right) = h$

$$\delta g(x) = \delta(x^2 - 2) = \left\{\left(x + \frac{1}{2}h\right)^2 - 2\right\} - \left\{\left(x - \frac{1}{2}h\right)^2 - 2\right\} = 2hx$$

$$\mu f(x) = \mu(x + 1) = \frac{1}{2}\left[\left\{\left(x + \frac{1}{2}h\right) + 1\right\} + \left\{\left(x - \frac{1}{2}h\right) + 1\right\}\right] = x + 1$$

$$\mu g(x) = \mu(x^2 - 2) = \frac{1}{2}\left[\left\{\left(x + \frac{1}{2}h\right)^2 - 2\right\} + \left\{\left(x - \frac{1}{2}h\right)^2 - 2\right\}\right]$$

$$= (x^2 - 2) + \frac{h^2}{4}, \quad \therefore \quad \delta(x^3 + x^2 - 2x - 2) = \delta(f(x)\, g(x))$$

$$= h\left\{\left(x^2 - 2\right) + \frac{h^2}{4}\right\} + 2hx\,(x + 1) = h\left(3x^2 + 2x - 2 + \frac{h^2}{4}\right)$$

EXERCISES

1. Prove the following identities:

(a) $1 + \frac{\delta^2}{2} = \left(1 - \delta^2_{\mu^2}\right)^{1/2}$

(b) $\Delta = \frac{1}{2}\,\delta^2 + \delta\sqrt{\left(1 + \frac{1}{4}\delta^2\right)}$

(c) $\Delta = \mu\delta + \frac{1}{2}\,\delta^2 = \frac{1}{2}\,\delta^2 + \delta\left(1 + \frac{\delta^2}{2}\right)^{1/2}$

(d) $1 + \delta^2_{\mu^2} = 1 + \delta^2 + \frac{1}{4}\delta^4$

(e) $\delta\left[\frac{f(x)}{g(x)}\right] = \dfrac{\mu g(x)\delta f(x) - \mu f(x)\delta g(x)}{g\left(x - \frac{h}{2}\right)g\left(x + \frac{h}{2}\right)}$

(f) $\mu\{f(x)\, g(x)\} = \{\mu f(x)\}\ \{\mu g(x)\} + \frac{1}{4}\ \{\delta f(x)\}\ \{\delta g(x)\}$

(g) $\mu\left\{\frac{f(x)}{g(x)}\right\} = \dfrac{\{\mu f(x)\}\{\mu g(x)\} - \frac{1}{4}\{\delta f(x)\}\{\delta g(x)\}}{f\left(x - \frac{h}{2}\right)g\left(x + \frac{h}{2}\right)}$

FINITE DIFFERENCE APPROXIMATIONS TO DERIVATIVES

If the values of x are equally spaced with h as the step-length, then by the Taylor's series $f(x + h)$ can be written as:

$$f(x+h) = f(x) + hf'(x) + \frac{h^2}{2!} f''(x) + \frac{h^3}{3!} f'''(x) + \ldots \tag{1}$$

Similarly,

$$f(x-h) = f(x) - hf'(x) + \frac{h^2}{2!} f''(x) - \frac{h^3}{3!} f'''(x) + \ldots \tag{2}$$

From expression (1), we get

$$f'(x) = \frac{1}{h}[f(x+h) - f(x)] - \frac{h}{2!} f''(x) - \frac{h^2}{3!} f'''(x) - \ldots \tag{3}$$

Neglecting the terms containing h, h^2, ... (3) yields

$$f'(x) \approx \frac{1}{h}[f(x+h) - f(x)]$$

$$= \frac{1}{h} \Delta f(x) \tag{4}$$

which is the first approximation of $f'(x)$ by its first forward difference at x.

Similarly, neglecting the terms containing h, h^2, ... we get

$$f'(x) \approx \frac{1}{h}[f(x) - f(x-h)]$$

$$= \frac{1}{h} \Delta f(x) \tag{5}$$

Expression (5) gives the approximation of $f'(x)$ by its first backward difference at x.

Next, subtracting (2) from (1), we obtain

$$f'(x) \approx \frac{1}{2h}[f(x+h) - f(x-h)] \tag{6}$$

On ignoring terms containing h^3, h^5, ... For step-length $p = 2h$, this expression reads

$$f'(x) = \frac{1}{p}\left[f\left(x+\frac{p}{2}\right) - f\left(x-\frac{p}{2}\right)\right]$$

$$= \frac{1}{p} \delta f(x) \tag{7}$$

This gives the approximation of $f'(x)$ by its first central difference at x.

Lastly, adding (1) and (2), we obtain

$$f''(x) \approx \frac{1}{h^2}\,[f(x + h) - 2f(x) + f(x - h)]; \tag{8}$$

Dropping the terms containing h^4, h^6,... this may be rewritten as:

$$f''(x) \approx \frac{1}{h^2}\,[\{f(x + h) - f(x)\} - \{f(x) - f(x - h)\}]$$

$$= \frac{1}{h^2}\left[\delta f\left(x+\frac{h}{2}\right)-\delta f\left(x-\frac{h}{2}\right)\right]$$

$$= \frac{1}{h^2}\,\delta^2 f(x) \tag{9}$$

This is the approximation of $f''(x)$ by its second central difference at x.

Differential operator D

It is often convenient to write $f'(x)$ as $Df(x)$, $f''(x)$ as $D^2 f(x)$, ..., $f^{(n)}(x)$ as $D^n f(x)$. That is,

$$Df(x) = f'(x),\ D^2f(x) = f''(x),\ \ldots\ D^n f(x) = f^{(n)}(x)\ \ldots \tag{10}$$

The symbol D is called the *first-order differential operator,* D^2 is called the *second-order differential operator*, and so on. In terms of these differential operators, formulas (4), (5), (7) and (9) read as follows:

$$Df(x) \approx \frac{1}{h}\,\Delta f(x) \approx \frac{1}{h}\,\nabla f(x) \approx \frac{1}{p}\,\delta f(x) \tag{11}$$

$$D^2 f(x) \approx \frac{1}{h^2}\,\delta^2 f(x) \tag{12}$$

Shift operator

Let us recall the following expression for the first forward difference of a function $f(x)$ for equally spaced values of x with step-length h:

$$\Delta f(x) = f(x + h) - f(x) \tag{1}$$

This may be rewritten as

$$f(x + h) = f(x) + \Delta f(x)$$

$$= (1 + \Delta)\, f(x) \tag{2}$$

We denote $(1 + \Delta)$ by E and call it as the (first order) shift operator.

Thus, $$E = 1 + \Delta \tag{3}$$

and $$Ef(x) = f(x + h) \tag{4}$$

Since Δ is a linear operator, it automatically follows that E is also a linear operator.

We define $E^2f(x)$, $E^3f(x)$, $E^4f(x)$,... by

$$E^2f(x) = E\{Ef(x)\} = Ef(x + h) = f(x + 2h)$$
$$E^3f(x) = E\{E^2f(x)\} = Ef(x + 2h) = f(x + 3h)$$
$$E^4f(x) = E\{E^3f(x)\} = Ef(x + 3h) = f(x + 4h)$$

and so on.

In general, $E^nf(x)$ is defined by

$$E^nf(x) = f(x + nh),\ n = 1, 2, 3,... \tag{5}$$

In (5), E^n is called the nth order shift order as it shifts the value of the function at x to the value at $x + nh$.

We can put formula (5) in an alternative form:

$$E^n y_m = y_{m+n} \tag{6}$$

Inverse shift operator

Since $\Delta f(x) = \nabla f(x + h)$, as seen by the right-hand side of (1),

we have

$$\Delta f(x) = \Delta f(x - h)$$
$$= f(x) - f(x - h)$$
$$f(x - h) = f(x) - \nabla f(x)$$
$$= (1 - \nabla)\, f(x) \tag{7}$$

Thus, the operator $1 - \nabla$ operating on $f(x)$ shifts $f(x)$ *backward* to its immediately *preceding value* $f(x - h)$. We denote this operator by E^{-1} (or $1/E$) and refer to it as the (first-order) *inverse shift operator*. Thus,

$$E^{-1} = 1 - \nabla \tag{8}$$

and

$$E^{-1} f(x) = f(x - h) \tag{9}$$

We immediately note that

$$E\{E^{-1} f(x)\} = Ef(x - h) = f(x)$$

and

$$E^{-1}\{Ef(x)\} = E^{-1}\{f(x + h)\} = f(x)$$

This means that the operator E^{-1} is actually the inverse of the operator E. Since ∇ is a linear operator, it follows from (8) that E^{-1} is also a linear operator.

The *nth-order inverse operator* E^{-n} is defined by an expression analogous to (5). That is,

$$E^{-n} f(x) = f(x - nh),\ n = 1, 2, 3,... \tag{10}$$

Thus, the operator E^{-n} shifts the value of the function at x to the value at $x - nh$.

The formula (10) can be put in the following *alternative form*:

$$E^{-n} y_m = y_{m-n} \tag{11}$$

It can be easily proved that for any two non-zero integers m and n,

$$E^m\, E^n\, f(x) = E^{m+n}\, f(x) = E^n\, E^m\, f(x) \tag{12}$$

The operators $E^{n/2}$ and $E^{-n/2}$

Now, $\quad E^n\, f(x) = f(x + nh);\; E^{-n}\, f(x) = f(x - nh)$

Sometimes we need an operator that shifts $f(x)$ to $f(x \pm nh/2)$; this operator is denoted by $E^{\pm n/2}$. Thus,

$$E^{n/2}\, f(x) = f\left(x + \frac{nh}{2}\right) \tag{13}$$

and

$$E^{-n/2}\, f(x) = f\left(x - \frac{nh}{2}\right) \tag{14}$$

In particular,

$$E^{1/2}\, f(x) = f\left(x + \frac{h}{2}\right) \tag{15}$$

$$E^{-1/2}\, f(x) = f\left(x - \frac{h}{2}\right) \tag{16}$$

We immediately note that the operators $E^{n/2}$ and $E^{-n/2}$ are inverse of each other. That is,

$$E^{n/2}\, E^{-n/2}\, f(x) = E^{-n/2}\, E^{n/2} f(x) = f(x) \tag{17}$$

Relation between E and D

We know the differential operators $D, D^2, \ldots D^n$

$Df(x) = f'(x),\; D^2 f(x) = f''(x)\; \ldots\; D^n\, f(x) = f^{(n)}(x).$

In this notation, the Taylor's expansion of $f(x + h)$ is

$$f(x + h) = f(x) + hD\, f(x) + \frac{h^2}{2!}\, D^2\, f(x) + \ldots + \frac{h^n}{n!}\, D^n\, f(x) + \ldots$$

$$= \left\{1 + hD + \frac{h^2}{2!} D^2 + \cdots + \frac{h^n}{n!} D^n + \cdots\right\} f(x) \tag{18}$$

The *series of operators* present in this expansion is analogous to the *exponential series*

$$1 + x + \frac{x^2}{2!} + \ldots + \frac{x^n}{n!} + \ldots$$

whose sum is e^x. In view of this analogy, the series of operators present in (18) is *symbolically* written as e^{hD}. That is,

$$1 + hD + \frac{h^2}{2!}\, D^2 + \ldots + \frac{h^n}{n!}\, D^n + \ldots = e^{hD} \tag{19}$$

Also, we observe that the left-hand side of (18) is equal to $Ef(x)$. Hence, (18) can be rewritten as

$$Ef(x) = e^{hD} f(x) \tag{20}$$

Thus, the operators E and e^{hD} are symbolically *equal*. That is,

$$E = e^{hD} \tag{21}$$

In view of this relation between E and D, hD is *symbolically* written as log E. That is,

$$hD = \log E \tag{22}$$

Relation between Δ and D

Since $E = 1 + \Delta$, the relation (22) can be rewritten as

$$hD = \log (1 + \Delta) \tag{23}$$

In view of the *logarithmic series*

$$\log (1 + x) = x - \frac{x^2}{2} + \frac{x^3}{3} - + \ldots,$$

the *operator* log $(1 + \Delta)$ is written *symbolically* as

$$\log (1 + \Delta) = \Delta - \frac{\Delta^2}{2} + \frac{\Delta^3}{3} - + \ldots \tag{24}$$

Then, (23) becomes

$$D = \frac{1}{h}\left[\Delta - \frac{\Delta^2}{2} + \frac{\Delta^3}{3} - + \cdots\right] \tag{25}$$

Example 22. Evaluate

(a) $E(x^2 - 3x + 1)$

(b) $(E^2 + 1)(x^3 + 2x^2 - 5x)$

(c) $E^{-1}(x^5 - x^3 + 1)$

(d) $E^{-3}(x^7 + x^5 + x^3 + x + 1)$

Solution:

(a) $E(x^2 - 3x + 1) = E(x^2) - 3E(x) + E(1)$
$= (x + h)^2 - 3(x + h) + 1$

(b) $(E^2 + 1)(x^3 + 2x^2 - 5x) = E^2(x^3 + 2x^2 - 5x) + (x^3 + 2x^2 - 5x)$
$= (x + 2h)^3 + 2(x + 2h)^2 - 5(x + 2h) + x^3 + 2x^2 - 5x$

(c) $E^{-1}(x^5 - x^3 + 1) = E^{-1}(x^5) - E^{-1}(x^3) + E^{-1}(1)$
$= (x - h)^5 - (x - h)^3 + 1$

(d) $E^{-3}(x^7 + x^5 + x^3 + 1) = (x - 3h)^7 + (x - 3h)^5 + (x - 3h)^3 + (x - 3h) + 1.$

Example 23. Evaluate

(a) E^3e^x and then E^ne^x.

where x varies by a constant interval h

(b) $(E - 1)(E - 2)x$

(c) $(E - 3)(E + 4)3^x$

(d) $(E + 2)(E - 1)(2^{x/h} + x)$

Solution:

(a) Here $Ee^x = e^{x+h}$,

$$\therefore \quad E^2e^x = EEe^x$$
$$= Ee^x + h$$
$$= e^{(x+h)+h}$$
$$= e^{x+2h}$$
$$E^3e^x = EE^2e^x = Ee^{x+2h}$$
$$= e^{(x+2h)+h} = e^{x+2h}$$

Proceeding in this way, we ultimately get $E^ne^x = e^{x+nh}$

(b) Here $Ex = (x + h)$

$$E_x^2 = EEx = E(x + h) = (x + h) + h = x + 2h$$
$$\therefore \quad (E - 1)(E - 2)x = (E^2 - 3E + 2)x = E_x^2 - 3Ex + 2x$$
$$= (x + 2h) - 3(x + h) + 2x$$
$$= x + 2h - 3x - 3h + 2x = -h$$

(c) $E3^x = 3^{x+h}$

$$\therefore \quad E^2\,3x = EE3^x = E3^{x+h} = 3^{(x+h)+h}$$
$$= 3^{x+2h}$$
$$\therefore \quad (E - 3)(E + 4)\cdot 3^x = (E^2 + E - 12)3^x$$
$$= E^23^x + E3^x - 12\,(3^x)$$
$$= 3^{x+2h} + 3^{x+h} - 12\,(3^x)$$
$$= 3^x\,(3^{2h} + 3^h - 12)$$
$$= 3^x\,(9^h + 3^h - 12)$$

(d) $(E - 2)(E - 1)(2^{x/h} + x)$

Here $E(x) = (x + h)$

$$E^2(x) = EEx = E(x + h) = (x + h) + h = x + 2h$$
$$E(2^{x/h}) = 2^{(x + h)}/h \text{ and } E^2\,(2^{x/h})$$
$$= E.\ E2^{x/h}$$
$$= E(2^{(x+h)/h})$$
$$= 2^{(x+h+h)/h} = 2^{(x+2h)/h}$$

Now $(E - 2)(E - 1)(2^{x/h} + x)$

$= (E^2 - 3E + 2)(2^{x/h} + x)$

$= (E^2 - 3E + 2)\,2^{x/h} + (E^2 - 3E + 2)x$

$= E^2\,(2^{x/h}) - 3E\,(2^{x/h}) + 2\,(2^{x/h}) + E^2x - 3Ex + 2x$

$= 2^{(x+2h)/h} - 3.2^{(x+h)/h} + 2^{(x+h)/h} + (x + 2h) - 3(x + h) + 2x$
$= 2^{(x+h)/h} [2^{h/n} - 3 + 1] + x + 2h - 3x - 3h + 2x$
$= -h$

Example 24. Expand the following:

(a) $(2E - 3)(5\Delta + 4)x^2$
(b) $\Delta^2 E^{-3} x^5$ and prove that $\Delta^3 y_3 = \Delta^3 y_6$

Solution:

(a) $(2E - 3)(5D + 4)x^2$

$$= (2E - 3)\{5\Delta x^2 + 4x^2\}$$
$$= (2E - 3)[5\{(x + h)^2 - x^2\} + hx^2]$$
$$= 2E\{5(x + h)^2 - x^2\} - 3\{5(x + h)^2 - x^2\}$$
$$= 2\{5(x + 2h)^2 - (x + h)^2\} - 3\{5(x + h)^2 - x^2\}$$

(b) $\Delta^2 E^{-3} x^5 = \Delta^2 (E^{-3}x^5) = \Delta^2 (x - 3h)^5$

part (i)

$$= \Delta\{\Delta(x - 3h)^5\}$$
$$= \Delta\{(x - 2h)^5 - (x - 3h)^5\}$$
$$= \{(x - h)^5 - (x - 2h)^5\} - \{(x - 2h)^5 - (x - 3h)^5\}$$

and part (ii)

$$\Delta^3 y_3 = (E - 1)^3 y_3$$
$$= (E - 1)(E - 1)(E - 1)y_3$$
$$= (E^3 - 3E^2 + 3E - 1)y_3$$
$$= y_6 - 3y_5 + 3y_4 - y_3 \qquad (1)$$

Next we have

$$\Delta^3 y_6 = (1 - E^{-1})^3$$
$$= (1 - E^{-1})(1 - E^{-1})(1 - E^{-1})y_6$$
$$= (1 - 3E^{-1} + 3E^{-2} - E^{-3})y_6$$
$$= y_6 - 3y_5 + 3y_4 - y_3 \qquad (2)$$

From (1) and (2), we get the required result.

Example 25. Evaluate

$$\frac{\Delta^2}{E} \sin(x + h) + \frac{\Delta^2 \sin(x + h)}{E \sin(x + h)}$$

Solution:

We have

$$\frac{\Delta^2}{E} \sin(x + h) + \frac{\Delta^2 \sin(x + h)}{E \sin(x + h)}$$

$$= E^{-1}(E - 1)^2 \sin(x + h) + \frac{(E - 1)^2 \sin(x + h)}{E \sin(x + h)}$$

$$= (E - 2 + E^{-1}) \sin(x + h) + \frac{(E^2 - 2E + 1)\sin(x+h)}{E \sin(x+h)}$$

$$= [\sin(x + 2h) - 2\sin(x + h) + \sin x]$$

$$+ \frac{\sin(x+3h) - 2\sin(x+2h) + \sin(x+h)}{\sin(x+2h)}$$

$$= 2\sin(x + h)(\cos h - 1) + \frac{2\sin(x+2h)(\cos h - 1)}{\sin(x+2h)}$$

$$= 2(\cos h - 1)[\sin(x + h) + 1]$$

Example 26. Prove the following:

(i) $\Delta = E - 1$ (ii) $\Delta = E\nabla = \nabla E = \delta E^{1/2}$

(iii) $\Delta^2 = E^2 - 2E + 1$ (iv) $\delta = E^{1/2} - E^{-1/2}$

(v) $E^{1/2} + E^{-1/2} = 2\mu$ (vi) $E^{\pm 1/2} = \mu \pm \dfrac{\delta}{2}$

(vii) $\Delta E^{-1/2} = \nabla E^{1/2} = \delta$ (viii) $E^{-1}\Delta = \Delta E^{-1} = \nabla$

(ix) $E - E^{-1} = 2\mu\delta$

Solution:

(i) We have

$$\begin{aligned}\Delta f(x) &= f(x + h) - f(x)\\ &= E[f(x)] - f(x)\\ &= (E - 1) f(x)\end{aligned}$$

Since $f(x)$ is arbitrary, so omitting it, we obtain

$$\Delta = E - 1 \text{ or } E = 1 + \Delta$$

(ii) $\Delta = E\nabla = \nabla E = \delta E^{1/2}$

We have
$$\begin{aligned}E\nabla[f(x)] &= E[f(x) - f(x - h)]\\ &= E[f(x) - E(f(x - h)]\\ &= f(x + h) - f(x) = \Delta f(x)\end{aligned}$$

$$\Rightarrow \qquad E\nabla = \Delta$$

Again $\nabla E[f(x)] = \nabla f(x + h) = f(x + h) - f(x) = \Delta f(x)$

$$\Rightarrow \qquad \nabla E = \Delta$$

Again $\delta E^{1/2}[f(x)]$

$$= \delta[f(x + h/2)] = f(x + h) - f(x) = \Delta f(x)$$

$$\Rightarrow \qquad \delta E^{1/2} = \Delta$$

(iii)
$$\begin{aligned}\Delta^2[f(x)] &= (E^{-1})^2 f(x)\\ &= (E - 1)\{(E - 1) f(x)\}\\ &= (E - 1)\{f(x + h) - f(x)\}\\ &= f(x + 2h) - f(x + h) - f(x + h) + f(x)\end{aligned}$$

$$= (E^2 - 2E + 1) f(x)$$
$$\Rightarrow \quad \Delta^2 = E^2 - 2E + 1$$

(iv) $E^{1/2} - E^{-1/2} = \delta$

$$(E^{1/2} - E^{-1/2}) f(x) = f\left(x + \frac{h}{2}\right) - f\left(x - \frac{h}{2}\right)$$
$$= \delta f(x)$$
$$\Rightarrow \quad E^{1/2} - E^{-1/2} = \delta$$

(v) $(E^{1/2} + E^{-1/2}) f(x)$

$$= f\left(x + \frac{h}{2}\right) + f\left(x - \frac{h}{2}\right)$$
$$= 2\mu f(x)$$
$$\Rightarrow \quad E^{1/2} + E^{-1/2} = 2\mu$$

(vi) Consider the results of (iv) and (v) and taking the sum and the difference, we get

i.e. $\quad (E^{1/2} - E^{-1/2}) + (E^{1/2} + E^{-1/2}) = \delta + 2\mu$

and $\quad (E^{1/2} - E^{-1/2}) - (E^{1/2} + E^{-1/2}) = \delta - 2\mu$

$\Rightarrow \quad 2E^{1/2} = \delta + 2\mu$

and $\quad 2E^{-1/2} = \delta - 2\mu$

The consolidated result is

$$E^{\pm 1/2} = \mu \pm \frac{\delta}{2}$$

(vii) $\Delta E^{-1/2} f(x)$

$$= \Delta \{E^{-1/2} f(x)\}$$
$$= \Delta\left\{f\left(x - \frac{h}{2}\right)\right\}$$
$$= f\left(x + \frac{h}{2}\right) - f\left(x - \frac{h}{2}\right)$$

$$\nabla E^{1/2} f(x) = \nabla\{E^{1/2} f(x)\}$$
$$= \nabla f\left(x + \frac{h}{2}\right)$$
$$= f\left(x + \frac{h}{2}\right) - f\left(x - \frac{h}{2}\right)$$

Thus,

$$\nabla E^{-1/2} f(x) = \nabla E^{1/2} f(x)$$
$$= f\left(x + \frac{h}{2}\right) - f\left(x - \frac{h}{2}\right)$$

$$= \delta f(x)$$

$$\Rightarrow \quad \Delta E^{-1/2} = \nabla E^{1/2} = \delta$$

(viii) From (ii), we get

$$E^{-1}\{E\nabla f(x)\} = E^{-1}\{\Delta f(x)\}$$

or $$\nabla f(x) = E^{-1}\,\Delta f(x)$$

On the other hand,

$$\Delta E^{-1}\,f(x) = \Delta f(x-h)$$
$$= f(x) - f(x-h)$$
$$= \Delta f(x)$$

Thus, $$\nabla f(x) = E^{-1}\,\Delta f(x)$$
$$= \Delta E^{-1}\,f(x)$$

$$\Rightarrow \quad \nabla = E^{-1}\Delta = \Delta E^{-1}$$

(ix) $$\mu\delta\,f(x) = \mu\,\{\delta f(x)\} = \mu\left\{f\left((x+\frac{h}{2}\right) - f\left(x-\frac{h}{2}\right)\right\}$$

$$= \frac{1}{2}\,\{f(x+h) + f(x)\} - \frac{1}{2}\,\{f(x) + f(x-h)\}$$

$$= \frac{1}{2}\,\{f(x+h) - f(x-h)\}$$

$$= \frac{1}{2}\,\{Ef(x) - E^{-1}\,f(x)\}$$

$$= \frac{1}{2}\,(E - E^{-1})\,f(x)$$

$$\Rightarrow \quad E - E^{-1} = 2\mu\delta$$

Example 27. Prove the following:

(a) $E^{-1} = e^{-hD}$

(b) $D = \dfrac{1}{h}\,\sinh^{-1}\,(\mu\delta)$

(c) $\delta = 2\sinh\left(\dfrac{hD}{2}\right)$

(d) $\mu = \cosh\left(\dfrac{hD}{2}\right)$

Solution:

(a) We can write $f(x) = E^{-1}Ef(x)$

$$= E^{-1}\{e^{hD}f(x)\}, \quad (\because \quad Ef(x) = e^{hD}f(x))$$

Hence, $\quad E^{-1}e^{hD} = 1$

or $$E^{-1} = \frac{1}{e^{hD}} = e^{-hD}.$$

Solution:

(b) $$\mu\delta = \frac{1}{2}(E - E^{-1})$$

$$= \frac{1}{2}(e^{hD} - e^{-hD})$$

$$= \sinh(hD)$$

$$\Rightarrow \quad D = \frac{1}{2}\sinh^{-1}(\mu\delta)$$

$$\mu\delta f(x) = \mu\{\delta f(x)\}$$

$$= \mu\left\{f\left(x + \frac{h}{2}\right) - f\left((x - \frac{h}{2}\right)\right\}$$

$$= \frac{1}{2}\{f(x + h) + f(x)\}$$

$$- \frac{1}{2}\{f(x) + f(x - h)\}$$

$$= \frac{1}{2}\{f(x + h) - f(x - h)\}$$

$$= \frac{1}{2}\{Ef(x) - E^{-1} f(x)\}$$

$$= \frac{1}{2}(E - E^{-1}) f(x)$$

$$\Rightarrow \quad \frac{E - E^{-1}}{2} = \mu\delta$$

Solution:

(c) $2 \sinh\left(\frac{hD}{2}\right)$

$$= (e^{hD/2} - e^{-hD/2}), \quad \because \sin\theta = \frac{(e^{\theta} - e^{-\theta})}{2}$$
$$= (e^{hD})^{1/2} - (e^{-hD})^{1/2}$$
$$= E^{1/2} - E^{-1/2} \quad (\because \quad E = e^{hD})$$
$$= \delta$$

$$\Rightarrow \quad \delta = 2\sinh\left(\frac{hD}{2}\right)$$

(d) We have

$$\cosh\left(\frac{hD}{2}\right) = \frac{1}{2}\,(e^{hD} + e^{-hD})$$

$$= \frac{1}{2}\,(E^{1/2} + E^{-1/2}) = \mu$$

Example 28. Prove(i) $\left(\frac{\Delta^2}{E}\right)u_n \neq \frac{\Delta^2 u_n}{E u_n}$

(ii) $\left\{\frac{\Delta^2}{E}e^x\right\}\left\{\frac{E\,e^x}{\Delta^2 e^x}\right\} = e^x$

Solution: (i)

Now
$$\left(\frac{\Delta^2}{E}\right)u_n = (\Delta^2 E^{-1})u_n$$

$$= [(E-1)^2\,E^{-1}]u_n$$
$$= [E - 2 + E^{-1}]u_n$$
$$= u_{n+2} - 2u_n + u_{n-2} \qquad (1)$$

$$\frac{\Delta^2 u_n}{E u_n} = \frac{(E-1)^2 u_n}{u_{n+h}} = \frac{(E^2 - 2E + 1)\,u_n}{u_{n+h}}$$

$$= \frac{u_{n+2h} - 2u_{n+h} + u_n}{u_{n+h}} \qquad (2)$$

From (1) and (2), the r.h.s of (1) and (2) are not equal.

$$\therefore \qquad \left(\frac{\Delta^2}{E}\right)u_n \neq \left(\frac{\Delta^2 u_n}{E u_n}\right)$$

Solution: (ii) We have,

$$Ee^x = e^{x+h} \qquad (1)$$
$$\Delta^2 e^x = (E-1)^2 e^x$$
$$= (E^2 - 2E + 1)e^x$$
$$= e^{x+2h} - 2e^{x+h} + e^x \qquad (2)$$

$$\frac{\Delta^2}{E}\,e^x = E^{-1}\,(\Delta^2 e^x)$$

$$= E^{-1}\,(e^{x+2h} - 2e^{x+h} + e^x)$$
$$= e^{x+h} - 2e^x + e^{x-h} \qquad (3)$$

and
$$\frac{Ee^x}{\Delta^2 e^x} = \frac{e^{x+h}}{e^{x+2h} - 2e^{x+h} + e^x}$$

$$= \frac{e^x}{e^{x+h} - 2e^x + e^{x-h}} \tag{4}$$

From (3) and (4), we get

$$\left\{\frac{\Delta^2}{E}e^x\right\}\left\{\frac{Ee^x}{\Delta^2 e}\right\} = e^x$$

Example 29. Estimate the missing term in the following table:

x :	0	1	2	3	4
$y = f(x)$:	1	2	6	?	51

Solution: We are given 4 values, so the third differences are constants and the fourth differences are zero

i.e. $\Delta^4 f(x) = 0$ for all values of x

$\Rightarrow$ $$(E - 1)^4 f(x) = 0$$

$\Rightarrow$ $$(E^4 - 4E^3 + 6E^2 - 4E + 1) f(x) = 0$$

$\Rightarrow$ $$E^4 f(x) - 4E^3 f(x) + 6E^2 f(x) - 4Ef(x) + f(x) = 0$$

$\Rightarrow$ $$f(x + 4) - 4 f(x + 3) + 6f(x + 2) - 4 f(x + 1) + f(x) = 0$$

where the interval of differencing is 1.

Now putting $x = 0$, we obtain

$$f(4) + 4f(3) + 6f(2) - 4 f(1) + f(0) = 0$$

$\Rightarrow$ $$51 - 4 f(3) + 36 - 8 + 1 = 0$$

$\Rightarrow$ $$f(3) = 20$$

Example 30. If $u_0 = 5$, $u_1, = 11$, $u_2 = 22$, $u_3 = 40$, $u_5 = 140$, then find u_4, given that the general term is represented by a fourth-degree polynomial.

Solution: By data, we are given 5 values of the function u_n, and u_n to be a polynomial of 4th degree and therefore its 4th differences are constant and higher differences are zero.

i.e $$\Delta^5 u_n = 0$$

i.e. $$(E - 1)^5 u_n = 0$$

$\Rightarrow$ $$(E^5 - 5E^4 + 10E^3 - 10E^2 + 5E - 1)u_n = 0$$

$\Rightarrow$ $$u_{n+5} - 5u_{n+4} + 10E_{n+3} - 10u_{n+2} + 5u_{n+1} - u_n = 0 \tag{1}$$

Taking $n = 0$ and substituting the given values of u_0, u_1, u_2, u_3, and u_5, we get

$$140 - 5u_4 + 400 - 220 + 55 - 5 = 0$$

$\Rightarrow u_n = 74$ is the required value.

Example 31. Find $\log_{10}^{7}$ and $\log_{10}^{11}$ from the data given $f(x) = \log_{10}^{x}$ for some values of x.

x	6	7	8	9	10	11	12
$f(x) = \log_{10}^{x}$	0.77815	?	0.90309	0.95424	1.0000	?	1.07918

Solution: Here, 5 values of x and the corresponding values of $y = f(x)$ are given

$\therefore$ $\quad \Delta^5 f(x) = 0 \quad \forall x$

i.e. $\quad (E - 1)^5 f(x) = 0$

$\Rightarrow \quad (E^5 - 5E^4 + 10E^3 - 10E^2 + 5E - 1) f(x) = 0 \qquad (1)$

Here $\quad x_0 = 6;\ x_1 = 7,\ x_2 = 8,\ x_3 = 9,\ x_4 = 10,$

$x_5 = 11,\ x_6 = 12$

and $\quad y_0 = 0.77815,\ y_5 = 0.90309, \ldots\ y_6 = 1.07918$

For $\quad f(x) = f(x_0) = y_0$

and $\quad f(x) = f(x_1) = y_1$

Expression (1) gives respectively,

$$y_5 - 5y_4 + 10y_3 - 10y_2 + 5y_1 - y_0 = 0$$
$$y_6 - 5y_5 + 10y_4 - 10y_3 + 5y_2 - y_1 = 0$$

Substituting the values of y_0, y_2, y_3, y_4 and y_6 in these equations, we get

$$5y_1 + y_5 = 5.26615$$
$$y_1 + 5y_5 = 6.05223$$

or solving, we get

$$y_1 = f(x_1) = f(_7) = \log_{10}^{7} = 0.84494$$
$$y_3 = f(x_5) = f(_{11}) = \log_{10}^{11} = 1.04146$$

EXERCISES

1. Evaluate the following:

   (i) $E(x + 1)^2$ (ii) $E(5x - 4)^{1/3}$

   (iii) $\dfrac{\Delta}{E} \sin x$ (iv) $E^2(x - 1)^3$

   (v) $(2E - 1)(3\Delta + 2)x^2$ (vi) $(x\Delta E)^2 x^2$

   (vii) $E^{-3}(x^3 + 2x - 1)$

2. Prove the following:

   (i) $\Delta^3 y_2 = \nabla^3 y_5$

   (ii) $\dfrac{\Delta^2}{E}(x^3) + \dfrac{\Delta^2 x^3}{E x^3} = 6h^2 \left[x + \dfrac{1}{(x+h)^2} \right]$

(iii) $\frac{\Delta^2}{E}\cos(x+h) + \frac{\Delta^2\cos(x+h)}{E\cos(x+h)} = -8\sin^2\left(\frac{h}{2}\right)\cos^2\left(\frac{x+h}{2}\right)$

3. Prove the identity (12).
4. Prove the following:

   (i) $E\{f(x)g(x)\} = \{Ef(x)\}\{Eg(x)\}$
   (ii) $E\{f(x)\}^n = \{Ef(x)\}^n$
   (iii) $E^m\{f_1(x)f_2(x)\ldots f_n(x)\} = \{E^m f_1(x)\} \times \{E^m f_2(x)\}\ldots\{E^m f_n(x)\}$

5. Prove the following identities:

   (i) $\Delta E = E\Delta$
   (ii) $\Delta = \delta E^{1/2}$
   (iii) $\nabla = \delta E^{-1/2}$
   (iv) $\delta = \Delta E^{-1/2} = \nabla E^{1/2}$
   (v) $(E^{1/2} + E^{-1/2})(1+\Delta)^{1/2} = 2 + \Delta$
   (vi) $(E+1)\delta = 2(E-1)\mu$
   (vii) $\mu = \frac{1}{2}\Delta(1+E^{-1})$
   (viii) $E^{1/2} = \left(1+\frac{\delta^2}{4}\right)^{1/2} + \frac{\delta}{2}$
   (ix) $E^r = \left(\mu+\frac{1}{2}\delta\right)^{2r}$
   (x) $\delta^n = \nabla^n E^{n/2} = \Delta^n E^{-n/2}$
   (xi) $\Delta^{2m+1} = E^m\left\{\mu\delta^{2m+1} + \frac{1}{2}\delta^{m+2}\right\}$
   (xii) $E^n y_r = (1+\Delta)^n y_r$

6. Prove the following identities:

   (i) $\nabla = 1 - e^{-hD}$
   (ii) $hD = \log(1-\nabla)^{-1} = 2\sinh^{-1}\left(\frac{\delta}{2}\right)$
   (iii) $\nabla^2 = h^2D^2 - h^3D^3 + \frac{7}{12}h^4D^4 - + \ldots$
   (iv) $\Delta^{-1} = \frac{1}{hD} - \frac{1}{2} + \frac{1}{12}hD - \frac{1}{720}(hD)^3 + \ldots$
   (v) $D^2 = \frac{1}{h^2}\left[\Delta^2 - \Delta^3 + \frac{11}{12}\Delta^4 - \frac{5}{6}\Delta^5 + \ldots\right]$

7. The operator δ^{-1} is defined by $\delta^{-1}\{\delta f(x)\} = f(x)$. Show that

$$\delta^{-1} = E^{1/2}\,\Delta^{-1} = E^{-1/2}\,\nabla^{-1}$$

8. If the operator M is defined by $Mf(x) = \frac{1}{2}\,[f(x+h) + f(x)]$, prove the following:

   (i) $M = \frac{1}{2}(E+1) = E - \frac{1}{2}\Delta$

   (ii) $\mu = ME^{-1/2}$

9. Find the missing values in the following tables:

(i)

x:	1	2	3	4	5
y:	2	?	40	83	150

(ii)

x:	1	2	3	4	5
$f(x)$:	15.75	17.9	?	22.75	43.2

(iii)

x:	4	5	6	7	8	9
y:	72	?	146	192	?	302

(iv)

x:	45	50	55	60	65
$f(x)$:	3.0	?	2.0	?	−2.4

(v)

x:	0	1	2	3	4	5	10	15
y:	5	?	?	?	?	27	179	611

10. If $y_1 = 40$, $y_3 = 45$, $y_5 = 54$, find y_2 and y_4.
11. If $y_0 = 3$, $y_1 = 12$, $y_2 = 81$, $y_3 = 200$, $y_4 = 100$, and $y_5 = 8$, find $\Delta^5 y_0$ without forming the differences table.
12. The first, third, fourth, sixth and seventh terms of a sequence are 0.135, 0.111, 0.100, 0.082 and 0.074. Find the second and fifth terms.

Answers:

1. (i) $(x+h+1)^2$ (ii) $\{5(x+h)-4\}^{1/3}$
   (iii) $2(\cosh - 1)\sin x$ (iv) $(x+2h-1)^3$
   (v) $2x^2 + 14hx + 19h^2$ (vi) $4h^2x^2 + 9h^3x$
   (vii) $(x-3h)^3 + 2(x-3h) - 1$.

9. (i) 15 (ii) 17.275 (iii) 106, 244
   (iv) 2.325, 2.025 (v) 6.2, 7.6, 11, 17

10. 42, 49 11. 755 12. 0.1230, 0.0904

Chapter **6**

Interpolation

INTRODUCTION

The process of finding a value of a function between two known values by a procedure other than the law which is given by the function itself. For example, in linear interpolation, the procedure is based on the assumption that the three points having these values of the function for ordinates lie on a straight line. Interpolation is the art of reading the values in between the given set of data. If we are interested to evaluate $f(x)$, $(= y)$ for a value of x outside the given range, then this is known as extrapolation. Some of the interpolation formulae can be utilised for extrapolation.

POLYNOMIAL INTERPOLATION

Let us assume the set of values of (x, y) as:

$$y = y_0, y_1 \ldots, y_n; x = x_0, x_1, \ldots x_n$$

We require the value of y corresponding to x_k which lies in $(x_r, x_r + 1)$.

The equation of line is $\dfrac{y - y_r}{x - x_r} = \dfrac{y_{r+1} - y_r}{x_{r+1} - x_r}$.

or
$$y_k = y_r + \left(\frac{y_{r+1} - y_r}{x_{r+1} - x_r}\right)(x_k - x_r)$$

gives the value of y at $x = x_k$

$$(x_r < x_k < x_r + 1)$$

Error in Polynomial Interpolation

If $y = f(x)$ is the exact curve and $y = p_n(x)$ is the interpolating polynomial curve,

$$\text{Error} = f(x) - P_n(x) = \left\{ \frac{(x-x_0)(x-x_1)\cdots(x-x_n)\, f_{(c)}^{(n+1)}}{(n+1)!} \right\}$$

For any x where $x_0 < x < x_n$ and $x_0 < c < x_n$.

Example 1. Find y at $x = 0.5$, $x = 0.75$, given the following table.

x:	0	1	2	5
y:	2	3	12	147

Solution:
$$y_k = y_r + \left(\frac{y_{r+1} - y_r}{x_{r+1} - x_r} \right)(x_k - x_r)$$

$$y\,(0.5) = 2 + \frac{(3-2)}{(1-0)}(0.5 - 0) = 2.5$$

$$y\,(0.75) = 2 + \frac{(3-2)}{(1-0)}(0.75 - 0) = 2.75$$

Interpolation with equal intervals (i.e. when the function-values are given at equidistant intervals).

A. Newton-Gregory Formula for Forward Interpolation

Let $x_i = x_0 + ih$, $i = 0, 1, 2, \ldots n$ and $f(x_i) = f_i$. Suppose $f(x)$ is to be found for a given x.

We define $k = \dfrac{x - x_0}{h}$ or $x = x_0 + kh$ (1)

Then $f(x) = f(x_0 + kh) = f_k = E^k f_0 = (1 + \Delta)^k f_0$

$$= \left[1 + \binom{k}{1}\Delta + \binom{k}{2}\Delta^2 + \cdots \right] f_0$$

or
$$f(x_0 + ih) = f_0 + \binom{k}{1}\Delta f_0 + \binom{k}{2}\Delta^2 f_0 + \ldots + \binom{k}{n}\Delta^n f_0 \qquad (2)$$

This is known as Newton-Gregory forward-difference interpolation formula.

B. Newton-Gregory Formula for Backward Interpolation

If we define $k = \dfrac{x_n - x}{h}$ or $x = x_n - kh$ (3)

then $f(x) = f(x_n - kh) = E^{-k} f_n$

$$= (1 - \nabla)^k f_n = \left[1 - \binom{k}{1}\nabla + \binom{k}{2}\nabla^2 - \cdots \right] f_n$$

or $$f(x_n - kh) = f_n - \binom{k}{1}\nabla f_n + \binom{k}{2}\nabla^2 f_n - \dots + (-1)^n \binom{k}{n}\nabla^n f_n \qquad (4)$$

This is known as Newton-Gregory backward interpolation formula.

Since the interpolation k is non-integer, the series in (2) and (4) are infinite series. However, with $(n + 1)$ entries in the table, only the powers of Δ or ∇ available to us are retained in (2) and (4).

Example 2. Derive Newton-Gregory forward difference interpolation formula from the Newton divided difference formula.

***Solution*:** Since $x_i = x_0 + ih$, we find the divided difference takes the simple form

$$f(x_0, x_1) = \frac{f_1 - f_0}{h} = \frac{1}{h}\Delta f_0$$

Similarly, $f(x_0, x_1, x_2) = \dfrac{1}{2!\,h^2}\Delta^2 f_0$

or in general $f(x_0, x_1, \dots x_n) = \dfrac{1}{n!\,h^n}\Delta^n f_0$

Hence, the Newton's divided difference formula takes the form

$$f(x) = f_0 + \frac{(x - x_0)}{h}\Delta f_0 + \frac{(x - x_0)(x - x_0 - h)}{2!\,h^2}\Delta^2 f_0 + \cdots$$

$$+ \frac{(x - x_0)(x - x_0 - h)\cdots(x - x_0 - \overline{(n-1)}\,h)}{n!\,h^n}\Delta^n f_0$$

Setting $x = x_0 + kh$, we get

$$f(x_0 + kh) = f_0 + k\Delta f_0 + \frac{k(k-1)}{2!}\Delta^2 f_0 + \cdots$$

$$\cdots + \frac{k(k-1)\cdots(k - \overline{n-1})}{n!}\Delta^n f_0$$

which is the Newton-Gregory forward difference interpolation formula.

The error term takes the form

$$E = \binom{k}{n+1} h^{n+1} f^{n+1}(\xi)$$

Example 3. Compute log (105.6) by Newton-Gregory forward difference formula given the values:

log (105) = 2.0211; log (106) = 2.0253;
log (107) = 2.0293; log (108) = 2.0334;
log (109) = 2.0374.

Solution: Since log (105.6) is near the beginning of the given table, we use Gregory-Newton interpolation formula.

$$f(x_0 + kh) = f_0 + \binom{k}{1}\Delta f_0 + \binom{k}{2}\Delta^2 f_0 + \ldots$$

where $$k = \frac{x - x_0}{h}$$

Let us prepare the finite differences table:

x_i	f_i	Δ	Δ^2	Δ^3	Δ^4
105	2.0211				
		0.0042			
106	2.0253		– 0.0002		
		0.0040		0.0003	
107	2.0293		0.0001		– 0.0005
		0.0041		– 0.0002	
108	2.0034		– 0.0001		
		0.0040			
109	2.0374				

We note that the required value (i.e. $x = 105.6$) is near the beginning of the table, and hence the value of 'x_0' is '105'.

Here $x_0 = 105$, $h = 1$; $x = 105.6$ and $k = \dfrac{x - x_0}{h}$

$$= \frac{105.6 - 105}{1}$$

$$= 0.6$$

$$f(105 + 0.6 \times 1) = f(105.6) = 2.0211 + 0.6\,(0.0042)$$

$$+ \frac{0.6(0.6 - 1)}{2!}(-0.0002)$$

$$+ \frac{0.6(0.6 - 1)\,(0.6 - 2)}{3!}(0.0003)$$

$$+ \frac{0.6(0.6 - 1)\,(0.6 - 2)\,(0.6 - 3)}{4!}(-0.0005)$$

$$= 2.0236776$$

Example 4. Given the following table, find $y(78)$.

x:	80	85	90	95	100
y:	5026	5674	6362	7088	7854

***Solution*:** It is an extrapolation problem. We have the Gregory-Newton forward interpolation formula as:

$$f(x_0 + kh) = f_0 + \binom{k}{1}\Delta f_0 + \binom{k}{2}\Delta^2 f_0 + \binom{k}{3}\Delta^3 f_0$$

where $$k = \frac{x - x_0}{h}$$

Let us prepare the finite differences table:

x_i	f_i	Δ	Δ^2	Δ^3	Δ^4
80	5026				
		648			
85	5674		40		
		688		–2	
90	6362		38		4
		726		2	
95	7088		40		
		766			
100	7854				

We note that the required value (i.e. $x = 78$) is near the beginning of the table and hence the value of x_0 is '80'.

Hence $x_0 = 80$, $h = 5$, $x = 78$ and $k = \dfrac{x - x_0}{h} = \dfrac{78 - 80}{5}$

$$= -0.4$$

$$f[80 - 0.4 \times 5] = f(78) = 5026 + (-0.4)(648)$$

$$+ \frac{(-0.4)(-0.4-1)}{2!}(40)$$

$$+ \frac{(-0.4)(-0.4-1)(-0.4-2)}{3!}(-2)$$

$$+ \frac{(-0.4)(-0.4-1)(-0.4-2)(-0.4-3)}{4!}(4)$$

$$= 4779.21$$

Example 5. Compute $f(1.2)$ by Newton-Gregory forward difference interpolation formula from the data:

x:	1.0	1.5	2.0	2.5	3.0
f:	9.00	32.75	79.00	155.25	269.00

***Solution*:**

x_i	f_i	Δf_i	$\Delta^2 f_i$	$\Delta^3 f_i$	$\Delta^4 f_i$
1.0	9.00				
		23.75			
1.5	32.75		22.5		
		46.25		7.5	
2.0	79.00		30.0		0
		76.25		7.5	
2.5	155.25		37.5		
		113.75			
3.0	269.00				

Here, $x_0 = 1.0, \quad h = 0.5, \quad x = 1.2$ and $k = \dfrac{x - x_0}{h} = 0.4$

$$\text{Hence, } f(1.2) = f_0 + \binom{0.4}{1}\Delta f_0 + \binom{0.4}{2}\Delta^2 f_0 + \binom{0.4}{3}\Delta^3 f_0$$

$$= 9.00 + (0.4)\,(23.75) + \frac{(0.4)\,(-0.6)}{2}(22.5)$$

$$+ \frac{(0.4)\,(-0.6)\,(-1.6)}{6}(7.5) = 16.28$$

Example 6. Use Gregory-Newton backward formula to find y when $x = 23.4$ and $x = 25.4$ from the data:

x:	19	20	21	22	23	24	25
y:	91	100.25	110	120.25	131	142.25	154

***Solution*:** Since the values of x are at finite differences, let us set up the finite differences table:

x_i	f_i	Δ	Δ^2	Δ^3
19	91.00			
		9.25		
20	100.25		0.50	
		9.75		0.0
21	110.0		0.50	
		10.25		0.0
22	120.25		0.50	
		10.75		0.0
23	131.00		0.50	
		11.25		0.0
24	142.25		0.50	
		11.75		
25	154.00			

From Gregory-Newton backward formula, we have

$$f(x_n - kh) = f_n - \binom{k}{1}\nabla f_n + \binom{k}{2}\nabla^2 f_n - \ldots + (-1)^n \binom{k}{n}\nabla^n f_n$$

and $$f(x_0 + kh) = f_0 + \binom{k}{1}\Delta f_0 + \binom{k}{2}\Delta^2 f_0 + \binom{k}{3}\Delta^3 f_0$$

(i) For interpolation, i.e. for $x = 23.4$

$$k = \frac{x - x_0}{h} = \frac{23.4 - 25}{1} = -1.6.$$

Hence,
$$\begin{aligned} f(x_0 + kh) &= f(25 + (-1.6) \times 1) \\ &= f(23.4) \\ &= 154.00 + (-1.6)(11.75) \\ &\quad + \frac{(-1.6+1)(-1.6)}{2!}(0.50) \\ &= 135.44 \end{aligned}$$

(ii) For extrapolation, i.e. for $x = 25.4$:

$$k = \frac{25.4 - 25}{1} = 0.4$$

Hence, $f[25 + (0.4) \times 1]$

$$\begin{aligned} &= 154.00 + (0.4)(11.75) \\ &\quad + \frac{(0.4+1)(0.4)}{2!}(0.50) \\ &= 158.84 \end{aligned}$$

Example 7. Compute $f(0.9)$ by Newton-Gregory backward differences interpolation formula from the data:

x:	0	0.2	0.4	0.6	0.8	1.0
f:	1.00	1.16	3.56	13.96	41.96	101.00

Solution:

x_i	f_i	∇f_i	$\nabla^2 f_i$	$\nabla^3 f_i$	$\nabla^4 f_i$	$\nabla^5 f_i$
0	1.00					
		0.16				
0.2	1.16		2.24			
		2.40		5.76		
0.4	3.56		8.00		4.84	
		10.40		9.60		0
0.6	13.96		17.60		4.84	
		28.00		13.44		
0.8	41.96		31.04			
		59.04				
1.0	101.00					

Here $x_n = 1.0$, $h = 0.2$, $x = 0.9$ and $k = \dfrac{x_n - x}{h} = 0.5$

$$\text{Hence, } f(0.9) = f_n - \binom{0.5}{1}\nabla f_n + \binom{0.5}{2}\nabla^2 f_n - \binom{0.5}{3}\nabla^2 f_n + \binom{0.5}{4}\nabla^4 f_n$$

$$= 101 - (0.5)\,(59.04) + \frac{(0.5)\,(-0.5)}{2}(31.04)$$

$$- \frac{(0.5)\,(-0.5)\,(-1.5)}{6}(13.44)$$

$$+ \frac{(0.5)\,(-0.5)\,(-1.5)\,(-2.5)}{24}(4.84)$$

$$= 66.57094$$

Example 8. Find the interpolating polynomial for the data

x:	–1	0	1	2
f:	4	2	2	4

Solution:

x_i	f_i	Δf_i	$\Delta^2 f_i$	$\Delta^3 f_i$
–1	4			
		–2		
0	2		2	
		0		0
1	2		2	
		2		
2	4			

Here $x_0 = -1$, $h = 1$, $k = x + 1$.
Hence,

$$f(x) = f_0 + \binom{x+1}{1}\Delta f_0 + \binom{x+1}{2}\Delta^2 f_0$$

$$= 4 + (x + 1)\,(-2) + \frac{(x+1)\,x}{2}2$$

$$= x^2 - x + 2.$$

EXERCISES

1. Estimate $f(0.3)$ given

x:	0.1	0.5	0.9	
f:	0.09983	0.47943	0.78333	(Ans: 0.71445)

2. Find $f(0.73)$ from the data

x:	0.4	0.6	0.8	1.0
f:	0.423	0.684	1.030	1.557

(Ans: 0.893)

3. Compute $f(0.158)$ given

x:	0.125	0.250	0.375	0.500
f:	0.79168	0.77334	0.74371	0.70413

(Ans: 0.78801)

4. Find $f\left(\frac{5}{2}\right)$ given

x:	–1	0	1	2
f:	1	1	1	–5

$\left(\text{Ans: } \frac{-7}{8}\right)$

5. Compute $f(0.5)$ for the data

x:	0	1	2	3	4
f:	1.000	2.718	7.389	20.086	54.598

(Ans: 1.467)

6. Find $f(3.17)$ from the table

x:	3.1	3.2	3.3	3.4	3.5
f:	0	0.6	1.0	1.2	1.3

(Ans: 0.438)

7. Find $f(0.636)$ from the data

x:	0.375	0.500	0.625	0.750
f:	0.74371	0.70413	0.65632	0.60228

(Ans: 0.65178)

8. Find the interpolating polynomial for the data.

x:	1	2	3	4	5
f:	11	18	37	74	135

(Ans: $x^3 + 10$)

9. Find the interpolating polynomial for the data.

x:	0	1	2	3	4	5
f:	1	3	19	85	261	631

(Ans: $x^4 + x + 1$).

10. Derive Newton-Gregory backward differences formula from the Newton's divided difference formula.

C. Central Difference Interpolation Formulae

Upto now we have discussed Newton's forward and backward interpolation formulae which are normally used for interpolation near the beginning and the end of the table respectively. Now, we shall develop central differences formulae which are ideally used for interpolation near the middle of the table.

We discuss below various central differences interpolation formulae.

Central differences are defined by the operator 'δ' as follows:

$$\delta y_{1/2} = y_1 - y_0;\ \delta y_{\frac{3}{2}} = y_2 - y_1,\ \ldots,\ \delta y_{n-\frac{1}{2}} = y_n - y_{n-1}$$

Higher order central differences are

$$\delta^2 y_1 = \delta y_{\frac{3}{2}} - \delta y_{\frac{1}{2}},\ \delta^2 y_2 = \delta y_{\frac{5}{2}} - \delta y_{\frac{3}{2}},\ \text{etc.}$$

It should be noted that the subscript of δy for any difference is the average of the subscripts of the two members of the difference. Differences on the same horizontal line have the same suffix. Odd order differences are known only for half values of the suffix (having fractional suffix) while even order differences for only integral values of the suffix. Observe that $\Delta y_0 = y_1 - y_0 = \nabla y_1 = \delta y_{\frac{1}{2}}$. Also $\Delta^3 y_2 = \nabla^3 y_5 = \delta^3 y_{\frac{7}{2}}$, etc. i.e. the same numbers occur in the same positions but identified in different ways.

With the above convention, the central differences table can be written as (we note that the set of values are written on alternative lines):

x	$f(x)$	δ	δ^2	δ^3	δ^4	δ^5
x_{-2}	y_{-2}					
		δy_{-2}				
			$\delta^2 y_{-2}$			
x_{-1}	y_{-1}					
		δy_{-1}		$\delta^3 y_{-2}$		
x_0	y_0	$\cdots$	$\delta^2 y_{-1}$	$\cdots$	$\delta^4 y_{-2}$	$\cdots$
		$\ddots\,\delta y_0$	$\delta^3 y_{-1}$			$\delta^5 y_{-2}$
x_1	y_1				$\delta^4 y_{-1}$	
		δy_1	$\ddots\,\delta^2 y_0$			
x_2	y_2			$\ddots\,\delta^3 y_0$	$\ddots$	
		δy_2	$\delta^2 y_1$			
x_3	y_3					

The various central difference values are computed as in the case of finite differences table as follows:

FIRST ORDER DIFFERENCES	$\delta y_0 = y_1 - y_0$ $\delta y_{-1} = y_0 - y_{-1}$ $\delta y_{-2} = y_{-1} - y_{-2}$
SECOND ORDER DIFFERENCES	$\delta^2 y_0 = \delta y_1 - \delta y_0 = y_2 - 2y_1 + y_0$ $\delta^2 y_{-1} = \delta y_0 - \delta y_{-1} = y_1 - 2y_0 + y_{-1}$ $\delta^2 y_{-2} = \delta y_{-1} - \delta y_{-2} = y_0 - 2y_{-1} + y_{-2}$
THIRD ORDER DIFFERENCES	$\delta^3 y_0 = \delta^2 y_1 - \delta^2 y_0 = y_3 - 3y_2 + 3y_1 - y_0$ $\delta^3 y_{-1} = \delta^2 y_0 - \delta^2 y_{-1} = y_2 - 3y_1 + 3y_0 - y\ -1$ $\delta^3 y_{-2} = \delta^2 y_{-1} - \delta^2 y_{-2} = y_1 - 3y_0 + 3y_{-1} - y_{-2}$ $\vdots$

Similarly, higher order differences can be formed.
(For example: $\delta^4 y_{-1} = \delta^3 y_0 - \delta^3 y_{-1} = y_3 - 4y_2 + 6y_1 - 4y_0 + y_{-1}$)

Interpolation formulae of Gauss, Stirling and Bessel, can be derived by rearranging the central differences table with the initial values x_0 and y_0 placed at the middle of the table as shown below:

$x_i = x_0 + ih$, for $i = 0, \pm 1, \pm 2, \ldots$
$y_i = f(x_i)$ and $\Delta y_i = y_{i+1} - y_i$, $\Delta^2_{yi} = \Delta y_{i+1} - \Delta y_i$, etc.

Gaussian interpolation formulae

Let $x_{-n}, x_{-(n-1)}, \ldots, x_{-1}, x_0, x_1, \ldots x_{n-1}, x_n$ be $(2n + 1)$ equally spaced points, where

$$\Delta x_i = x_{i+1} - x_i = h = \text{constant}$$

with $i = -n, -(n-1), \ldots, -1, 0, 1, \ldots, (n-1), n$. Let $P(x)$ be a polynomial of degree $2n$, given by $P(x) = a_0 + a_1(x - x_0) + \ldots + a_{2n}(x - x_{-(n-1)}) \ldots (x - x_{-1})(x - x_0) \ldots, \ldots (x - x_{n-1})(x - x_n)$.

If $P(x_i) = y_i = f(x_i)$ for $i = 0, \pm 1, \pm 2, \ldots \pm n$, then $a_0 = y_0$, $a_1 = \dfrac{\Delta y_0}{h}$,

$$a_2 = \frac{\Delta^2 y_{-1}}{2!h^2}, \ldots, a_{2n} = \frac{\Delta^{2n} y_{-n}}{(2n)!h^{2n}}.$$

Introducing $k = \dfrac{x - x_0}{h}$, we get Gauss' (forward) first interpolation formula

$$P(x) = y_0 + k\Delta y_0 + \frac{k(k-1)}{2!}\Delta^2 y_0 + \frac{(k+1)k(k-1)}{3!}\Delta^2 y_{-1} + \frac{(k+1)k(k-1)(k-2)}{4!}\Delta^2 y_{-2} + \ldots + \frac{(k+n-1)\cdots(k-n)}{(2n)!}\Delta^{2n} y_{-n} \quad (1)$$

Similarly, Gaussian (backward) second interpolation formula is

$$P(x) = y_0 + k\Delta y_{-1} + \frac{(k+1)k}{2!}\Delta^2 y_{-1} + \frac{(k+1)(k-1)}{3!}\Delta^3 y_{-2}$$

$$+ \frac{(k+2)(k+1)k(k-1)}{4!}\Delta^4 y_{-2} + \dots$$

$$+ \frac{(k+n)(k+n-1)\cdots(k-n+1)}{(2n)!}\Delta^{2n} y_{-n} \quad (2)$$

Note: In order to shift the initial value from (x_0, y_0) to (x_1, y_1) replace k by $k-1$ and increase the indices in r.h.s. of (2) by (1). This yields another Gaussian interpolation formula

$$P(x) = y_1 + (k-1)\Delta y_0 + \frac{k(k-1)}{2!}\Delta^2 y_0 + \frac{k(k-1)(k-2)}{3!}\Delta^3 y_{-1}$$

$$+ \frac{(k+1)k(k-1)(k-2)}{4!}\Delta^4 y_{-1} + \dots$$

$$+ \frac{(k+n-1)\cdots(k-n)}{(2n)!}\Delta^{2n} y_{-(n-1)} \quad (3)$$

Stirling's interpolation formula

It is the arithmetic mean of formulas (1) and (2)

$$P(x) = y_0 + k\frac{\Delta y_{-1} + \Delta y_0}{2} + \frac{k^2}{2}\Delta^2 y_{-1} + \frac{k(k^2-1)}{3!}\frac{\Delta^3 y_{-2} + \Delta^3 y_{-1}}{2}$$

$$+ \frac{k^2(k^2-1)}{4!}\Delta^4 y_{-2}$$

$$+ \frac{k(k^2-1)(k^2-2^2)}{5!}\frac{\Delta^5 y_{-3} + \Delta^5 y_{-2}}{2} + \dots \quad (4)$$

Bessel's interpolation formula

It is the arithmetic mean of (1) and (3).

$$P(x) = y_0 + k\Delta y_0 + \frac{k(k-1)}{2}\frac{\Delta^2 y - 1 + \Delta^2 y_0}{2} + \frac{\left(k - \frac{1}{2}\right)k(k-1)}{3!}\Delta^3 y_{-1}$$

$$+ \frac{k(k-1)(k+1)(k-2)}{4!}\frac{\Delta^4 y_{-2} + \Delta^4 y_{-1}}{2} + \dots$$

Note: We use Stirling's formula for $|k| \leq 0.25$ and Bessel's formula for $(0.25 \leq k \leq 0.75)$ Stirling's formula consists of the even differences along the central line and mean of the differences just above and below the central line.

$$\begin{array}{ccccccc} & \Delta y_{-1} & & \Delta^3 y_{-2} & & \Delta^5 y_{-3} & \\ -y_0 & --- & \Delta^2 y_{-1} & --- & \Delta^4 y_{-2} & --- & \Delta^6 y_{-3} - \text{central line.} \\ & \Delta y_0 & & \Delta^5 y_{-1} & & \Delta^5 y_{-2} & \end{array}$$

Bessel's formula consists of means of even differences along the central line and odd differences below the central line

$$\begin{array}{ccccccc} y_0 & --- & \Delta^2 y_{-1} & --- & \Delta^4 y_{-2} & --- & \Delta^6 y_{-3} -- \text{central line} \\ & \Delta y_0 & & \Delta^3 y_{-1} & & \Delta^5 y_{-2} & \\ & & \Delta^2 y_0 & & \Delta^4 y_{-1} & & \Delta^6 y_{-2} \end{array}$$

Example 9. With the help of Stirling's formula, compute $f(1.22)$ from the following data:

x:	1.0	1.1	1.2	1.3	1.4
$f(x)$:	0.841	0.891	0.932	0.963	0.985

***Solution*:** Here, let origin be taken at $x_0 = 1.2$, also $h = 0.1$, then

$$k = \frac{x - x_0}{h} = \frac{1.22 - 1.2}{0.1} = 0.2$$

If $|k| \leq \frac{1}{4}$, we use Stirling's formula.

Since $k = 0$: 2 lies in between $-\frac{1}{4}$ and $\frac{1}{4}$, Stirling's formula is

$$y = y_0 + k\left(\frac{\Delta y_{-1} + \Delta y_0}{2}\right) + \frac{k^2}{2!}\Delta^2 y_{-1} + \frac{k(k^2-1)}{3!}\left(\frac{\Delta^3 y_{-2} + \Delta^3 y_{-1}}{2}\right)$$

$$+ \frac{k^2(k^2-1)}{4!}\Delta^4 y_{-2} + \ldots.$$

Substituting the values from the table,

$$y(1.22) = 0.932 + 0.2\ \frac{(0.041 + 0.031)}{2} + \frac{(0.2)^2}{2!}(-0.01)$$

$$+ \frac{(0.2)\left[(0.2)^2 - 1\right]}{6}\ \frac{(-0.001 + 0.001)}{2} + \frac{(0.2)^2\left[(0.2)^2 - 1\right]}{24}(0.002)$$

$$y(1.22) = 0.9389968.$$

x	k	y	Δy	$\Delta^2 y$	$\Delta^3 y$	$\Delta^4 y$
1.0	−2	$y_{-2} = 0.841$				
			$\Delta y_{-2} = 0.05$			
1.1	−1	$y_{-1} = 0.891$		$\Delta^2 y_{-2} = -0.009$		
			$\Delta y_{-1} = 0.041$		$\Delta^3 y_{-2} = -0.001$	
1.2	0	$y_0 = 0.932$		$\Delta^2 y_{-1} = -0.01$		$\Delta^4 y_{-2} = 0.002$
			$\Delta y_0 = 0.031$		$\Delta^3 y_{-1} = 0.001$	
1.3	1	$y_1 = 0.963$		$\Delta^2 y_0 = -0.009$		
			$\Delta y_1 = 0.022$			
1.4	2	$y_2 = 0.985$				

Example 10. Find the value of $f(x)$ at $x = 1.95$ using Bessel's formula for the data

x:	1.7	1.8	1.9	2.0	2.1	2.2	2.3
$f(x)$:	2.979	3.144	3.283	3.391	3.463	3.997	4.491

***Solution*:** Here, the origin at $x_0 = 2.0$, given $h = 0.1$ at $k = \dfrac{x - x_0}{h} = \dfrac{1.95 - 2.0}{0.1}$

$= -0.5$, since $k = -0.5$ lies between $-\dfrac{1}{4}$ and $\dfrac{3}{4}$, Bessel's formula is applicable,

Preparation of the central differences table

x	k	y	Δy	$\Delta^2 y$	$\Delta^3 y$	$\Delta^4 y$	$\Delta^5 y$
1.7	–3	2.979					
			0.165				
1.8	–2	3.144		−0.026			
			−0.0139		−0.005		
1.9	–1	3.283		−0.031		0	
			−0.108		−0.005		0.503
2.0	0	3.391		−0.036		0.503	
			+0.072		0.498		1.503
2.1	1	3.463		0.462		−1.0	
			0.53		−0.502		
2.2	2	3.997		−0.94			
			−0.494				
2.3	3	4.491					

The Bessel's formula is

$$y = y_0 + k\Delta y_0 + \frac{k(k-1)}{2!}\left(\frac{\Delta^2 y_{-1} + \Delta^2 y_0}{2}\right) + \frac{\left(k - \frac{1}{2}\right)k(k-1)}{3!}\Delta^3 y_{-1}$$

$$+ \frac{(k+1)(k)(k-1)(k-2)}{4!}\left(\frac{\Delta^4 y_{-2} + \Delta^4 y_{-1}}{2}\right)$$

Substituting the values from the table (underlined)

$$y(1.95) = 3.391 + (-0.5)\ (0.072)$$

$$+ \frac{(-0.5)(-0.5-1)}{2}\ \frac{(-0.036 + 0.462)}{2}$$

$$+ \frac{\left(-0.5 - \frac{1}{2}\right)(-0.5)(-0.5-1)}{6}(0.498)$$

$$+\frac{(-0.5+1)(0.5)(-0.5-1)(-0.5-2)}{24}\frac{(0.503-1.0)}{2}$$

$y(1.95) = 3.362917.$

Example 11. Using Bessel's formula prove that

$$y_{\frac{1}{2}} = \frac{1}{2}(y_0 + y_1) - \frac{1}{16}(\Delta^2 y_{-1} + \Delta^2 y_0)$$

assuming that third differences are constant.

Solution: Bessel's formula with differences constant is:

$$y = \frac{1}{2}(y_0 + y_1) + \left(k - \frac{1}{2}\right)\Delta y_0 + \frac{k(k-1)}{2!}\left[\frac{\Delta^2 y_{-1} + \Delta^2 y_0}{2}\right]$$

$$+\frac{\left(k - \frac{1}{2}\right)k(k-1)}{3!}\Delta^3 y_{-1}$$

where $$k = \frac{x - x_0}{h}$$

Let $x_0 = 0$ and $h = 1$ $\therefore$ $k = x$

Put $x = \frac{1}{2}$. Then

$$y_{\frac{1}{2}} = \frac{1}{2}(y_0 + y_1) - \frac{1}{16}(\Delta^2 y_{-1} + \Delta^2 y_0)$$

Example 12. Use Gauss' forward formula to find the value of $f(32)$ with the help of the data:

x:	25	30	35	40	45
$f(x)$:	2734.4	2686.6	2412.4	2341.3	2138.8

***Solution*:** Let $x_0 = 35$, $k = 5$ and $x = 32$

$$\therefore \quad k = \frac{x - x_0}{h} = \frac{32 - 35}{5} = \frac{-3}{5} = -0.6$$

The central differences table is

k	x	$y = f(x)$	Δy	$\Delta^2 y$	$\Delta^3 y$	$\Delta^4 y$
−2	25	2734.4				
			−47.8			
−1	30	2686.6		−226.4		
			−274.2		429.5	
0	35	2412.4		203.1		−764.0
			−71.1		−334.5	
1	40	2341.3		−131.4		
			−202.5			
2	45	2138.8				

Gauss' forward formula is

$$y = y_0 + k\Delta y_0 + \frac{k(k-1)}{2!}\Delta^2 y_{-1} + \frac{k(k^2-1)}{3!}\Delta^3 y_{-1} + \frac{k(k^2-1)(k-2)}{4!}\Delta^4 y_{-2}$$

$$\therefore \quad y = 2412.4 - 0.6(-71.1) + \frac{(-0.6)(-1.6)}{2!}(203.1)$$

$$+ \frac{(-0.6)(-0.64)}{6}(-334.5) + \frac{(-0.6)(-0.64)(-2.6)}{24}(-764)$$

$$= 2412.4 + 42.66 + 97.488 - 21.408 + 31.7824$$

$$= 2562.9$$

Example 13. Using Gauss' backward formula, obtain $f(25)$ given that $f(14) = 1.1461$, $f(18) = 1.2553$, $f(22) = 1.3424$, $f(26) = 1.4150$, $f(30) = 1.4771$.

***Solution*:** Let $x_0 = 22$, $h = 4$ and $x = 25$

$$\therefore \quad k = \frac{x - x_0}{h} = \frac{25 - 22}{4} = \frac{3}{4} = 0.75$$

k	x	$y = f(x)$	Δy	$\Delta^2 y$	$\Delta^3 y$	$\Delta^4 y$
−2	14					
		1.1461				
−1	18		0.1092			
		1.2553		−0.0221		
0	22		0.0871		0.0076	
		1.3424		−0.0145		−0.0036
1	26		0.0726		0.0040	
		1.4150		−0.0105		
2	30		0.0621			
		1.4771				

Gauss' backward formula is

$$y = y_0 + k\Delta y_{-1} + \frac{(k+1)k}{2!}\Delta^2 y_{-1} + \frac{(k^2-1)k}{3!}\Delta^3 y_{-2}$$

$$\therefore \qquad + \frac{(k+2)(k^2-1)k}{4!}\Delta^3 y_{-2}$$

$$f(25) = 1.3424 + (0.75)(0.0871) + \frac{(1.75)(0.75)}{2}(-0.0145)$$

$$+ \frac{(-0.4375)(0.75)}{6}(0.0076)$$

$$+ \frac{(2.75)(-0.4375)(0.75)}{24}(-0.0036)$$

$$= 1.3424 + 0.0653 - 0.0095 - 0.0004 + 0.0001$$

$$= 1.3979$$

EXERCISES

1. Using Gauss' forward interpolation formula, find $f(25)$ from the data:

   $f(20) = 14, f(24) = 32, f(28) = 35, f(32) = 40$

2. Using Gauss' forward formula, evaluate y_{30} given that:

   $y_{21} = 18.4708,\quad y_{25} = 17.8144,\quad y_{29} = 17.1070,$
   $y_{33} = 16.3432,\quad y_{37} = 15.5154$

3. Using Gauss' forward formula, find $f(41)$ given that:

x :	30	35	40	45	50
$f(x)$:	3678.2	2995.1	2400.1	1876.2	1416.3

4. Using Gauss' forward formula, find $f(53)$ given that:

x :	40	50	55	60	65	70
$f(x)$:	1.6021	1.6990	1.7404	1.7781	1.8129	1.8451

5. Using Gauss' backward formula, find $f(18)$ given that:

x :	4	7	10	13	16	19
$f(x)$:	0.6020	0.8451	1.0000	1.1139	1.2041	1.2788

6. Using Gauss' forward formula, find y_{32} given that

x :	15°	20°	25°	30°	35°	40°
$f(x)$:	0.2618	0.3491	0.4363	0.5236	0.6109	0.6981

7. Using Gauss' forward formula, find y_{13} given that:

x :	4	6	8	10	12	14	16
y_x :	4.250	6.167	8.125	10.100	12.083	14.071	16.062

8. Use Stirling's formula to find y_{28} given that:

x :	20	25	30	35	40
y_x :	49225	48316	47236	45926	44306

9. Use Stirling's formula to find $f(1.22)$ from the data:

x :	1.0	1.1	1.2	1.3	1.4	1.5	1.6
$f(x)$:	0.8415	0.8912	0.9320	0.9636	0.9854	0.9975	0.9996

10. Use Stirling's formula to find $f(3.32)$ from the data:

x :	3.0	3.1	3.2	3.3	3.4	3.5	3.6	3.7
$f(x)$:	20.086	22.198	24.532	27.113	29.964	33.115	36.598	40.447

11. Use Bessel's formula to find $f(1.22)$ from the table:

x :	0	0.5	1.0	1.5	2.0	2.5	3.0
$f(x)$:	0	0.1915	0.3413	0.4332	0.4772	0.4938	0.4986

12. Use Bessel's formula to find $y_{0.41}$ from the data:

x :	0	0.1	0.2	0.3	0.4	0.5	0.6	0.7	0.8
y_x :	1.000	0.995	0.980	0.955	0.921	0.878	0.825	0.765	0.697

13. Apply Bessel's formula to find $y_{2.73}$ given that :

$y_{2.5} = 0.4938,\quad y_{2.6} = 0.4953,\quad y_{2.7} = 0.4965$
$y_{2.8} = 0.4974,\quad y_{2.9} = 0.4981,\quad y_{3.0} = 0.4987$

14. Find the value of y if $x = 3.75$ using Bessel's formula given:

x :	2.5	3.0	3.5	4.0	4.5	5.0
y :	24.145	22.043	20.225	18.644	17.262	16.047

Answers:

1. 33.49
2. 16.9216
3. 2289.99
4. 1.7243
5. 1.2553
6. 0.5585
7. 13.077
8. 47691.83
9. 0.9390
10. 27.660
11. 0.3887
12. 0.917
13. 0.4968
14. 19.407

LAGRANGE'S INTERPOLATION FORMULA

(Usually for both equal and unequal intervals)

Let $x_0, x_1, \ldots x_n$ be distinct values and a function $f(x)$ satisfy the condition $f(x_i) = f_i$. The interpolating polynomial is given by

$$P(x) = \sum_{i=0}^{n} f_i L_i(x) \tag{1}$$

where $L_i(x)$ are polynomials of degree n such that

$$L_i(x_j) = \delta_{ij} = 0 \text{ if } j \neq i$$
$$= 1 \text{ if } j = i \tag{2}$$

Indeed $P(x_j) = \sum_{i=0}^{n} f_i L_i(x_j) = \sum_{i=0}^{n} f_i \delta_{ij} = f_j$.

To find $L_i(x)$, called the Lagrange coefficient function, we first note that $L_i(x)$ vanishes at x_{ij}, $j \neq i$.

Hence, $L_i(x) = c_i(x - x_0) \ldots (x - x_{i-1})(x - x_{j+1}) \ldots (x - x_n)$.

Since $L_i(x_i) = 1$, we get

$$c_i = \frac{1}{(x_i - x_0) \cdots (x_i - x_{i-1})(x_i - x_{i+1}) \cdots (x_i - x_n)}$$

Hence, $$L_i(x) = \frac{(x - x_0) \cdots (x - x_{i-1})(x - x_{i+1}) \cdots (x - x_n)}{(x_i - x_0) \cdots (x_i - x_{i-1})(x_i - x_{i+1}) \cdots (x_i - x_n)} \tag{3}$$

The function

$P(x) = \sum_{i=0}^{n} f_i L_i(x)$ is called the Lagrange's interpolating polynomial.

Example 14. Use Lagrange's interpolation formula to fit a polynomial to the following data:

x:	–1	0	1	3
f:	2	1	0	–1

Solution: Here $x_0 = -1$; $x_1 = 0$, $x_2 = 1$, $x_3 = 3$

$$L_0(x) = \frac{(x-0)(x-1)(x-3)}{(-1-0)(-1-1)(-1-3)} = -\frac{1}{8}x(x-1)(x-3)$$

$$L_1(x) = \frac{(x-0)(x-1)(x-3)}{(-1-0)(-1-1)(-1-3)} = \frac{1}{3}(x^2-1)(x-3)$$

$L_2(x)$ is not needed since $f_2 = 0$

$$L_3(x) = \frac{(x+1)(x-0)(x-1)}{(3+1)(3-0)(3-1)} = \frac{1}{24}x(x^2-1)$$

Hence, $P(x) = 2L_0 + L_1 - L_3 = \dfrac{1}{24}(x^3 - 25x + 24)$.

Example 15. Compute $f(4)$ from the data:

x :	1	2	3	5
f :	0	7	26	124

Solution:

$$f(4) = P(4) = (4-1)(4-2)(4-3)(4-5)$$

$$\left[\frac{7}{(4-2)(2-1)(2-3)(2-5)} + \frac{26}{(4-3)(3-1)(3-2)(3-5)} + \frac{124}{(4-5)(5-1)(5-2)(5-3)}\right]$$

$$= 63.$$

Example 16. The function $f(x) = e^x$ is interpolated at the points $x = -1, 0, 1$ and $\sqrt{e}$ is computed. Find the bound for the error and compare it with the true error.

x :	-1	0	1
y :	$\frac{1}{e}$	1	e

Solution:

$$L_0(x) = \frac{(x-0)(x-1)}{(-1-0)(-1-1)} = \frac{1}{2}(x^2 - x)$$

$$L_1(x) = \frac{(x+1)(x-1)}{(0+1)(1-0)} = 1 - x^2$$

$$L_2(x) = \frac{(x+1)(x-0)}{(1+1)(1-0)} = \frac{1}{2}(x^2 + x)$$

Hence,

$$P(x) = \frac{1}{e}L_0 + L_1 + eL_2$$

$$= \frac{1}{2e}(x^2 - x) + 1 - x^2 + \frac{e}{2}(x^2 + x)$$

$\sqrt{e}$ is approximated as $P\left(\frac{1}{2}\right) = -\frac{1}{8e} + \frac{3}{4} + \frac{3e}{8}$

$= 1.723$

Error: $\frac{W\left(\frac{1}{2}\right)e^{\xi}}{3!} = \frac{e^{\xi}}{32} < \frac{e}{32} = 0.085$

True error $= P\left(\frac{1}{2}\right) - \sqrt{e}$

$= 1.723 - 1.649$

$= 0.074$

Example 17. Compute $f(5)$ and $f(6)$ from the data:

x :	1	2	3	7
$f(x)$:	2	4	8	128

explain why the results differ from those obtained by $f(x) = 2^x$.

Solution: $x_0 = 1, x_1 = 2, x_2 = 3, x_3 = 7;$

$f(x_0) = 2, f(x_1) = 4, f(x_2) = 8, f(x_3) = 128.$

$$f(x) = \frac{(x-2)(x-3)(x-7)}{(1-2)(1-3)(1-7)}f(x_0) + \frac{(x-1)(x-3)(x-7)}{(2-1)(2-3)(2-7)}f(x_1)$$

$$+ \frac{(x-1)(x-2)(x-7)}{(3-1)(3-2)(3-7)}f(x_2) + \frac{(x-1)(x-2)(x-3)}{(7-1)(7-2)(7-3)}f(x_3)$$

$$= \frac{(x-2)(x-3)(x-7)}{-12}(2) + \frac{(x-1)(x-3)(x-7)}{5}(4)$$

$$+ \frac{(x-1)(x-2)(x-7)}{-8}(8) + \frac{(x-1)(x-2)(x-3)}{120}(128)$$

$$\therefore \quad f(5) = \frac{3.2.(-2)}{-12}(2) + \frac{4.2.(-2)}{5}(4) + \frac{4.3.(-2)}{-8}(8) + \frac{4.3.2}{120}(128)$$

$$= 2 - 12.8 + 24 + 25.6 = 38.8$$

$$f(6) = \frac{4.3.(-1)}{-12}(2) + \frac{5.3.(-1)}{5}(4) + \frac{5.4.(-1)}{-8}(8) + \frac{5.4.3}{120}(128)$$

$$= 2 - 12 + 20 + 64 = 74$$

But, actual values of $f(5)$ and $f(6)$ are

$$f(5) = 2^5 = 32$$
$$f(6) = 2^6 = 64$$

The difference in values of $f(5)$ and $f(6)$ are due to the assumption of $f(x)$ as a polynomial, when it is an exponential function of the from: 2^x.

EXERCISES

1. Use Lagrange's interpolation formula to solve the following problems

(i) Given (1, 2) (3, 5), (7,12), (13, 20) find (4, ?)

$$\left(\text{Hint: } -\frac{27}{144}(2) + \frac{81}{80}(5) + \frac{27}{144}(12) - \frac{9}{720}(20)\right)$$

(Ans: 6.6875)

(ii) Fit polynomial of third degree and find y (0.2) from the following data:

x_i :	0	0.1	0.3	0.5
y_i :	−0.5	0	0.2	1

$$\left(\text{Ans: } -0.15004,\ \frac{125}{3}x^3 - 30x^2 + \frac{73}{12}x - 0.5\right)$$

(iii) Fit a polynomial of 3rd degree

x :	0	1	3	4
y :	−12	0	6	12

(Ans: $x^3 - 7x^2 + 18x - 12$)

(iv) Determine $P(v = 21)$ given the following:

v :	10	15	22.5	33.75	50.625	75.937
P :	0.3	0.675	1.519	3.417	7.689	17.3

(Ans: 1.323)

(v) Compute $f(9)$ from the following data:

x :	5	7	11	13	17
$f(x)$:	150	392	1452	2366	5202

$$\left(\text{Ans: } -\frac{50}{3} + \frac{3136}{15} + \frac{3872}{3} - \frac{2366}{3} + \frac{578}{5} = 810\right)$$

(vi) Compute $f(27)$ from the data:

x:	14	17	31	35
$f(x)$:	68.7	64.0	44.0	39.1

(Ans: 49.3)

(vii) Given (300, 2.4771), (304, 2.4829), (305, 2.4843), (307, 2.4871), find (301,2).

(Ans: 2.4786)

INVERSE INTERPOLATION USING LAGRANGE'S INTERPOLATION FORMULA

So far, given a set of values of x and y, we were finding the values of y corresponding to some $x = x_i$ (which is not given in the table). In the inverse interpolation, for a given value of y, the corresponding value of x is to be determined. For example, to find the value of the root x for which the function $y = f(x)$ become zero is an inverse interpolation problem. Inverse interpolation problem is similar to direct interpolation since the roles of x and y are interchanged. Thus, Lagrange's interpolation formula for inverse interpolation takes the form given by

$$x = \sum_{i=0}^{n} \frac{(y-y_1)(y-y_2)\cdots(y-y_{i-1})(y-y_{i+1})\cdots(y-y_n)}{(y_i-y_1)(y_i-y_2)\cdots(y_i-y_{i-1})(y_i-y_{i+1})\cdots(y_i-y_n)} x_i$$

$$x = \frac{(y-y_1)(y-y_2)\cdots(y-y_n)}{(y_0-y_1)(y_0-y_2)\cdots(y_0-y_n)} x_0 + \frac{(y-y_0)(y-y_2)\cdots(y-y_n)}{(y_1-y_0)(y_1-y_2)\cdots(y_1-y_n)} x_1 + \ldots$$

$$+ \frac{(y-y_1)(y-y_2)\cdots(y-y_{i-1})(y-y_{i+1})\cdots(y-y_n)}{(y_i-y_1)(y_i-y_2)\cdots(y_i-y_{i-1})(y_i-y_{i+1})\cdots(y_i-y_n)} x_i \ldots$$

$$+ \frac{(y-y_0)(y-y_1)\cdots(y-y_{n-1})}{(y_n-y_0)(y_n-y_1)\cdots(y_n-y_{n-1})} x_n \qquad (1)$$

Example 18. Obtain a value of x when $y = 19$, by Lagrange's method, given that for $x = 0, 1, 2$, $y = 0, 1, 20$ respectively.

***Solution*:**

For a given $x_0 = 0$, $x_1 = 1$, $x_2 = 2$,

$y_0 = 0$, $y_1 = 1$, $y_2 = 20$

Lagrange's inverse interpolation formula gives

$$x = \frac{(y-y_1)(y-y_2)\cdots(y-y_n)}{(y_0-y_1)(y_0-y_2)\cdots(y_0-y_n)} \cdot x_0$$

$$+ \frac{(y-y_0)(y-y_2)\cdots(y-y_n)}{(y_1-y_0)(y_1-y_2)\cdots(y_1-y_n)} \cdot x_1 + \ldots$$

$$+ \frac{(y-y_0)(y-y_1)\cdots(y-y_{n-1})}{(y_n-y_0)(y_n-y_1)\cdots(y_n-y_{n-1})} \cdot x_n$$

i.e.,
$$x = \frac{(19-1)(19-20)}{(0-1)(0-20)} \times 0 + \frac{(19-0)(19-20)}{(1-0)(1-20)} \times 1$$
$$+ \frac{(19-0)(19-1)}{(20-0)(20-1)} \times 2$$
$$= 0 + 1 + 1.8 = 2.8$$

Example 19. Find the value of x when $y = 15$ from the given data;

x:	5	6	9	11
y:	12	13	14	16

Solution: Using Lagrange's inverse interpolation formula, we have

$$x = \frac{(15-13)(15-14)(15-16)}{(12-13)(12-14)(12-16)} \times 5$$
$$+ \frac{(15-12)(15-14)(15-16)}{(13-12)(13-14)(13-16)} \times 6$$
$$+ \frac{(15-12)(15-13)(15-16)}{(14-12)(14-13)(14-16)} \times 9$$
$$+ \frac{(15-12)(15-13)(15-14)}{(16-12)(16-13)(16-14)} \times 11$$
$$= \frac{5}{4} - 6 + \frac{27}{2} + \frac{11}{4} = \frac{46}{4} = 11.5$$

Example 20. Find the value of x, when $y = 8$ by inverse interpolation using Lagrange's formula

x:	–2	–1	1	2
y:	–7	2	0	11

Solution:

$$x = \frac{(y-2)(y-0)(y-11)}{(-7-2)(-7-0)(-7-11)}(-2)$$
$$+ \frac{(y+7)(y-0)(y-11)}{(2+7)(2-0)(2-11)}(-1) + \frac{(y+7)(y-2)(y-11)}{(0+7)(0-2)(0-11)}(1)$$
$$+ \frac{(y+7)(y-2)(y-0)}{(11+7)(11-2)(11-0)}(2)$$

put $y = 8$, then

$$x = \frac{6(8)(-3)}{(-9)(-7)(-18)}(-2) + \frac{(15)(8)(-3)}{(9)(2)(-9)}(-1)$$

$$+ \frac{(15)(6)(-3)}{7(-2)(-11)}(1) + \frac{(15)(6)8}{(18)(9)(11)}\,2$$

$$x = -\frac{8}{21} - \frac{20}{9} - \frac{135}{77} + \frac{80}{99} + = -3.5483.$$

EXERCISES

(1) Determine x (7) from the data:

x :	1	3	4
y :	4	12	19

(2) Find x (12) using Lagrange's technique

x :	1.2	2.1	2.8	4.1	4.9	6.2
y :	4.2	6.8	9.8	13.4	15.5	19.6

(Ans: 3.55)

(3) Find x corresponding to $y = 10$

x :	10	15	17	20
y :	3	7	11	17

(Ans: 16.641)

(4) Use Lagrange's formula to find the number whose logarithm is 0.30500 having given log 1 = 0, log 2 = 0.30103, log 3 = 0.47712 and log 4 = 0.60206. (Ans: 2.018)

(5) Given $f(0) = 16.35 = y_0$

$f(5) = 14.88 = y_1$

$f(10) = 13.5 = y_2$

$f(15) = 12.46 = y_3$

find x for $y = 14.00$ (Ans: = 8.34)

(6) Apply Lagrange's formula inversely to find a root of the equation $f(x) = 0$ when $f(30) = -30$, $f(34) = -13$, $f(38) = 3$, $f(42) = 18$.

(Ans: 37.230378)

(7) Let $y(x) = \frac{2}{\sqrt{11}}\int_0^x e^{-t^2}\,dt$ be the probability integral. Given

x :	0.46	0.47	0.48	0.49
$y(x)$:	0.4846555	0.4937452	0.5027498	0.5116683

Determine for what value of x, $y(x) = \frac{1}{2}$. (Ans: 0.476937)

NEWTON'S DIVIDED DIFFERENCE INTERPOLATION

Before we obtain formulae for interpolation when the data is at unequal intervals, we shall define divided difference and study its properties.

Suppose $f(x)$ be given at $(n+1)$ distinct points $x_0, x_1, \ldots x_n$. We define the first divided difference of $f(x)$ as

$$f(x_0, x_1) = \frac{f(x_1) - f(x_0)}{x_1 - x_0}$$

The second divided difference is defined as

$$f(x_0, x_1, x_2) = \frac{f(x_1, x_2) - f(x_0, x_1)}{x_1 - x_0}$$

and in general, the nth divided difference is defined by

$$f(x_0, x_1, x_2, \ldots, x_n) = \frac{f(x_1, \cdots x_n) - f(x_0, \cdots x_{n-1})}{x_n - x_0}$$

We observe that the divided differences are symmetric in all their arguments. From the defined equations, we have

$$f(x) = f(x_0) + (x - x_0) f(x, x_0)$$
$$f(x, x_0) = f(x_0, x_1) + (x - x_1) f(x, x_0, x_1)$$
$$f(x, x_0, x_1) = f(x_0, x_1, x_2) + (x - x_2) f(x, x_0, x_1, x_2)$$
$$\cdots \quad \cdots \quad \cdots \quad \cdots \quad \cdots \quad \cdots \quad \cdots$$
$$f(x, x_0, \ldots, x_{n-1}) = f(x_0, x_1, \ldots x_{n-1}) + (x - x_{n-1}) - f(x, x_0 \ldots x_n).$$

Multiplying the 2nd equation by $(x - x_0)$, the third equation by $(x - x_0)(x - x_1)$, and so on, and finally adding, we get the Newton's divided difference formula $f(x) = P(x) + E(x)$,

where $P(x) = f(x_0) + (x - x_0) f(x_0, x_1) + (x - x_0)(x - x_1) f(x_0, x_1, x_2) + \ldots$
$+ (x - x_0)(x - x_1) \ldots (x - x_{n-1}) f(x_0, x_1, \ldots, x_n)$

is the interpolating polynomial and

$E(x) = W(x) f(x, x_0, x_1, \ldots, x_n)$ is the error term.

By actual substitution, we find $P(x_i) = f(x_i)$. Comparing the error term $E(x)$ with that in Lagrange's interpolation formula, we get

$$f(x, x_0, \ldots, x_n) = \frac{f_{(\xi)}^{(n+1)}}{(n+1)!}$$

Example 21. Using Newton's divided difference interpolation, find the polynomial satisfying the data.

x :	−1	0	1	3
f :	2	1	0	−1

Solution: Divided differences table is

x	f	1st order	2nd order	3rd order
−1	2			
		−1		
0	1		0	
		−1		$\frac{1}{24}$
1	0		$\frac{1}{6}$	
		$\frac{-1}{2}$		
3	−1			

The first divided differences are

$$\frac{1-2}{0-(-1)} = -1, \quad \frac{0-1}{1-0} = -1, \quad \frac{-1-0}{3-1} = -\frac{1}{2}$$

The second divided differences are

$$\frac{-1-(-1)}{1-(-1)} = 0, \quad \frac{-\frac{1}{2}-(-1)}{3-0} = \frac{1}{6}$$

The third divided difference is $\dfrac{\frac{1}{6}-0}{3-(-1)} = \dfrac{1}{24}$

Hence, $f(x_0, x_1) = -1$, $f(x_0, x_1, x_2) = 0$, $f(x_0, x_1, x_2, x_3) = \dfrac{1}{24}$

and $$P(x) = 2 + (x + 1)(-1) + (x + 1)\,x(x - 1)\left(\frac{1}{24}\right)$$

$$= \frac{x^3 - 25x + 24}{24}$$

Example 22. Find $f(1)$ and $f(9)$ from the following data:

x:	-1	0	2	5	10
f:	-2	-1	7	124	999

Solution: The divided differences table is

x	f	1st order	2nd order	3rd order	4th order
–1	–2				
		1			
0	–1		1		
		4		1	
2	7		7		0
		39		1	
5	124		17		
		175			
10	999				

Hence,

$$f(1) = P(1) = -2 + (1 + 1)(1) + (1 + 1)(1 - 0)(1) + (1 + 1)(1 - 0)(1 - 2)(1) = 0$$

$$f(9) = P(9) = -2 + (9 + 1)(1) + (9 + 1)(9 - 0)(1) + (9 + 1)(9 - 0)(9 - 2)(1) = 728.$$

Note: It can be easily seen that computation of $P(x)$ for a given x involves a total of $n(n + 1)$ multiplications and divisions. Hence, this method is superior to the Lagrange's method which involves $2n(n + 1)$ multiplications and divisions.

EXERCISES

1. Show that

(i) $f(x_0, x_1) = f(x_1, 0) = \dfrac{f(x_0)}{x_0 - x_1} + \dfrac{f(x_1)}{x_1 - x_0}$

(ii) $f(x_0, x_1, x_2) = \dfrac{f(x_0)}{(x_0 - x_1)(x_0 - x_2)} + \dfrac{f(x_1)}{(x_1 - x_0)(x_1 - x_2)} + \dfrac{f(x_0)}{(x_2 - x_0)(x_3 - x_1)}$

2. Through the limiting process, show that

   (a) $f(x, x) = f'(x)$

   (b) $f(x, x, x) = \dfrac{f''(x)}{2!}$

   (c) The nth order divided difference,

$$f(x, x, x, \ldots x) = \frac{f_{(x)}^{(n)}}{n!}$$

3. Find the interpolating polynomials for the data

(a)

x :	1	2	3	4
f :	1	1	1	k

$\left(\text{Ans: } 1 + \frac{1}{6}(x-1)(x-2)(x-3)(k-1)\right)$

(b)

x :	1	2	3	4
	1	2	3	k

$\left(\text{Ans: } x + \frac{1}{6}(x-1)(x-2)(x-3)(k-4)\right)$

(c)

x :	–2	0	2	3
f :	0	1	$\frac{2}{5}$	–1

$\left(\text{Ans: } \frac{1}{30}(30 + 7x - 6x^2 - x^3)\right)$

4. Determine $f(1)$ from the data

x :	–1	0	2	5
f :	7	10	22	235

(Ans: 11)

5. Determine $f(3)$ from the table

x :	0	1	2	4	5	6
f :	1	14	15	5	6	9

(Ans: 8)

6. Find $f(1)$ if

x :	–4	–1	0	2	5
f :	1245	33	5	9	1335

(Ans: –5)

Chapter 7

Numerical Differentiation

INTRODUCTION

In this chapter, we consider differentiation of functions given in the form of tables for equally spaced abscissa. As the name 'numerical differentiation' suggests, here we study the technique of applying the power of differentiation to a set of data (x_i, y_i), $i = 0,1, \ldots, n$, given in numerical form about two quantities x and y. For this, it is essential to establish a relation $y = f(x)$, based on the given set of data, where $f(x)$ is a differentiable function based on the position of x, we have the following interpolation formulae to represent the function $f(x)$.

NEWTON'S FORWARD FORMULA

Let $x_i = x_0 + ih$, $i = 0,1,2, \ldots, n$, and $f(x_i) = f_i$,
Then for any real K,

$$f(x) = f(x_0 + ih) = E^k f_0 = (1 + \Delta)^k f_0 \tag{1}$$

which is the forward difference interpolation formula. Noting that $dx = hdk$, we get from (1), the derivative

$$f'(x) = \frac{1}{h}\frac{d}{dk}(1 + \Delta)^k f_0 \tag{2}$$

Similarly, we get

$$f''(x) = \frac{1}{h^2}\frac{d^2}{dk^2}(1 + \Delta)^k f_0 \tag{3}$$

or in general, the pth derivative

$$f^{(p)}(x) = \frac{1}{h^p}\frac{d^p}{dx^p}(1 + \Delta)^k f_0 \tag{4}$$

This formula can also be obtained as the following:

$$Ef_0 = f_1 = f(x_0 + h) = f_0 + \frac{h}{1!} f_0' + \frac{h^2}{2!} f_0'' + \ldots$$

$$= \left(1 + \frac{hD}{1!} + \frac{h^2 D^2}{2!} + \cdots\right) f_0 = e^{hD} f_0$$

Showing that

$$E = e^{hD} \text{ or } D = \frac{1}{h} \log (1 + \Delta) \qquad (5)$$

Now $$f^{(p)}(x) = D^p f_k = \frac{1}{h^p} [\log (1 + \Delta)]^p (1 + \Delta)^k f_0$$

$$= \frac{1}{h^p} \frac{d^p}{dk^p} (1 + \Delta)^k f_0$$

Using the well-known formula $\frac{d^p}{dx^p} a^x = a^x (\log a)^p$

NEWTON'S BACKWARD FORMULA

Using $x = x_n - kh$, we get $f(x) = f(x_n - kh) = E^{-k} f_n = (1 - \nabla)^k f_n$
Noting that $dx = -hdk$, we obtain from (6) the derivatives (6)

$$f'(x) = -\frac{1}{h} \frac{d}{dk} (1 - \nabla)^k f_n \qquad (7)$$

$$f''(x) = \frac{1}{h^2} \frac{d^2}{dk^2} (1 - \nabla)^k f_n \qquad (8)$$

and $$f^{(p)}(x) = \frac{(-1)^p}{h^p} \frac{d^p}{dx^p} (1 - \nabla)^k f_n \qquad (9)$$

Alternatively, since $D = \frac{1}{h} \log E = -\frac{1}{h} \log (1 - \nabla)$, we have

$$f^{(p)}(x) = D^p f_{n-k} = D^p E^{-k} f_n$$

$$= \frac{(-1)^p}{h^p} [\log (1 - \nabla)]^p (1 - \nabla)^k f_n$$

which is same as (9).

Expanding the r.h.s member of (7) and (9), we get

$$f'(x) = \frac{1}{h}\left[\nabla f_n - \frac{(2k-1)}{2}\nabla^2 f_n + \frac{(3k^2-6k+2)}{6}\nabla^3 f\right.$$

$$\left. - \frac{(2k^3-9k^2+11k-3)}{12}\nabla^4 f_n + \cdots\right] \tag{10}$$

$$f''(x) = \frac{1}{h^2}\left[\nabla^2 f_n - (k-1)\nabla^3 f_n + \frac{(6k^2-18k+11)}{12}\nabla^4 f_n + \cdots\right] \tag{11}$$

In particular for $k = 0$,

$$f'(x_n) = \frac{1}{h}\left[\nabla f_n + \frac{1}{2}\nabla^2 f_n + \frac{1}{3}\nabla^3 f_n + \cdots\right] \tag{12}$$

or $\quad f'(x_n) = -\frac{1}{h}\log(1 - \nabla) f_n$

$$f''(x_n) = \frac{1}{h^2}\left[\nabla^2 f_n + \nabla^3 f_n + \frac{11}{12}\nabla^4 f_n + \cdots\right] \tag{13}$$

Note: The formula from (6) to (13) can be easily written down from those with forward-differences (1) to (4), (6)–(9) by changing $D \to -D$, $\Delta \to -\nabla$ and $f_0 \to f_n$.

COMPUTATION OF DERIVATIVES

In actual computation of derivatives, we expand $(1 + \Delta)^k$ in (2)–(4) by binomial series and then perform differentiation.

Thus we get

$$f'(x) = \frac{1}{h}\left[\nabla f_0 + \frac{(2k-1)}{2}\nabla^2 f_0 + \frac{(3k^2-6k+2)}{6}\nabla^3 f_0\right.$$

$$\left. + \frac{(2k^3-9k^2+11k-3)}{12}\nabla^4 f_0 + \cdots\right] \tag{14}$$

$$f''(x) = \frac{1}{h^2}\left[\nabla^2 f_0 + (k-1)\nabla^3 f_0 + \frac{(6k^2-18k+11)}{12}\nabla^4 f_0 + \cdots\right] \tag{15}$$

In particular for $x = x_0$, i.e. $k = 0$

$$f'(x_0) = \frac{1}{h}\left[\nabla f_0 - \frac{1}{2}\nabla^2 f_0 + \frac{1}{3}\nabla^3 f_0 - \frac{1}{4}\nabla^4 f_0 + \cdots\right] \tag{16}$$

or $$f'(x_0) = \frac{1}{h}\log(1 + \Delta) f_0$$

$$f''(x_0) = \frac{1}{h^2}\left[\nabla^2 f_0 - \Delta^3 f_0 + \frac{11}{12}\Delta^4 f_0 - \cdots\right] \tag{17}$$

Notice that the Series (6)–(9) are infinite series. However with $(n + 1)$ entries in the table of values, only the power of Δ upto Δ^n are available for use and this causes truncation error in computing the derivatives.

Example 1. Find f' and f'' at $x = 1.0$, 1.2 and 2.9 from the following table:

x:	1.0	1.5	2.0	2.5	3.0
f:	27.00	106.75	324.00	783.75	1621.00

Solution: The finite differences table is

x	f	Δf	$\Delta^2 f$	$\Delta^3 f$	$\Delta^4 f$
1.0	27.00				
		79.75			
1.5	106.75		137.5	105	
		217.25			30
2.0	324.00		242.5	105	
		459.75			
2.5	783.75		377.5		
		837.25			
3.0	1621.00				

Hence $\Delta f_0 = 79.75$, $\Delta^2 f_0 = 137.5$, $\Delta^3 f_0 = 105$, $\Delta^4 f_0 = 30$.
Using (6), we get

$$f'(x) = 2\left[79.75 + \frac{(2k-1)}{2}137.5 + \frac{(3k^2 - 6k + 2)}{6}105 + \frac{(2k^3 - 9k^2 + 11k - 3)}{12}30\right]$$

Since $k = \dfrac{x - x_0}{h} = 2(x - 1)$, we have $k = 0$ for $x = 1$,
$k = 0.4$ for $x = 1.2$ and $k = 3.8$ for $x = 2.9$.
Using three values of k, we get $f'(1) = 77$, $f'(1.2) = 135.24$ $f'(2.9) = 1948.12$

Similarly (7) takes the form

$$f''(x) = 4\left[137.5 + (k-1)\,105 + \frac{(6k^2 - 18k + 11)}{12}30\right]$$

which gives $f''(1) = 240$, $f''(1.2) = 345.6$, $f''(2.9) = 2018.4$.

Example 2. Compute f' and f'' at

(a) $x = 16$
(b) $x = 15$
(c) $x = 24$
(d) $x = 25$

from the following table

x :	15	17	19	21	23	25
$f(x) = \sqrt{x}$:	3.873	4.123	4.359	4.583	4.796	5.0.

***Solution*:** The finite differences table is

x	$y = \sqrt{x}$	Δy	$\Delta^2 y$	$\Delta^3 y$	$\Delta^4 y$	$\Delta^5 y$
15	3.873					
		0.250				
17	4.123		−0.014			
		0.236		0.002		
19	4.359		−0.012		−0.001	
		0.224		0.001		0.002
21	4.583		−0.011		0.001	
		0.213		0.002		
23	4.796		−0.009			
		0.204				
25	5.000					

(a) To find f', f'' at $x = 16, 15$, we use the forward difference results.

$$f'(x) = \frac{1}{h}\left[\Delta f_0 + \frac{2k-1}{2}\Delta^2 f_0 + \frac{3k^2 - 6k + 2}{6}\Delta^3 f_0 + \frac{4k^3 - 18k^2 + 22k - 6}{24}\Delta^4 f_0\right]$$

Here $k = \dfrac{x - x_0}{h}$, $x_0 = 15$, $h = 2$ and $x = 16$, then

$$k = \frac{16-15}{2} = \frac{1}{2} = 0.5$$

Also, $\Delta f_0 = 0.250$, $\Delta^2 f_0 = -0.014$, $\Delta^3 f_0 = 0.0002$, $\Delta^4 f_0 = -0.001$

$$\therefore \quad f'(16) = \frac{1}{2}\left[0.250 + 0 + \frac{3(0.5)^2 - 6(0.5) + 2}{6}(.002) + \frac{4(0.5)^3 - 18(0.5)^2 + 22(0.5) - 6}{24}(-0.001)\right]$$

$$f'(16) = 0.1249375 \text{ (Exact value: 0.125)}$$

Similarly

$$f''(16) = \frac{1}{2^2}[-0.014 + (0.5 - 1)(0.002)] + \frac{6(0.5)^2 - 18(0.5) + 11(0.001)}{12}$$

$$f''(16) = -0.0038229 \text{ (Exact value} = -0.00390625)$$

(b) When $x = 15$, then $k = 0$

$$f'(15) = \frac{1}{2}\left[0.250 - \frac{1}{2}(-0.014) + \frac{1}{3}(0.002) - \frac{1}{4}(-0.001)\right]$$

$$f'(15) = 0.128958 \text{ (Exact value 0.12909)}$$

$$f''(15) = \frac{1}{4}\left[-0.014 - 0.002 + \frac{11}{12}(-0.001)\right]$$

$$f''(15) = -0.004229 \text{ (Exact value: } -0.0043033).$$

We use Newton's backward formula for (c) and (d)

Since $x = 24$; $x = 25$ are at the end of the table.

(c) Here $k = \dfrac{x - x_n}{h}$, $x = 24$, $x_n = 25$, $h = 2$, $k = \dfrac{24 - 25}{2} = -0.5$, $k = -0.5$.

Here $\nabla f_n = 0.204$, $\nabla^2 f_n = -0.009$, $\nabla^3 f_n = 0.002$, $\nabla^4 f_n = 0.001$.

$$f'(24) = \frac{1}{2}\left[0.204 + \frac{1}{2}(2(0.5) + 1)(-0.009) + \frac{3(-0.5)^3 + 6(-0.5) + 2}{6}(0.002)\right]$$

$$= 0.09727 \text{ (Exact value: 0.10206)}$$

$$f''(24) = \frac{1}{2^2}\left[-0.009 + (-0.5 + 1)(0.002) + \frac{6(-0.5)^2 + 18(-0.5) + 11}{12}(0.001)\right]$$

$$= -0.00242708 \text{ (Exact value: } -0.002126293)$$

(d) When $x = 25$ which is a tabulated value, $k = 0$, then

$$f'(25) = \frac{1}{2}\left[0.204 + \frac{1}{2}(-0.009) + \frac{1}{3}(0.002) + \frac{1}{4}(0.001)\right]$$

$$= 0.100208 \text{ (Exact value: 0.100)}$$

$$f''(25) = \frac{1}{2^2}\left[-0.009 + 0.002 + \frac{11}{12}(0.001)\right]$$

$$= -0.00225 \text{ (Exact value: } -0.002)$$

If $f = \sqrt{x}$, $f'(x) = \dfrac{1}{2\sqrt{x}}$, $f''(x) = \dfrac{-1}{4x^{3/2}}$.

STIRLING'S FORMULA

Example 3. A rod is rotating in a plane. The following table gives the angle θ (radians) through which the rod has turned for various values of time t (seconds).

t:	0	0.2	0.4	0.6	0.8	1.0
θ:	0	0.12	0.49	1.12	2.02	3.20

Calculate the angular velocity and the angular acceleration of the rod when $t = 0.6$ second.

Solution: Here $h = 0.2$, to find the angular velocity and angular acceleration at $t = 0.6$ seconds, we have to find $\left(\dfrac{d\theta}{dt}\right)_{t=0.6}$ and $\left(\dfrac{d^2\theta}{dt^2}\right)_{t=0.6}$

Since $t = 0.6$ is near to the middle value of t we use Stirling's formula Choosing the origin at $t_0 = 0.6$ we have

$$k = \frac{t - t_0}{h} = \frac{0.6 - 0.6}{0.2} = 0.$$

Stirling's formula for derivative:

$$\left(\frac{d\theta}{dt}\right)_{t=0.6} = \left(\frac{d\theta}{dt}\right)_{k=0} = \frac{1}{h}\left[\left(\frac{\Delta\theta_0 + \Delta\theta_{-1}}{2}\right) - \frac{1}{6}\left(\frac{\Delta^3\theta_{-1} + \Delta^3\theta_{-2}}{2}\right)\right.$$

$$\left. + \frac{1}{30}\left(\frac{\Delta^5\theta_{-2} + \Delta^5\theta_{-3}}{2}\right) + \cdots\right]$$

The central differences table

t	p	θ	$\Delta\theta$	$\Delta^2\theta$	$\Delta^3\theta$	$\Delta^4\theta$	$\Delta^5\theta$
0	−3	0					
			0.12				
0.2	−2	0.12		0.25			
			0.37		0.01		
0.4	−1	0.49		0.26		0	
			0.63		0.01		0
0.6	0	1.12		0.27		0	
			0.90		0.01		
0.8	1	2.02		0.28			
			1.18				
1.0	2	3.20					

$$\left(\frac{d\theta}{dt}\right)_{k=0} = \frac{1}{0.2}\left[\left(\frac{0.90+0.63}{2}\right)-\frac{1}{6}\left(\frac{0.01-0.01}{2}\right)\right]$$

$$= 5\left(\frac{1.53}{2}\right)$$

$$= 3.825 \text{ radians/sec}$$

$$\left(\frac{d^2\theta}{dt^2}\right)_{t=0.6} = \left(\frac{d^2\theta}{dt^2}\right)_{k=0}$$

$$= \frac{1}{h^2}\left[\Delta^2\theta_{-1} - \frac{1}{12}\Delta^4\theta_{-2} + \cdots\right]$$

$$= 25\left[0.27 - \frac{1}{12}(0)\right]$$

$$= 6.75 \text{ radians/sec}^2$$

Example 4. Find the gradient of the road at the middle point of the elevation above a datum line of seven points of the road which are given below:

x:	0	300	600	900	1200	1500	1800
y:	135	149	157	183	201	205	193

Solution:

We are asked here to find the gradient of the road at the middle point of the elevation we shall find $\left(\frac{dy}{dt}\right)_{x=x_0}$ by Stirling's formula.

Here $h = 300$. Choose the origin at $x = x_0 = 900$ we get $k = 0$.

Stirling's formula for derivative is

$$\left(\frac{dy}{dx}\right)_{x=x_0} = \frac{1}{h}\left[\left(\frac{\Delta y_0 + \Delta y_{-1}}{2}\right) - \frac{1}{6}\left(\frac{\Delta^2 y_{-1} + \Delta^3 y_{-2}}{2}\right) + \frac{1}{30}\left(\frac{\Delta^5 y_{-2} + \Delta^5 y_{-3}}{2}\right) - + \cdots\right]$$

The central differences table is:

x	k	y	Δy	$\Delta^2 y$	$\Delta^3 y$	$\Delta^4 y$	$\Delta^5 y$	$\Delta^6 y$
0	−3	135						
			14					
300	−2	149		−6				
			08		24			
600	−1	157		18		−50		
			26		−26		70	
900	0	183		−8		20		−86
			18		−6		−16	
1200	1	201		−14		4		
			4		−2			
1500	2	205		−16				
			−12					
1800	3	193						

$$\left(\frac{dy}{dt}\right)_{x=x_{900}} = \frac{1}{300}\left[\left(\frac{18+26}{2}\right) - \frac{1}{6}\left(\frac{-6+(-26)}{2}\right) + \frac{1}{30}\left(\frac{-16+70}{2}\right) - \cdots\right]$$

$$= 0.0852$$

Example 5. Calculate $y'(4.16)$ from the table

x :	4.10	4.12	4.14	4.16	4.18	4.20	4.22
$f(x)$:	0.147	0.165	0.178	0.189	0.195	0.214	0.234

***Solution*:** The central differences table is

x	$y = f(x)$	Δy	$\Delta^2 y$	$\Delta^3 y$	$\Delta^4 y$	$\Delta^5 y$	$\Delta^6 y$
4.10	0.147						
		0.018					
4.12	0.165		−0.005				
		0.013		0.003			
4.14	0.178		−0.003		−0.006		
		0.011		−0.003		0.027	
4.16	0.189		−0.005		0.021		−0.078
		0.006		0.018		−0.051	
4.18	0.195		0.013		−0.030		
		−0.019		−0.012			
4.20	0.214		0.001				
		−0.020					
4.22	0.234						

Let $x_0 = 4.16$, $h = 0.02$

$$\therefore \qquad k = \frac{x - x_0}{h}$$

$$\therefore \qquad \frac{dk}{dx} = \frac{1}{h}$$

By Gauss' forward formula

$$y = y_0 + k\Delta y_0 + \frac{k(k-1)}{2!}\Delta^2 y_{-1} + \frac{k(k^2-1)}{3!}\,.\,\Delta^3 y_{-1}$$

$$+ \frac{k(k^2-1)(k-2)}{4!}\Delta^4 y_{-2} + \frac{k(k^2-1)(k^2-2^2)}{5!}\Delta^5 y_{-2} + \cdots$$

$$\therefore \qquad \frac{dy}{dx} = \frac{dy}{dk}\cdot\frac{dk}{dx}$$

$$= \frac{1}{h}\left\{\Delta y_0 + \frac{2k-1}{2!}\Delta^2 y_{-1} + \frac{3k^2-1}{3!}\Delta^3 y_{-1}\right.$$

$$\left. + \frac{4k^3 - 6k^2 - 2k + 2}{4!}\Delta^4 y_{-2} + \cdots\right\}$$

At $x = 4.16$, we have $k = 0$

$$\therefore \quad \frac{dy}{dx} = \frac{1}{h}\left[\Delta y_0 - \frac{1}{2}\Delta^2 y_{-1} - \frac{1}{6}\Delta^3 y_{-1} + \frac{2}{24}\Delta^4 y_{-2} + \frac{4}{5!}\Delta^5 y_{-2} - \frac{12}{6!}\Delta^6 y_{-3}\right]$$

$$= \frac{1}{0.02}\left[0.006 - \frac{1}{2}(-0.005) - \frac{1}{6}(0.018)\right.$$

$$\left. + \frac{1}{12}(0.021) + \frac{1}{30}(-0.051) - \frac{1}{60}(-0.078)\right]$$

$$= \frac{1}{0.02}[0.0060 + 0.0025 - 0.0030 + 0.0018 - 0.0017 + 0.0013]$$

$$= \frac{1}{0.02}[0.0069] = 0.345$$

EXERCISES

(1) Find $\frac{dy}{dx}$ at $x = 1$ from the following table using Newton's-Gregory forward formula:

x :	1	2	3	4	5	6
y :	1986	2955	3894	4794	5646	6442

(2) Find $\frac{dy}{dx}$ and $\frac{d^2y}{dx^2}$ at $x = 30$ from the table:

x :	30	32	34	36	38	40	42
y :	5.4772	5.6568	5.8310	6.0000	6.1644	6.3245	6.4807

(3) Find $\frac{dy}{dx}$ at $x = 5$ using central differences formula from the table:

x :	2	3	4	5	6	7	8
$f(x)$:	0.3010	0.4771	0.6020	0.6990	0.7781	0.8451	0.9031

(4) Find $\frac{dy}{dx}$ at $x = 1$ using central differences table:

x :	0.7	0.8	0.9	1.0	1.1
y :	0.6442	0.7174	0.7833	0.8415	0.8912
x :	1.2	1.3			
y :	0.9320	0.9636			

(5) Use central differences formula to find $f'(15)$, $f''(15)$ from the table:

x :	15	17	19	21	23	25
$f(x)$:	3.873	4.123	4.359	4.583	4.796	5.000

(6) Using central differences formula find $f'(1)$ from the data:

x :	1.00	1.05	1.10	1.15	1.20	1.25	1.30
y :	1.000	1.025	1.049	1.072	1.095	1.118	1.140

(7) From the values in the table given below find the value of sec 31° using numerical differentiation.

x (degree) :	31	32	33	34
tan x :	0.6008	0.6249	0.6494	0.6745

(8) The variation in the steam pressure p with temperature θ has been found to be as follows:

θ:	90°	95°	100°	105°	110°	115°	120°
p:	1463	1765	2116	2524	2994	3534	4152

Find $\frac{dp}{d\theta}$ and $\frac{d^2 p}{d\theta^2}$ at $\theta = 105°$

(9) A rod is rotating in a plane. The angle θ (radians) through which the rod has turned in time t are tabulated as follows:

t:	0	0.2	0.4	0.6	0.8	1.0	1.2
θ:	0	0.12	0.49	1.12	2.02	3.20	4.67

Find (i) the angular velocity of the rod and
(ii) its angular acceleration, when $t = 0.6$ sec.

(10) Determine $f'(4)$ from the table

x:	1	2	4	8	10
$f(x)$:	0	1	5	21	27

(11) From the following data find $f'(10)$:

x:	3	5	11	27	34
$f(x)$:	– 13	23	899	17315	35606

(12) The following table gives the velocity v of a body at time t. Find its acceleration at $t = 1.1$.

t:	1.0	1.1	1.2	1.3	1.4
v:	43.1	47.7	52.1	56.4	60.8

(13) The following data gives the correspondence between pressure p and specific volume V of a heated system:

V:	2	4	6	8	10
p:	10.5	42.7	25.3	16.7	13

Find the rate of change of pressure w.r.t volume when $v = 2$.

(14) The temperature of a cooling material observed w.r.t time t is as follows:

t:	1	3	5	7	9
θ:	85.3	74.5	67.0	60.5	54.3

Find the rate of cooling when $t = 8$.

(15) Find the first and second derivatives of the function $y = f(x)$ at $x = 1.1$ from the following table:

x :	1	1.2	1.4	1.6	1.8	2.0
$f(x)$:	0.00	0.1280	0.5440	1.2960	2.4320	4.0000

(16) Using Gauss' forward formula evaluate $f'(7.50)$ from the table:

x :	7.47	7.48	7.49	7.50	7.51	7.52	7.53
$f(x)$:	0.193	0.195	0.198	0.201	0.203	0.206	0.208

(17) From Stirling's formula prove that

$$\frac{d}{dx}(y_x) = \frac{2}{3}(y_{x+1} - y_{x+1}) - \frac{1}{12}(y_{x+1} - y_{x-2}) \text{ nearly.}$$

Answers:

(1) 980 (approx)
(2) 0.0913; – 0.0015
(3) 0.8794
(4) 0.5403
(5) 0.1289; – 0.0004
(6) 0.5002
(7) 1.17
(8) 87.53 and 2.48
(9) 3.81 rad/sec; 6.73 rad/sec^2
(10) 2.83
(11) 2.33
(15) 0.630 and 6.60
(16) 0.23

Chapter 8

Numerical Integration

INTRODUCTION

The process of evaluating a definite integral from a set of tabulated values of the integrand $f(x)$, which is not known explicitly is called numerical integration.

In this chapter, we are concerned with the numerical evaluation of the integrals of the type:

$$\int_a^b f(x)\,dx$$

In elementary courses on integration, we have studied various methods of evaluation of the integrals. The methods are such that the integration can be performed if the integrand (that is, $f(x)$) is in simple form for integration or in a form specified by some rules. In these cases, basically, we have

$$\int_a^b f(x)\,dx = [F(x)]_a^b$$

i.e., the integration of $f(x)$ results with another function.

For example, $$\int e^x dx = e^x + c = F(x, c)$$

or $$\int x^2 dx = \frac{x^3}{3} + c = F(x, c)$$

Such type of evaluation of integrals are said to be in 'closed form'.

A few examples, where in the closed form is not possible or solutions are difficult, are given below:

(1) $\int_0^1 e^{-x^2} \cdot dx$, for which no closed form solution is possible. The value of this integral in provided with in the form of a table.

(2) $\int \frac{e^x - e^{-x}}{x}$, (a closed form solution is not possible).

(3) Given the following data, measured for the velocity of an object at various times

Time :	0.1	0.2	0.3	1.0	1.2
Velocity :	2.4443	3.446	4.004	10.230	12.006

In order to compute the distance travelled by the object we have to evaluate the integral of the velocity w.r.t. 't' in the form:

$$V = \frac{ds}{dt} \Rightarrow S = \int V \cdot dt + \text{(constant)}$$

and we note that V is known through the above table only.

NEWTON'S-COTE'S FORMULAE

Suppose $I = \int_{x_0}^{x_n} f(x)\, dx$, taking values $f(x_i) = f_i$,

$$x_i = x_0 + ih,\ i = 0,1,2, \ldots, n$$

If $x = x_0 + kh,\ 0 \le k \le n$, then

$$f(x_k) = f_k = E^k f_0 = (1 + \Delta)^k f_0$$

Hence
$$I = \int_{x_0}^{x_n} f(x)\, dx = h\int_0^n (1 + \Delta)^k f_0 dk \qquad (1)$$

(**Note:** $k = \frac{x - x_0}{h}$ or $x = x_0 + kh \Rightarrow dx = hdk$)

This is the basic integration formula. By specifying the value of n in (1) we obtain various integration formulae, called Newton-Cote's formulae. It should be noted that with $(n + 1)$ entries in a table, we have the power of Δ only upto Δ^n for use in (1).

Trapezoidal Rule

Let $n = 1$ in (1) and retaining terms upto Δ in the expansion of $(1 + \Delta)^k$. We get

$$\int_{x_0}^{x_1} f(x)\,dx = h\int_0^1 (1 + k\Delta)\, f_0\, dk \;=\; \frac{h}{2}[2f_0 + \Delta f_0]$$

Since $\Delta f_0 = f_1 - f_0$, this gives the trapezoidal rule,

$$\int_{x_0}^{x_1} f(x)\,dx \;=\; \frac{h}{2}[f_0 + f_1] \tag{2}$$

Error in Trapezoidal Rule

The loading term dropped in the expansion of $(1 + \Delta)^k$ gives the error

$$h\int_0^1 \frac{k\,(k-1)}{2!}\,\Delta^2 f_0\, dk \;=\; -\,\frac{h}{12}\Delta^2 f_0$$

Since $\Delta \approx hD$ for small h,

$$\text{Error} = -\,\frac{h^3}{12} f''(\xi),\; x_0 < \xi < x_1 \tag{3}$$

Here ξ is introduced to take into account the effect of the remaining terms dropped.

For a general interval (2) gives

$$\int_a^b f(x)\,dx \;=\; \frac{(b-a)}{2}\,[f(a) + f(b)] \tag{4}$$

using (4) for each of the sub-interval $[x_i, x_{i+1}]$, $i = 0, 1, 2, \ldots n-1$

We get the composite formula

$$\int_{x_0}^{x_n} f(x)\,dx \;=\; \frac{h}{2}\,[f_0 + 2\,(f_1 + f_2 + \ldots + f_{n-1}) + f_n] \tag{5}$$

with global error given by

$$\text{Error} = -\,\frac{(x_n - x_0)}{12} h^2 f''(\xi),\; x_0 < \xi < x_n \tag{6}$$

Simpson's $\frac{1}{3}$ Rule

Setting $n = 2$ in (1) and retain terms upto Δ^2.

we get, $$\int_{x_0}^{x_2} f(x)dx = h\int_0^2 \left[1+k\Delta+\frac{k(k-1)}{2!}\Delta^2\right] f_0\, dk$$

$$= \frac{h}{3}[6 + 6\Delta + \Delta^2] f_0$$

Now, using $\Delta f_0 = f_1 - f_0$; $\Delta^2 f_0 = f_2 - 2f_1 + f_0$

We obtain the Simpson's $\frac{1}{3}$ rule

$$\int_{x_0}^{x_2} f(x)\,dx = \frac{h}{3}\ [f_0 + 4f_1 + f_2] \qquad (7)$$

Simpson's $\frac{3}{8}$ Rule

Setting $n = 3$ in (1) and retain terms upto Δ^3.

We get

$$\int_{x_0}^{x_3} f(x)dx = h\int_0^3 \left[1+k\Delta+\frac{k(k-1)}{2!}\Delta^2+\frac{k(k-1)(k-2)}{3!}\Delta^3\right] f_0\, dk$$

$$= \frac{3h}{8}[8 + 12\Delta + 6\Delta^2 + \Delta^3] f_0$$

Now, using $\Delta f_0 = f_1 - f_0$, $\Delta^2 f_0 = f_2 - 2f_1, + f_0$,
$\Delta^3 f_0 = f_3 - 3f_2, + 3f_1 - f_0$, we obtain the

Or, Simpson's $\frac{3}{8}$ Rule given by:

$$\int_{x_0}^{x_3} f(x)dx = \frac{3h}{8}\ [f_0 + 3f_1, + 3f_2 + f_3] \qquad (8)$$

Weddle's Rule

Putting $n = 6$ in (1) and retain term upto Δ^6.

We get,

$$\int_{x_0}^{x_6} f(x)dx = h\int_0^6 \left[1+\binom{k}{1}\Delta+\binom{k}{2}\Delta^2+\binom{k}{3}\Delta^3+\binom{k}{4}\Delta^4 + \binom{k}{5}\Delta^5+\binom{k}{6}\Delta^6\right] f_0 dk$$

$$= h\left[6+18\Delta+27\Delta^2+24\Delta^3+\frac{123}{10}\Delta^4+\frac{33}{10}\Delta^5+\frac{41}{140}\Delta^6\right] f_0$$

Adding the small quantity $\frac{h\Delta^6}{140}$ on the right side, we obtain.

$$\int_{x_0}^{x_6} f(x)dx = \frac{3h}{10}[20 + 60\Delta + 90\Delta^2 + 80\Delta^3 + 41\Delta^4 + 11\Delta^5 + \Delta^6]f_0.$$

Now, using $\Delta f_0 = f_1 - f_0$, etc...,
We get the Weddle's rule.

$$\int_{x_0}^{x_6} f(x)dx = \frac{3h}{10}[f_0 + 5f_1 + f_2 + 6f_3 + f_4 + 5f_5 + f_6]... \quad (9)$$

Error in Simpson's $\frac{1}{3}$ rule

For this we look for the leading term dropped and find $h\int_0^2 \frac{k(k-1)(k-2)}{3!}$ $\Delta^3 f_0\, dk = 0$. Hence, the error is obtained from the next term as

$$h\int_0^2 \frac{k(k-1)(k-2)(k-3)}{4!}\Delta^4 f_0 dk = -\frac{h}{90}\Delta^4 f_0 \text{ giving}$$

$$\text{Error} = -\frac{h^5}{90} f^{iv}(\xi),\ (x_0 < \xi < x_2) \quad (1)$$

For a general interval, $\int_{x_0}^{x_2} f(x)dx = \frac{h}{3}[f_0 + 4f_1 + f_2]$ takes the form

$$\int_a^b f(x)dx = \frac{(b-a)}{6}\left[f(a) + 4f\left(\frac{a+b}{2}\right) + f(b)\right] \tag{2}$$

Using this for each of the sub-intervals $[x_i, x_{i+2}]$.
$i = 0, 2, 4, \ldots(2x - 2)$. We obtain the composite formula

$$\int_{x_0}^{x_{2n}} f(x)dx = \frac{h}{3}[f_0 + 4(f_1 + f_3 + \ldots + f_{2n-1}) +$$

$$2(f_2 + f_4 + \ldots + f_{2n-2}) + f_{2n}] \tag{3}$$

With the global error given by

$$\text{Error} = -\frac{(x_{2n} - x_0)\,h^4}{180} f^{\text{iv}}(\xi),\ (x_0 < \xi < \xi_{2n}) \tag{4}$$

Error in Simpson's $\frac{3}{8}$ rule

The leading term dropped in the expansion of $(1 + \Delta)^k$ is

$$h\int_0^3 \frac{k(k-1)(k-2)(k-3)}{4!}\Delta^4 f_0\, dk$$

$$= -\frac{3h}{80}\Delta^4 f_0$$

Hence

$$\text{Error} = -\frac{3h^5}{80} f^{\text{iv}}(\xi),\ x_0 < \xi < x_3$$

GLOBAL ERROR

For a general interval,

$$I = \int_{x_0}^{x_3} f(x)dx = \frac{3h}{8}[f_0 + 3f_1 + 3f_2 + f_3]$$

takes the form:

$$\int_a^b f(x)\,dx = \frac{(b-a)}{8}\left[f(a) + 3f\left(\frac{2a+b}{3}\right) + 3f\left(\frac{a+2b}{3}\right) + f(b)\right]$$

Using each of the subintervals $[x_i, x_{i+3}]$ (1)
$i = 0, 3, 6, \ldots (3n-3)$, we obtain the composite formula

$$\int_{x_0}^{x_{3n}} f(x)dx = \frac{3h}{8}[f_0 + 3(f_1 + f_2 + f_4 + f_5 + \ldots + f_{3n-2} + f_{2n-1})$$

$$+ 2(f_3 + f_6 + \ldots + f_{3n-3}) + f_{3n}] \quad (2)$$

With global error given by

$$\text{Error} = -\frac{(x_{3n} - x_0)h^4}{80}, \; f^{iv}(\xi), \; x_0 < \xi < x_{3n} \quad (3)$$

Example 1. Let a function $f(x)$ be given by the data:

x :	1.0	1.5	2.0	2.5	3.0
f :	1.000	2.875	7.000	14.125	25.000

Evaluate $I = \int_1^3 f(x)dx$ by (i) trapezoidal rule and (ii) Simpson's $\frac{1}{3}$ rule

Solution: Here $h = 0.5$ and $n = 4$.

(i) By trapezoidal rule,

$$\int_1^3 f(x)dx = \frac{h}{2}[f_0 + 2\,(f_1 + f_2 + \ldots + f_{n-1}) + f_n]$$

$$I = \frac{0.5}{2}[1.000 + 2\,(2.875 + 7.000 + 14.125) + 25.000]$$

$$= 18.5$$

(ii) By Simpson's $\frac{1}{3}$ rule, we have

$$\int_{x_0}^{x_{2n}} f(x)dx = \frac{h}{3}\,[f_0 + 4\,(f_1 + f_3 + \ldots + f_{2n-1}) +$$

$$2(f_2 + f_4 + \ldots + f_{2n-2}) + f_{2n}]$$

$$I = \frac{0.5}{3}\,[1.000 + 4(2.875 + 14.125) + 2(7.000) + 25.000]$$

$$= 18.0$$

Example 2. Evaluate $\int_0^1 e^x dx$ approximately in steps of 0.5 using trapezoidal rule.

Solution:

Let $y = f(x)\ e^x$ and $h = 0.2$, $n = 5$

Then the values of x and $f(x)$ are as follows:

	x_0	x_1	x_2	x_3	x_4	x_5
x :	0	0.2	0.4	0.6	0.8	1.0
$f(x)$:	1	1.2214	1.4918	1.8221	2.2255	2.7183
	y_0	y_1	y_2	y_3	y_4	y_5

By trapezoidal rule,

$$\int_0^1 e^x dx = h\left[\frac{1}{2}(y_0 + y_5) + (y_1 + y_2 + y_3 + y_4)\right]$$

$$= 0.2\left[\frac{1}{2}(1 + 2.7183) + (1.2114 + 1.4918 + 1.822 + 2.2255)\right]$$

$$= 0.2\ [1.8592 + 6.7508]$$

$$= 1.7220 \text{ (approx.)}$$

Example 3. Evaluate $I = \int_0^1 \frac{dx}{1+x^2}$ with $h = \frac{1}{6}$ by

(i) Trapezoidal rule

(ii) Simpson's $\frac{1}{3}$ rule

(iii) Simpson's $\frac{3}{8}$ rule

(iv) Weddle's rule

Solution: (i) Trapezoidal rule, using

$$\int_{x_0}^{x_n} f(x)\,dx = \frac{h}{2}[\,f_0 + 2\,(\,f_1 + f_2 + \ldots + f_{n-1}) + f_n]$$

We get

$$I = \frac{1}{12}\left[\frac{1}{1+0}+2\left\{\frac{1}{1+\left(\frac{1}{6}\right)^2}+\frac{1}{1+\left(\frac{2}{6}\right)^2}+\frac{1}{1+\left(\frac{3}{6}\right)^2}\right.\right.$$

$$\left.\left.+\frac{1}{1+\left(\frac{4}{6}\right)^2}+\frac{1}{1+\left(\frac{5}{6}\right)^2}\right\}+\frac{1}{1+1}\right]$$

$$= \frac{1}{12}\left[1+72\left(\frac{1}{37}+\frac{1}{40}+\frac{1}{45}+\frac{1}{52}+\frac{1}{61}\right)+\frac{1}{2}\right]$$

$$= 0.7842$$

(ii) Using,

$$\int_{x_0}^{x_{2n}} f(x)dx = \frac{h}{3}[f_0 + 4(f_1 + f_2 + \ldots + f_{n-1})$$

$$+ 2(f_2 + f_4 + \ldots + f_{2n-2}) + f_{2n}]$$

We get

$$I = \frac{1}{18}\left[1+4\left\{\frac{1}{1+\left(\frac{1}{6}\right)^2}+\frac{1}{1+\left(\frac{3}{6}\right)^2}+\frac{1}{1+\left(\frac{5}{6}\right)^2}\right\}\right.$$

$$\left.+2\left\{\frac{1}{1+\left(\frac{2}{6}\right)^2}+\frac{1}{1+\left(\frac{4}{6}\right)^2}\right\}+\frac{1}{2}\right]$$

$$= \frac{1}{18}\left[1+144\left(\frac{1}{37}+\frac{1}{45}+\frac{1}{61}\right)+72\left(\frac{1}{40}+\frac{1}{52}\right)+\frac{1}{2}\right]$$

$$= 0.7854 \left(\text{Exact value of } \frac{\pi}{4}\right)$$

(iii) Using

$$\int_{x_0}^{x_3} f(x)dx = \frac{3h}{18}[f_0 + 3f_1 + 3f_2 + f_3]$$

We get,

$$I = \frac{1}{16}\left[1 + \frac{3}{1+\left(\frac{1}{6}\right)^2} + \frac{3}{1+\left(\frac{2}{6}\right)^2} + \frac{2}{1+\left(\frac{3}{6}\right)^2} + \frac{3}{1+\left(\frac{4}{6}\right)^2} + \frac{3}{1+\left(\frac{5}{6}\right)^2} + \frac{1}{2}\right]$$

$$= 0.7854$$

(iv) Using,

$$\int_{x_0}^{x_6} f(x)\,dx = \frac{3h}{10}[f_0 + 5f_1 + f_2 + 6f_3 + f_4 + 5f_5 + f_6]$$

We get

$$I = \frac{1}{20}\left[1 + \frac{5}{1+\left(\frac{1}{6}\right)^2} + \frac{1}{1+\left(\frac{2}{6}\right)^2} + \frac{6}{1+\left(\frac{3}{6}\right)^2} + \frac{1}{1+\left(\frac{4}{6}\right)^2} + \frac{5}{1+\left(\frac{5}{6}\right)^2} + \frac{1}{2}\right]$$

$$= 0.7854$$

Example 4. Using trapezoidal rule integrate the function defined by the following between 0 and 7:

x:	0	1	2	4	7
f:	1	–5	–5	37	295

***Solution*:** We fiind here abscissas are not equally spaced. Hence *we do not* use

$$\int_{x_0}^{x_n} f(x)\,dx = \frac{h}{2}[f_0 + 2\,(f_1 + f_2 + \dots + f_{n-1}) + f_n]$$

Instead we use

$$\int_{x_0}^{x_1} f(x)\,dx = \frac{h}{2}[f_0 + f_1], \text{ repeatedly.}$$

thus

$$I = \frac{1}{2}(1-5) + \frac{1}{2}(-5-5) + \frac{2}{2}(-5+37) + \frac{3}{2}(37+295)$$

$$= 523$$

Example 5. Evaluate $\int_{-1}^{1} (x)dx$ with two sub-intervals by

(i) Trapezoidal rule and

(ii) Simpson's $\frac{1}{3}$ rule

(i) By trapezoidal rule, $I = \frac{1}{2}[1 + 1] = 1$ (exact value)

(ii) By Simpson's $\frac{1}{3}$ rule, $I = \frac{1}{3}[1 + 0 + 1] = \frac{2}{3}$.

Example 6. Apply the combination of trapezoidal rule and Simpson's $\frac{1}{3}$ rule to find

$$I = \int_0^2 f(x)dx, \text{ where } f(x) \text{ in given by}$$

x :	0	1.0	1.5	2.0
$f(x)$:	1.1	2.4	5.7	8.1

Solution:

$$I = \int_0^{1.0} f(x)dx + \int_{1.0}^{2.0} f(x)dx$$

$$= \frac{1}{2}[1.1 + 2.4] + \frac{0.5}{3}[2.4 + 4(5.7) + 8.1]$$

$$= 7.35$$

Example 7. Evaluate the following integral by using the trapezoidal rule:

$$I = \int_0^2 \frac{1}{1+x}dx. \text{ With increment value of 0.25?}$$

Solution: Given:

Increment $= 0.25 = h$

Lower limit $= 0$

Upper limit $= 2$

i.e,

x :	0	0.25	0.5	0.75	1.0	1.25	1.5	1.75	2.0
$1 + x$:	1	1.25	1.5	1.75	2.0	2.25	2.5	2.75	3.0
$\left(\frac{1}{1+x}\right)$:	1	0.8	0.67	0.57	0.5	0.44	0.4	0.36	0.33

$$\text{Trapezoidal rule} = I = \text{Interval}\left[\frac{\text{Initial}}{2} + \sum_{n=1}^{n-1} y_i + \frac{\text{last value}}{2}\right]$$

$$= 0.25\left[\frac{1}{2} + 0.8 + 0.67 + 0.57 + 0.5 + 0.44 + 0.4 + 0.36 + \frac{0.33}{2}\right]$$

$$= 1.101$$

Example 8. Evaluate the following integral by using Simpson's $\frac{1}{3}$ rule. Take number of panels $n = 4$.

$$I = \int_0^4 \log_e f(5 - 4\cos x)\,dx$$

Solution:

x	0	$\pi/4$	$\pi/2$	$3\pi/4$	π
$\cos x$	1	0.707	0	−0.707	−1
$-4\cos x$	−4	−2.828	0	2.828	4
$5 - 4\cos x$	1	2.172	5	7.828	9
$\log_e(5 - 4\cos x)$	0	0.775	1.609	2.057	2.197

$$\text{Simpson's rule} = I = \frac{1}{3}\left(\frac{\pi}{4}\right)[0 + 4(0.7756 + 2.057) + 2(1.609) + 2.197]$$

$$= 4{\cdot}384$$

Example 9. In the table given below the values of $f(x)$ are given for values of x from 0.2 to 0.8

x :	0.2	0.3	0.4	0.5	0.6	0.7	0.8
$f(x)$:	2.03	2.87	2.14	3.65	3.01	2.71	2.05

Determine the area bounded by the curve between $x = 0.8$ and the x-axis.

***Solution*:** Here, $h = 0.1$;

$$\text{Trapezoidal rule} = \text{Interval}\left[\frac{\text{Initial value}}{2} + \sum_{n=1}^{n-1} y_i + \frac{\text{last value}}{2}\right]$$

Using the trapezoidal rule.

$$= 0.1\left[\frac{2.03}{2} + 2.87 + 3.14 + 3.65 + 3.01 + 2.71 + \frac{2.05}{2}\right]$$

$$= 1.742 \text{ unit}^2.$$

Using Simpson's rule.

$$= \frac{0.1}{3}[2.03 + 4(2.87 + 3.65 + 2.71) + 2(3.14 + 3.01) + 2.05]$$

$$= 1.707 \text{ unit}^2.$$

Example 10. Applying Simpson's $\frac{3}{8}$ rule to evaluate the following.

$$\int_0^{\pi/2} e^{\sin x} dx$$

Solution:

Let $f(x) = e^{\sin x}$ and $h = \frac{\pi}{6}$; $n = 3$

$$x_0 = 0 \qquad x_1 = \frac{\pi}{6} \qquad x_2 = \frac{2\pi}{6} = \frac{\pi}{3}, \qquad x_3 = \frac{\pi}{2}$$

$$f_0 = e^{\sin 0} = 1, \qquad f_1 = e^{\sin\frac{\pi}{6}}, \qquad f_2 = e^{\sin\frac{\pi}{3}}, \qquad f_3 = e^{\sin\frac{\pi}{2}} = e$$

$$I = \int_0^{\pi/2} f(x)\,dx = \frac{3h}{8}[(f_0 + f_3) + 3(f_1 + f_2)]$$

$$= \frac{3}{8} \times \frac{\pi}{6}\left[\left(e^{\sin_0} + e^{\sin\frac{\pi}{2}}\right) + 3\left(e^{\sin\frac{\pi}{6}} + e^{\sin\frac{\pi}{3}}\right)\right]$$

$$= 0.1964\ [1 + 2.72 + 3(1.64924 + 2.38)]$$

$$= 0.1964 \times 15.808 = 3.1046$$

Example 11. Use Simpson's $\frac{3}{8}$ rule to obtain the approximate value of

$$\int_0^{0.3} (2x - x^2)^{1/2} dx.$$

Solution: Let $f(x) = (2x - x^2)^{1/2}$; $h = 0.05$

$\therefore \quad n = 6$

$x_0 = 0$, $\quad x_1 = 0.05$, $\quad x_2 = 0.10$, $\quad x_3 = 0.15$,

$x_4 = 0{\cdot}20$, $\quad x_5 = 0.25$, $\quad x_6 = 0.30$

$f_0 = 0$, $\quad f_1 = 0.3122$, $\quad f_2 = 0.4359$, $\quad f_3 = 0.5268$,

$f_4 = 0.6000$, $\quad f_5 = 0.6614$, $\quad f_6 = 0.7141$

By Simpson's $\frac{3}{8}$ rule, we get

$$I = \int_0^{0.3} (2x - x^2)^{\frac{1}{2}} dx = \frac{3h}{8}[(f_0 + f_6) + 3(f_1 + f_2 + f_4 + f_5) + 2(f_3)]$$

$$= \frac{3(0.05)}{8}[0.7141 + 3(2.0095) + 2(0.5268)]$$

$$= 0.1462$$

Example 12. Applying Simpson's $\frac{3}{8}$ rule, to evaluate the following:

$$I = \int_{.2}^{1.4} \left(\sin x - \log_e^x + e^x\right) dx$$

Solution: Let $f(x) = \sin x - \log_e^x + e^x$; $h = 0.1$ and $n = 12$.
i.e. divide the range (0.2, 1.4) into 12 equal parts of width 0.1

$$I = \frac{3}{8} \times 0.1 \times [(3.02950 + 4.70418) + 3(2.84935 + 2.79753$$

$$+ 2.89759 + \ldots) + 2(2.82130 + 3.16604 + 3.80008)]$$

$$= \frac{0.3}{4}[7.73368 + 3 \times 26.90750 + 2 \times 9.78742]$$

$$= 4.05116$$

While, actual value of integral = 4.05095

$\therefore$ Error = 4.05095 – 4.05116

= –0.00021.

Example 13. Find $\int_0^1 \frac{dx}{1+x^2}$ by using Simpson's $\frac{1}{3}$ rule and Simpson's $\frac{3}{8}$ rule.

Hence, obtain the approximate value of π in each case.

Solution: Divide the range (0, 1) into 6 equal parts each of width $\frac{1-0}{6} = \frac{1}{6}$;

$$\left\{\left[\because h = \frac{(b-a)}{n}\right] = \frac{(1-0)}{6} = \frac{1}{6}\right\}$$

Let $f(x) = \frac{1}{1+x^2}$; $h = \frac{1}{6}$; $n = 6$

$$x_0 = 0, \quad x_1 = \frac{1}{6}, \quad x_2 = \frac{2}{6}, \quad x_3 = \frac{3}{6},$$

$$x_4 = \frac{4}{6}, \quad x_5 = \frac{5}{6}, \quad x_6 = 1$$

$$f_0 = 1.0000000 \quad f_1 = \frac{36}{37}, \quad f_2 = \frac{36}{40}, \quad f_3 = \frac{36}{45},$$

$$f_4 = \frac{36}{52}, \quad f_5 = \frac{36}{61}, \quad f_6 = \frac{1}{2}$$

By Simpson's $\frac{1}{3}$ rule, we get

$$\int_0^1 \frac{dx}{1+x^2} = \frac{h}{3}[f_0 + f_6 + 4(f_1 + f_3 + f_5) + 2(f_2 + f_4)]$$

$$= \frac{(1/6)}{3}[1.0000 + 4(0.9729 + 0.8000 + 0.5901)$$

$$+ 2(0.9000 + 0.6923) + 0.5000]$$

$$= 0.7853$$

$$= 0.7853979, \text{ (Correct to 7 decimal places)} \tag{1}$$

By Simpson's $\frac{3}{8}$ rule, we get

$$I = \int_0^1 \frac{dx}{1+x^2} = \frac{3h}{8}[f_0 + f_6 + 3(f_1 + f_2 + f_4 + f_5) + 2f_3]$$

$$= \frac{1}{16}[1.500 + 3(3.1554) + 2(0.8000)]$$

$$= \frac{1}{16}[12.566334] = 0.7853958$$

$$\text{(Correct to 7 places of decimal)} \tag{2}$$

But $$\int_0^1 \frac{dx}{1+x^2} = [\tan^{-1} n]_0^1$$

$$= \tan^{-1} 1 - \tan^{-1} 0$$

$$= \pi/4 \tag{3}$$

Now from (1) and (3), we get

$$\frac{\pi}{4} = 0.7853979$$

or $\pi = 3.1415916.$

From (2) and (3), we get

$$\pi/4 = 0.7853958 \quad \text{or} \quad \pi = 3.1415916$$

Example 14. Find an approximate value of $\log_2$ applying the Simpson's $\frac{1}{3}$ rule to the integral $\int_0^1 \frac{dx}{1+x}$ using eleven ordinates.

Solution: Given; $\left\{\begin{array}{l} \text{11 ordinates} \\ \therefore \quad n = 10 \end{array}\right\}$

Accordingly we divide the range (0, 1) into 10 equal intervals with

$$h = \frac{(b-a)}{n} = \frac{(1-0)}{10} = \frac{1}{10};\ h = \frac{1}{10}$$

$$x_0 = 0, \quad x_1 = \frac{1}{10}, \quad x_2 = \frac{2}{10}, \quad x_3 = \frac{3}{10}, \quad x_4 = \frac{4}{10}, \quad x_5 = \frac{5}{10},$$

$$x_6 = \frac{6}{10}, \quad x_7 = \frac{7}{10}, \quad x_8 = \frac{8}{10}, \quad x_9 = \frac{9}{10}, \quad x_{10} = \frac{10}{10} = 1$$

$f_0 = 1,$ $\quad f_1 = 0.9999,$ $\quad f_2 = 0.8166,$
$f_3 = 0.7692,$ $\quad f_4 = 0.7142,$ $\quad f_5 = 0.6660,$
$f_6 = 0.6250,$ $\quad f_7 = 0.5852,$ $\quad f_8 = 0.5555,$
$f_9 = 0.5263,$ $\quad f_{10} = 0.5000$

Hence by Simpson's $\frac{1}{3}$ rule, we find

$$\int_0^1 \frac{dx}{1+x} = \frac{h}{3}[f_0 + 4(f_1 + f_3 + f_5 + f_7 + f_9)$$

$$+ 2(f_2 + f_4 + f_6 + f_8) + f_{10}]$$

$$= 0.7063$$

Now, $$\int_0^1 \frac{dx}{1+x} = (\log(1+x))_0^1 = \log_2 - \log_1 = 0.7063$$

$$\Rightarrow \log^2 = 0.7063$$

Example 15. Use Simpson's $\frac{3}{8}$ rule to obtain approximate value of

$\int_0^{0.3} (1-8x^3)^{1/2}$ considering 3 equal intervals.

Solution: Given: $n = 3$, $h = \frac{(b-a)}{n} = \frac{(0.3-0)}{3} = 0{\cdot}1$,

Hence, the table is

$x_0 = 0,$	$x_1 = 0.1,$	$x_2 = 0.2,$	$x_3 = 0.3$
$f_0 = 1,$	$f_1 = 0.9960,$	$f_2 = 0.9674,$	$f_3 = 0.8853$

∴ We have

$$I = \frac{3(0.1)}{8}[f_2 + 3f_1 + 3f_2 + f_3]$$

$$\approx 0.29158$$

Example 16. Evaluate $\int_0^1 \frac{\sin x}{x} dx$ with eleven ordinates using Simpson's $\frac{1}{3}$ rule.

Solution: Given:

$$n = 10$$

$$h = \frac{(b-a)}{h} = \frac{(1-0)}{10} = \frac{1}{10} = 0{\cdot}1$$

∴ $$h = 01.$$

Divide the range (0, 1) into 10 equal intervals

$x_0 = 0,$	$x_1 = \frac{1}{10},$	$x_2 = \frac{2}{10},$	$x_3 = \frac{3}{10},$
$x_4 = \frac{4}{10},$	$x_5 = \frac{5}{10},$	$x_6 = \frac{6}{10},$	$x_7 = \frac{7}{10},$
$x_8 = \frac{8}{10},$	$x_9 = \frac{9}{10},$	$x_{10} = \frac{10}{10} = 1$	
$f_0 = 1,$	$f_1 = 0.998,$	$f_2 = 0.993,$	$f_3 = 0.985,$
$f_4 = 0.974,$	$f_5 = 0.959,$	$f_6 = 0.961,$	$f_7 = 0.920,$
$f_8 = 0.892,$	$f_9 = 0.870,$	$f_{10} = 0.841$	

We note that when $x = 0$, $\frac{\sin x}{x} \to \frac{0}{0} \to$ indeterminate form. Hence we have to use L' Hospital rule for evaluation at this point

$$\therefore \qquad \lim_{x \to 0} \frac{\sin x}{x} = \lim_{x \to 0} \frac{\cos x}{1} = 1$$

Hence, by Simpson's $\frac{1}{3}$ rule, we find

$$\int_0^1 \frac{\sin x}{x} dx = \frac{h}{3} [f_0 + 4(f_1 + f_3 + f_5 + f_7 + f_9)$$
$$+ 2(f_2 + f_4 + f_6 + f_8) + f_{10}]$$
$$= 0.9459$$

Example 17. Evaluate $\int_1^4 e^{1/x} dx$ taking $h = 1$. Using Simpson's $\frac{3}{8}$ rule,

Solution: $x_0 = 1$, $x_1 = 2$, $x_2 = 3$, $x_3 = 4$
$f_0 = 2.718$, $f_1 = 1.649$, $f_2 = 1.396$, $f_3 = 1.284$

Simpson's $\frac{3}{8}$ rule requires a minimum of 3 equal intervals and hence, from the given values of x and f we have;

$$I = \frac{3h}{8} [f_0 + 3f_1 + 3f_2 + f_3]$$
$$= \frac{3 \times 1}{8} [2.718 + 3(1.649) + 3(1.396) + 1.284]$$
$$= 4.926.$$

Example 18. Evaluate $\int_0^1 e^x dx$ by Simpson's $\frac{1}{3}$ rule choosing h small enough of guarantee five decimal accuracy. How large h can be?

Solution: The global error formulation given by

$$\text{error} = -\left(\frac{1-0}{180}\right) h^4 e^{\xi}, \; 0 < \xi < 1$$

$$|\text{error}| < \frac{h^4 e}{180}.$$

For 5-decimal places of accuracy, $\frac{h^4 e}{180} < 5 < 10^{-6}$.

Hence $h < \frac{1}{10}\left(\frac{9}{e}\right)^{1/4} = 0.315$.

With $h = 0.2$, we get

$$I = \frac{0.1}{3}[1 + 4(e^{.1} + e^{.3} + e^{.5} + e^{.7} + e^{.9})$$
$$+ 2(e^{.2} + e^{.4} + e^{.6} + e^{.8}) + e]$$
$$= 1.718283. \text{ (exact value is 1.718282)}$$

Example 19. Use Simpson's $\frac{1}{3}$ rule with $h = 0.5$ to evaluate $\int_0^1 \frac{dx}{1+x}$. Find the error bound and the actual error.

Solution: $I = \frac{1}{6}\left[1 + \frac{4}{1.5} + \frac{1}{2}\right] = 0.6944.$

The error formula

$$\text{Error} = -\frac{h^5}{90} f^{iv}(\xi),\ x_0 < \xi < x_2$$

given with $f(x) = \frac{1}{1+x}$

$$\text{Error} = -\frac{h^5}{90} f^{iv}(\xi) = -\left(\frac{1}{2}\right)^5 \frac{1}{90} \frac{4!}{(1+\xi)^5}$$

$$= -\frac{1}{120} \frac{1}{(1+\xi)^5}$$

Hence $-0.0083 <$ Error < -0.0003.

Actual error = In 2 – 0.6944 = 0.6931 – 0.6944
= – 0.0013.

The error in Simpson's $\frac{1}{3}$ rule and Simpson's $\frac{3}{8}$ rule are both 0 (h^5). Hence, Simpson's $\frac{1}{3}$ formula is generally preferred. Comparing the error formulae of trapezoidal rule and Simpson's $\frac{1}{3}$ rule it is generally seen that the latter gives more accurate values than the former. It should be noted that for function with discontinuous derivatives. This may not be true. (see Ex: 5)

Example 20. Using Weddle's rule with 7 ordinates evaluate the integral

$$\int_{1.2}^{2.4} \frac{1}{\log_{10} x} dx$$

***Solution*:** Given: $n = 6$, (7 ordinates), $f(x) = \dfrac{1}{\log_{10} x} dx$

Hence $h = \dfrac{(b-a)}{n} = \dfrac{(2.4-1.2)}{6} = 0.2$

and the table is given by

$x_0 = 1.2$, $x_1 = 1.4$, $x_2 = 1.6$, $x_3 = 1.8$,
$x_4 = 2.0$, $x_5 = 2.2$, $x_6 = 2.4$

$(f_0 = 12.63)$, $(f_1 = 6.84)$, $(f_2 = 4.84)$, $(f_3 = 3.92)$,
$(f_4 = 3.32)$, $(f_5 = 2.92)$, $(f_6 = 2.63)$.

The value of the integral by Weddle's rule is

$$I = \frac{3h}{10}[f_0 + 5f_1 + f_2 + 6f_3 + f_4 + 5f_5 + f_6]$$

$$= \frac{3\times 0.2}{10}[12.53 + 5(6.84) + 4.84 + 6(3.92) + 3.32$$

$$+ 5(2.92) + 2.63]$$

$$= 5.744.$$

Example 21. Find by Weddle's rule the value of the integral

$$I = \int_{4}^{5.2} \log e^x \, dx$$

Considering 7 ordinates.

***Solution*:** Given: 7 ordinates

Hence, $n = 6$ equal intervals

We find $h = (b - a)/n = (5.2 - 4)/6 = 1.2/6 = 0.2$

Let us prepare the table with step length 0.2 starting from $x = 4$ to $x = 5.2$

$(x_0 = 4.2)$, $(x_1 = 4.2)$, $(x_2 = 4.4)$, $(x_3 = 4.6)$,
$(x_4 = 4.8)$, $(x_5 = 5.0)$, $(x_6 = 5.2)$.

$$f_0 = 1.38629463,$$
$$f_1 = 1.43508453,$$
$$f_2 = 1.48160545,$$
$$f_3 = 1.52605630,$$
$$f_4 = 1.56861522,$$
$$f_5 = 1.60943791,$$
$$f_6 = 1.64865863.$$

Weddle's formula requires 7 ordinates or 6 equal intervals and hence, the formula is used only once and after substituting the above values we have:

$$I = \int_4^{5.2} \log e^x \, dx$$

$$= \frac{3h}{10} [f_0 + 5f_1 + f_2 + 6f3 + f_4 + 5f_5 + f_6]$$

$$= 1.9274$$

Example 22. Find the value of $\int_0^{0.6} \frac{dx}{1+x^2}$ using Weddle's rule.

***Solution*:** Let $f(x) = \dfrac{1}{1+x^2}$; $n = 6$, $x_0 = 0$

$$h = \frac{(b-a)}{n} = \frac{0.6-0}{6} = 0.1$$

$\therefore \qquad h = 0.1$

Then

$x_0 = 0,\quad x_1 = 0.1,\quad x_2 = 0.2,\quad x_3 = 0.3,$
$x_4 = 0.4,\quad x_5 = 0.5,\quad x_6 = 0.6$

$(f_0 = 1),\quad (f_1 = 0.9901),\quad (f_2 = 0.9615),\quad (f_3 = 0.9174),$
$(f_4 = 0.8621),\quad (f_5 = 0.8000),\quad (f_6 = 0.7353).$

By Weddle's rule, we get

$$\int_0^{0.6} \frac{dx}{1+x^2} = \frac{3h}{10} [f_0 + 5f_1 + f_2 + 6f_3 + f_4 + 5f_5 + f_6]$$

$$= \frac{3(0.1)}{10} [1 + 4.9505 + 0.9615 + 5.5044 + 0.8621$$
$$+ 4.0000 + 0.7352]$$

$$= \frac{0.3}{10}[18.0138] = 0.5404$$

Example 23. Evaluate $\int_{0.2}^{1.4} e^{2x}\,dx$ using Weddle's rule with seven ordinates.

***Solution*:** Let $x_0 = 0.2$, $x_6 = 1.4$ and $f(x) = e^{2x}$, $n = 6$

$$\therefore \qquad h = \frac{(b-a)}{n} = \frac{(1.4-0.2)}{6} = \frac{1.2}{6} = 0.2$$

Then

$x_0 = 0.2,\quad x_1 = 0.4,\quad x_2 = 0.6,\quad x_3 = 0.8,$
$x_4 = 0.1,\quad x_5 = 1.2,\quad x_6 = 1.4$

$f_0 = 1.4918,\quad f_1 = 2.2255,\quad f_2 = 3.3201,\quad f_3 = 4.9530,$
$f_4 = 7.3890,\quad f_5 = 11.0232,\quad f_6 = 16.4446.$

By Weddle's rule, we get

$$\int_{0.2}^{1.4} e^{2x}\,dx = \frac{3h}{10}[f_0 + 5f_1 + f_2 + 6f_3 + f_4 + 5f_5 + f_6]$$

$$= \frac{3(0.2)}{10}[1.4918 + 11.1275 + 3.3201 + 29.7180$$
$$+ 7.3890 + 55.1160 + 16.4446]$$

$$= \frac{0.6}{10}[124.6070] = 7.4764.$$

EXERCISES

1. The function $f(x)$ is given by the table

x:	1.6	1.8	2.0	2.2	2.4	2.6	2.8
f:	4.95	6.05	7.39	9.02	11.02	13.46	16.44

Find $\int_{1.6}^{2.8} f(x)\,dx$ by trapezoidal rule, Simpson's $\frac{1}{3}$ rule, Simpson's $\frac{3}{8}$ rule and Weddle's rule.

2. In an experiment, a quantity f was measured as follows:

x:	20	21	22	23	24	25	26
f:	95.90	96.85	97.77	98.68	99.56	100.41	101.24

Compute $\int_{20}^{26} f(x)\,dx$ by Simpson's $\frac{1}{3}$ rule and Weddle's rule.

3. The velocity of a particle moving in a straight line is given as

t :	0.0	0.1	0.2	0.3	0.4	0.5	0.6	0.7	0.8
v :	1.534	1.668	1.811	1.971	2.151	2.352	2.577	2.828	3.107

Find the distance travelled in 0.8 second. (Here t is in seconds and v is in metres/sec.) Use trapezoidal rule with $h = 0.1$, $h = 0.2$ and $h = 0.4$.

4. A rocket is launched from the ground. Its acceleration during the first 80 seconds is given in the following table:

t(sec) :	0	10	20	30	40	50	60	70	80
a(m/sec.^{2}) :	30	31.63	33.44	35.47	37.75	40.33	43.29	46.69	50.67

Find the velocity at $t = 80$ by Simpson's rule.

5. Apply a combination of the two Simpson's rules to evaluate the integral from $x = 3$ to $x = 6.5$. Note that the number of sub-intervals is odd here.

x :	3.0	3.5	4.0	4.5	5.0	5.5	6.0	6.5
f :	–0.33906	–0.13738	–0.06604	–0.23106	–0.32758	–0.34144	–0.27668	–0.15384

6. Evaluate $\int_0^{0.4} \cos x\,dx$ by trapezoidal rule with $h = 0.1$.

7. Evaluate $\int_1^5 \frac{dx}{x}$ by trapezoidal rule with $h = 1$.

8. Evaluate $\int_0^1 \frac{x\,dx}{1+x}$ by Simpson's $\frac{1}{3}$ rule with $h = 0.1$.

9. Evaluate $\int_1^9 \sqrt{6x-5}\,dx$ by trapezoidal rule with $h = 1$.

10. Evaluate $\int_0^2 e^{-x^2}\,dx$ by Simpson's $\frac{1}{3}$ rule with $h = 0.5$.

11. Evaluate $\int_0^{1.2} \ln(1+x^2)\,dx$ by trapezoidal rule and by Simpson's $\frac{1}{3}$ rule with $h = 0.2$.

12. Evaluate $\int_0^1 e^{x^2}\,dx$ by Simpson's $\frac{1}{3}$ rule with $h = 0.1$.

13. Evaluate $\int_0^1 e^{-x^2}\,dx$ by trapezoidal rule with $h = 0.1$.

14. Evaluate $\int_4^{\pi/2} \frac{\sin x}{x}\,dx$ by Simpson's $\frac{1}{3}$ rule with $h = \frac{\pi}{24}$.

15. Evaluate $\int_0^{\pi/2} \sin^{10} x\,dx$ by Simpson's $\frac{1}{3}$ rule with $h = \frac{\pi}{8}$.

16. Evaluate $\int_0^{\pi/2} \frac{\cos x}{1+x}\,dx$ by Simpson's $\frac{3}{8}$ rule with $h = \frac{\pi}{12}$.

17. Evaluate $\int_0^{\pi} \frac{dx}{x+\cos x}$ by Simpson's $\frac{1}{3}$ rule with $h = \frac{\pi}{4}$ and with $h = \frac{\pi}{8}$.

18. Evaluate $\int_0^1 \sin x^2\,dx$ with $h = 0.1$ by trapezoidal rule and Simpson's $\frac{1}{3}$ rule.

19. Evaluate $\int_0^1 \cos x^2\,dx$ with $h = 0.1$ by trapezoidal rule and Simpson's $\frac{1}{3}$ rule.

20. Evaluate $\int_0^1 x^x\,dx$ by trapezoidal rule with $h = 1$, Simpson's $\frac{1}{3}$ rule with $h = \frac{1}{2}$, Simpson's $\frac{3}{8}$ rule $h = \frac{1}{3}$ and Weddle's rule with $h = \frac{1}{6}$.

21. Evaluate $\int_4^{5.2} \ln x\,dx$ with $h = 0.2$ by Simpson's $\frac{1}{3}$ rule and Weddle's rule.

22. Evaluate $\int_{0.2}^{1.4} (\sin x - \ln x + e^x)\, dx$ with $h = 0.1$ by Simpson's $\frac{1}{3}$ rule and Weddle's rule.

23. To compute $\int_{1.8}^{3.4} e^x\, dx$ by trapezoidal rule with five decimal accuracy (error < 0.000005), how small h must and be?

24. Show that the difference between the values of $\int_{x_0}^{x_6} f(x)\, dx$ obtained by Weddle's rule and Simpson's $\frac{1}{3}$ rule with six sub-intervals is $\frac{1}{30}(\Delta^4 + \Delta^5 + \Delta^6)\, f_0$.

25. Obtain the following Newton-Cote's formulae with $n = 4$, $n = 5$ and $n = 6$.

$$\int_{x_0}^{x_4} f(x)\,dx = \frac{2h}{45}[7f_0 + 32f_1 + 12f_2 + 32f_3 + 7f_4].$$

$$\text{Error} = -\frac{8h^7}{945} f^{\text{vi}}(\xi).$$

$$\int_{x_0}^{x_5} f(x)\,dx = \frac{5h}{288}[19f_0 + 75f_1 + 50f_2 + 50f_3 + 75f_4 + + 19f_5].$$

$$\text{Error} = -\int_{x_0}^{x_5} f(x)\,dx \frac{275}{12096} h^7 f^{\text{vi}}(\xi).$$

$$\int_{x_0}^{x_6} f(x)\,dx = \frac{h}{140}[41f_0 + 216f_1 + 27f_2 + 272f_3 + 27f_4 + 216f_5 + 41f_6].$$

$$\text{Error} = -\frac{9h^9}{1400} f^{\text{viii}}(\xi).$$

Chapter **9**

Numerical Solution of Ordinary Differential Equations

INTRODUCTION

In many practical problems, we come across differential equations having initial conditions and boundary conditions. An ordinary differential equation is one which involves the differential coefficients of functions of one variable. In this chapter, we shall discuss some methods to find the numerical solution of such equations. We know that the general solution of an equation of order m consists of m arbitrary constants which can be determined by given initial conditions. Thus, the numerical solution of an ordinary differential equation of the first order (say) with numerical coefficients and given initial conditions can be found to a desired degree of accuracy.

INITIAL VALUE PROBLEMS

An ordinary differential equation of the order n is a relation of the form

$$F(x, y, y' y'', \ldots y^{(n)}) = 0 \tag{1}$$

where $y = y(x)$ and $y^{(i)} = \dfrac{d^i y}{dx^i}$.

Its general solution contains n-arbitrary constants.

To determine the constants, we need to impose n subsidiary conditions.

If $y, y', y'', \ldots y^{(n-1)}$ are prescribed at a single value of x, say x_0, as

$$\begin{aligned} y(x_0) &= y_0 \\ y'(x_0) &= y_0' \\ y^{(n-1)}(x_0) &= y_0^{(n-1)} \ldots \end{aligned} \tag{2}$$

then these conditions are called the initial conditions and the d.e. together with the d.c. is called an initial value problem.

Our primary concern in this chapter is to solve numerically the first–order equation

$$y' = f(x,y),\ y(x_0) = y_0 \tag{3}$$

We also indicate the numerical methods of solving simultaneous first-order equations

$$\left.\begin{aligned} y' &= f(x, y, z), y(x_0) = y_0 \\ z' &= g(x, y, z), z(x_0) = z_0 \end{aligned}\right\} \tag{4}$$

and the second-order equation

$$y'' = g(x, y, y'),\ y(x_0) = y_0,\ y'(x_0) = y'_0 \tag{5}$$

If we get $y' = z$, then (5) takes the form (h).

PICARD'S ITERATION METHOD

To solve the initial value problem

$$y' = f(x, y),\ y(x_0) = y_0 \tag{1}$$

We integrate the equation from x_0 to x and obtain the integral equations

$$y(x) = y_0 + \int_{x_0}^{x} f(t, y(t))\, dt$$

which yields the iteration formula

$$y_{k+1}(x) = y_0 + \int_{x_0}^{x} f(t, y_k(t))\, dt,\ k = 0, 1, 2, \ldots \tag{2}$$

Starting with $y_0(x) = y_0$

We find successive approximations $y_1(x), y_2(x), \ldots$ under certain conditions as $f(x, y)$, the sequence $\{y_k(x)\}$ converges to the true solution of (1).

This method can easily be extended to a system of equations of first-order as a single-second-order equation.

The iteration formula for

$$y' = f(x, y, z),\ y(x_0) = y_0$$
$$z' = g(x, y, z),\ z(x_0) = z_0$$

is $$y_{k+1}(x) = y_0 + \int_{x_0}^{x} f(t, y_k(t), z_k(t))\, dt$$

$$z_{k+1}(x) = z_0 + \int_{x_0}^{x} g(t, y_k(t), z_k(t))\,dt$$

The iteration formula for

$$y'' = g(x, y, y'), \; y(x_0) = y_0$$
$$y'(x_0) = y'_0$$

is
$$y_{k+1}(x) = y_0 + (x - x_0)y'_0 + \int_{x_0}^{x} (x - t)\, g(t, y_k(t), y'_k(t))\,dt$$

Normally we start with

$$y_0(x) = y_0$$

and $z_0(x) = z_0$ in the above iteration formula.

Example 1. Solve $y' = x^2 + y^2$,
$y(0) = 1.$
$y_0(x) = 1.$

Solution:

$$y_{k+1}(x) = 1 + \int_{x_0}^{x} [t^2 + y_k^2(t)]\,dt, \; k = 0, 1, 2, \ldots$$

$$y_1(x) = 1 + \int_{x_0}^{x} (t^2 + 1)\,dt$$

$$= 1 + x + \frac{x^3}{3}$$

$$y_2(x) = 1 + \int_{0}^{x} \left[t^2 + \left(1 + t + \frac{t^3}{3} \right)^2 \right] dt$$

$$= 1 + x + x^2 + \frac{2}{3}x^3 + \frac{x^4}{6} + \frac{2}{15}x^5 + \frac{x^7}{63}$$

which is an approximate solution.

Example 2. Using Picard's method of successive approximations find the solution of the equations

$$\frac{dy}{dx} = 1 + xy$$

subject to the condition $y = 0$ when $x = 0$ upon third approximation and obtain y when $x = 0.2$.

Solution:

Given $\quad f(x, y) = 1 + xy$

$$x_0 = 0,\ y_0 = 0$$

We have
$$y = y_0 + \int_{x_0}^{x} f(x, y_0)\,dx \tag{1}$$

Let $y = y_0 = 0$, as r.h.s. of (1).
Then 1st approximate to y is

$$y_1 = 0 + \int_{x_0}^{x} f(x, 0)\,dx$$

$$= \int_0^x 1dx = [x]_0^x = x$$

Now take $y = y_1 = x$ on r.h.s. of (1),
We get the second approximate

$$y_2 = \int_0^x f(x, y_1)\,dx = \int_0^x (1 + x^2)\,dx = x + \frac{x^3}{3}$$

Again, taking $y = y_2 = x + \dfrac{x^3}{3}$ on r.h.s. (1), we get

$$y_3 = \int_0^x f(x, y_2)\,dx = \int_0^x \left[1 + x\left(x + \frac{x^3}{3}\right)\right]dx$$

$$= x + \frac{x^3}{3} + \frac{x^5}{15}$$

When $x = 0.2$, then

$$y_3 = 0.2 + \frac{(0.2)^3}{3} + \frac{(0.2)^5}{15}$$

$$= 0.202687$$

Example 3. Use Picard's method to approximate y when $x = 0.2$ given that $y = 1$ when $x = 0$ and $\dfrac{dy}{dx} = x - y$.

Solution: Here $f(x, y) = x - y$, $x_0 = 0$, $y_0 = 1$.

We have first approximation,

$$y^{(1)} = y_0 + \int_0^x f(x, y_0)\, dx$$

$$= 1 + \int_0^x (x - 1)\, dx = 1 - x + \frac{x^2}{2}$$

Second approximation,

$$y^{(2)} = y_0 + \int_0^x f(x, y^{(1)})\, dx = 1 + \int_0^x (x - y^{(1)})\, dx$$

$$= 1 + \int_0^x \left(x - 1 + x - \frac{x^2}{2}\right) dx = 1 - x + x^2 - \frac{x^3}{6}$$

Third approximation,

$$y^{(3)} = y_0 + \int_0^x f(x, y^{(2)})\, dx = 1 + \int_0^x (x - y^{(2)})\, dx$$

$$= 1 + \int_0^x \left(x - 1 + x - x^2 + \frac{x^3}{6}\right) dx$$

$$= 1 - x + x^2 - \frac{x^3}{3} + \frac{x^4}{24}$$

Fourth approximation,

$$y^{(4)} = y_0 + \int_0^x f(x, y^{(3)})\, dx$$

$$= 1 + \int_0^x (x - y^{(3)})\, dx$$

$$= 1 + \int_0^x \left(x - 1 + x - x^2 + \frac{x^3}{3} - \frac{x^4}{24}\right) dx$$

$$= 1 - x + x^2 - \frac{x^3}{3} + \frac{x^4}{12} - \frac{x^5}{120}$$

Fifth approximation,

$$y^{(5)} = y_0 + \int_0^x f(x, y^{(4)})\, dx$$

$$= 1 + \int_0^x (x - y^{(4)})\, dx$$

$$= 1 + \int_0^x \left(x - 1 + x - x^2 + \frac{x^3}{3} - \frac{x^4}{12} + \frac{x^5}{120} \right) dx$$

$$= 1 - x + x^2 - \frac{x^3}{3} + \frac{x^4}{12} - \frac{x^5}{60} + \frac{x^6}{720}$$

When $x = 0.2$, we get

$$y^{(1)} = 0.92,\ y^{(2)} = 0.83867,\ y^{(3)} = 0.83740$$

$$y^{(4)} = 0.83746 \text{ and } y^{(5)} = 0.83746.$$

Thus, $\quad y = 0.837$ when $x = 0.2$.

Example 4. Use Picard's method to approximate the value of y when $x = 0.1$, given that $y = 1$ when $x = 0$ and $\frac{dy}{dx} = 3x + y^2$.

Solution: Here $f(x, y) = 3x + y^2$, $x_0 = 0$, $y_0 = 1$

We have first approximation,

$$y^{(1)} = y_0 + \int_{x_0}^x f(x, y_0)\, dx = 1 + \int_0^x (3x + y^2)\, dx$$

$$= 1 + \int_0^x (3x + 1)\, dx = 1 + x + 3.\frac{x^2}{2}$$

Second approximation,

$$y^{(2)} = y_0 + \int_0^x f(x, y^{(1)})\, dx$$

$$= 1 + \int_0^x [3x + \{y^{(1)}\}^2]\, dx$$

$$= 1 + \int_0^x \left[3x + \left(\frac{3}{2} + x^2 + x + 1 \right)^2 \right] dx$$

$$= 1 + \int_0^x \left[\frac{9}{4}x^4 + 3x^3 + 4x^2 + 5x + 1 \right] dx$$

$$= 1 + x + \frac{5}{2}x^2 + \frac{4}{3}x^3 + \frac{3}{4}x^4 + \frac{9}{20}x^5$$

Third approximation,

$$y^{(3)} = y_0 + \int_0^x f(x, y^{(2)})\, dx$$

$$= 1 + \int_0^x [3x + \{y^{(2)}\}^2]\, dx$$

$$= 1 + \int_0^x \left[\frac{81}{400}x^{10} + \frac{27}{40}x^9 + \frac{141}{80}x^8 + \frac{17}{4}x^7 \right.$$

$$\left. + \frac{1157}{180}x^6 + \frac{136}{15}x^5 + \frac{125}{12}x^4 + \frac{23}{3}x^3 + 6x^2 + 5x + 1 \right] dx$$

$$= 1 + x + \frac{5}{2}x^2 + 2x^3 + \frac{23}{12}x^4 + \frac{25}{12}x^5 + \frac{68}{45}x^6$$

$$+ \frac{1157}{1260}x^7 + \frac{17}{32}x^8 + \frac{47}{240}x^9 + \frac{27}{400}x^{10} + \frac{81}{4400}x^{11}.$$

When $x = 0.1$, we have

$$y^{(1)} = 1.115,\ y^{(2)} = 1.1264,\ y^{(3)} = 1.12721.$$

Thus, $\quad y = 1.127$ when $x = 0.1$.

Example 5. Approximate y and z by using Picard's method for the particular solution of

$$\frac{dy}{dx} = x + z, \quad \frac{dz}{dx} = x - y^2$$

given that $y = 2$, $z = 1$, when $x = 0$.

Solution: Let $\phi(x, y, z) = x + z$

and $f(x, y, z) = x - y^2$

Here $x_0 = 0, y_0 = 2, z_0 = 1$

We have $\dfrac{dy}{dx} = \phi(x, y, z)$

$$\Rightarrow y = y_0 + \int_{x_0}^{x} \phi(x, y, z)\, dx, \tag{1}$$

Also $\dfrac{dz}{dx} = f(x, y, z)$

$$\Rightarrow z = z_0 + \int_{x_0}^{x} f(x, y, z)\, dx, \tag{2}$$

We have first approximation of y,

$$y^{(1)} = y_0 + \int_{x_0}^{x} \phi(x, y_0, z_0)\, dx = 2 + \int_{0}^{x} (x + z_0)\, dx$$

$$= 2 + \int_{0}^{x} (1 + x)\, dx = 2 + x + \frac{1}{2}x^2$$

Again first approximation of z is

$$z^{(1)} = z_0 + \int_{x_0}^{x} f(x, y_0, z_0)\, dx$$

$$= 1 + \int_{0}^{x} (x - y_0^2)\, dx$$

$$= 1 + \int_{0}^{x} (x - 4)\, dx = 1 - 4x + \frac{1}{2}x^2$$

Second approximation of y,

$$y^{(2)} = y_0 + \int_{x_0}^{x} \phi(x, y^{(1)}, z^{(1)})\, dx$$

$$= 2 + \int_0^x (x + z^{(1)})\, dx$$

$$= 2 + \int_0^x \left(x+1-4x+\frac{1}{2}x^2 \right) dx$$

$$= 2 + x - (3/2)x^2 + (1/6)x^3$$

Second approximation of z,

$$z^{(2)} = z_0 + \int_{x_0}^x f(x, y^{(1)}, z^{(1)})\, dx$$

$$= 1 + \int_0^x [x - \{y^{(1)}\}^2]\, dx$$

$$= 1 + \int_0^x \left[x - \left(2+x+\frac{1}{2}x^2 \right)^2 \right] dx$$

$$= 1 + \int_0^x \left[x - \left(4+x^2+\frac{1}{4}x^4+4x+x^3+2x^2 \right) \right] dx$$

$$= 1 - 4x - (3/2)x^2 - x^3 - (x^4/4) - (x^5/20)$$

Example 6. Solve $y' = 1 + xyz$, $y(0) = 0$

$$z' = x + y + z,\ z(0) = 1$$

Solution:

$$y_{k+1}(x) = \int_0^x [1 + t y_k(t) z_k(t)]\, dt$$

$$z_{k+1}(x) = 1 + \int_0^x [t + y_k(t) + z_k(t)]\, dt$$

With $y_0(x) = 0$, $z_0(x) = 1$, we get

$$y_1(x) = \int_0^x 1\, dt = x$$

$$z_1(x) = 1 + \int_0^x (t + 1)\, dt = 1 + x + \frac{1}{2}x^2$$

$$y_2(x) = \int_0^x \left[1+t^2\left(1+t+\frac{t^2}{2}\right)\right]dt$$

$$= x + \frac{1}{3}x^3 + \frac{1}{4}x^4 + \frac{1}{10}x^5$$

$$z_2(x) = 1 + \int_0^x \left(t+t+1+t+\frac{t^2}{2}\right)dt$$

$$= 1 + x + \frac{3x^2}{2} + \frac{1}{6}x^3$$

Example 7. Solve

$$y'' = y + xy^1,$$

$$y(0) = 1,\ y^1(0) = 0$$

Solution:

$$y_{k+1}(x) = 1 + \int_0^x (x-t)\ [y_k(t) + ty'_k(t)]dt$$

With $y_0(x) = 1, y'_0(x) = 0$, we get

$$y_1(x) = 1 + \int_0^x (x-t)\,dt$$

$$= 1 + \frac{1}{2}x^2,\ y'_1(x) = x$$

$$y_2(x) = 1 + \int_0^x (x-t)\left[1+\frac{1}{2}t^2+t^2\right]dt$$

$$= 1 + \frac{1}{2}x^2 + \frac{1}{8}x^4$$

$$y'_2(x) = x + \frac{1}{2}x^3$$

$$y_3(x) = 1 + \int_0^x (x-t)\left[1+\frac{1}{2}t^2+\frac{1}{8}t^4+t^2+\frac{1}{2}t^4\right]$$

$$= 1 + \frac{1}{2}x^2 + \frac{1}{8}x^4 + \frac{1}{48}x^6$$

which is an approximate solution.

Example 8. Use Picard's method to approximate y when $x = 0.1$ given that $y = 1$ when $x = 0$ and $\dfrac{dy}{dx} = \dfrac{y-x}{y+x}$.

Solution: Here first approximation $f(x, y) = \dfrac{y-x}{y+x}$, $x_0 = 0$, $y_0 = 1$

$$y_1 = y_0 + \int_{x_0}^{x} f(x, y_0)\,dx$$

$$= 1 + \int_0^x \frac{y_0 - x}{y_0 + x}\,dx$$

$$= 1 + \int_0^x \frac{1-x}{1+x}\,dx$$

$$= 1 + \int_0^x \left[\frac{2}{1+x} - 1\right] dx$$

$$= 1 + [2\log(1+x) - x]_0^1$$

$$= 1 - x + 2\log(1+x)$$

Second approximation,

$$y_2 = y_0 + \int_{x_0}^{x} f(x, y_1)\,dx$$

$$= 1 + \int_0^x \frac{y_1 - x}{y_1 + x}\,dx$$

$$= 1 + \int_0^x \frac{1 - 2x + 2\log(1+x)}{1 + 2\log(1+x)}\,dx$$

$$= 1 + \int_0^x \left[1 - \frac{2x}{1 + 2\log(1+x)}\right] dx$$

$$= 1 + x - 2\int_0^x \frac{x}{1 + 2\log(1+x)}\,dx$$

which is difficult to integrate. Sometimes we get the complicated integrals and it is the practical difficulty associated with this method. Here, in this example only first approximation can be obtained and so it gives the approximate value for y for $x = 0.1$.

Thus, when $x = 0.1$,

$$y_1 = 1 - 0.1 + 2 \log (1.1)$$
$$= 0.9828$$

EXERCISES

1. Solve $y' = y - x$, $y(0) = 2$.
2. Solve $y' - x^2y = x$, $y(0) = 1$.
3. Solve $y' = x^2 + y^2$, $y(0) = 0$.
4. Solve $y' = x + yz$, $y(0) = 1$, $z' = x^2 - y^2$, $z(0) = 0.5$.
5. Solve $y'' + xy = x^2$, $y(0) = 1$, $y'(0) = 0$.

EULER'S METHOD

Suppose we want to solve the differential equation

$$\frac{dy}{dx} = f(x, y);\ y(x_0) = y_0 \tag{1}$$

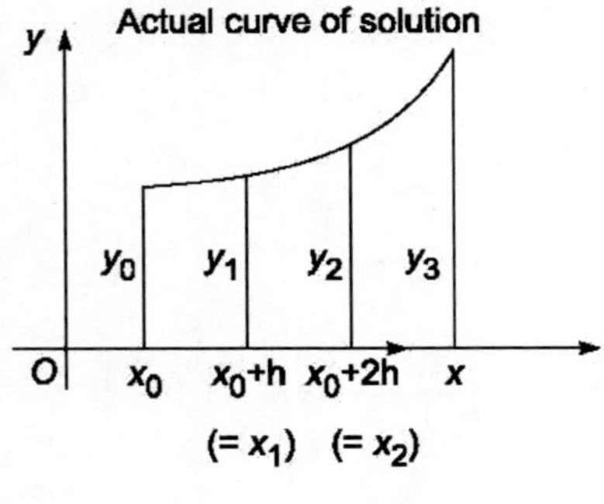

Let x_0, $x_1 (= x_0 + h)$, $x_2 (= x_0 + 2h)$, ... be equidistant values of x. The actual curve of solution is as shown. We want to find the approximate value y_1 corresponding to $x = x_1$.

In Euler's method, we use the property that, in a small interval, a curve is nearly a straight line. In the interval x_0 to x_1 of x, we approximate the curve by the tangent at the point (x_0, y_0). The equation of the tangent at (x_0, y_0) is

$$y - y_0 = \left(\frac{dy}{dx}\right)_{(x_0, y_0)} \cdot (x - x_0) = f(x_0, y_0) \cdot (x - x_0) \qquad \left[\because \frac{dy}{dx} = f(x, y)\right]$$

$$y = y_0 + f(x_0, y_0)\ (x - x_0)$$

This is the value of y corresponding to a point on the tangent. Since the curve is approximate by the tangent in the interval (x_0, x_1), the value of y on the curve corresponding to $x = x_1$ is given by the above value of y in (1) approximately.

Hence, putting $x = x_1$

$$y_1 = y_0 + f(x_0, y_0) \cdot (x_1 - x_0)$$
$$= y_0 + h \cdot f(x_0, y_0)$$

Similarly, approximating the curve in the next interval (x_1, x_2) by a line through (x_1, y_1) whose slope is $f(x_1, y_1)$, we get,

$$y_2 = y_1 + h\, f(x_1, y_1)$$
$$y_3 = y_2 + h\, f(x_2, y_2)$$

In general, $y_{m+1} = y_{m+h}\, f(x_m, y_m)$

$$m = 0, 1, 2, 3, \ldots$$

This is referred to as the Euler's algorithm.

The above can also be put in the form $y(x + h) = y(x) + h\, f(x, y)$.

Note: The Euler's method is a single step integration which depends on Taylor's series expansion of the function $y(x)$ about some starting point, say y_0.

Example 9. Using Euler's method, compute y for $x = 0.1$ and 0.2 choosing $h = 0.1$ from $y^1 = y - \dfrac{2x}{y}$, $y(0) = 1$.

***Solution*:** Since $y_0 = 1$

$$y_1 = 1 + (0.1)\ [1] = 1.1$$

$$y_2 = 1.1 + (0.1)\left[1.1 - \frac{2(0.1)}{1.1}\right]$$

$$= 1.19$$

$$y_{m+1} = y_{m+h}\, f(x_m, y_m)$$

Hence, $y(0.1) = 1.1$

and $y(0.2) = 1.19$

The true value correct to two decimal places are 1.10 and 1.18 respectively.

Example 10. Given $\dfrac{dy}{dx} = \dfrac{y - x}{y + x}$, with $y = 1$ for $x = 0$. Find y approximately for $x = 0.1$ by Euler's method. (Five steps)

Solution: Here, it is required the value of $x = 0.1$ from $x = 0$ in five steps. We break up the intervals 0 to 0.1 into five sub-intervals by introducing the points x_1, x_2, x_3, x_4, x_5. Let $h = 0.02$. We shall find the values of y at $x = 0.02, 0.06, 0.08$ and 0.1 successively. Thus, we have $x_0 = 0$, $y_0 = 1$, $h = 0.02$, $f(x, y) = \frac{y - x}{y + x}$, using $y_{n+1} = y_n + h\, f(x_n, y_n)$, we get

$$y_1 = y_0 + h\, f(x_0, y_0) = 1 + (0.02)\left[\frac{1-0}{1+0}\right] = 1.02$$

$$y_2 = y_1 + h\, f(x_1, y_1) = 1.02 + (0.02)\frac{1.02 - 0.02}{1.02 + 0.02} = 1.0392$$

$$y_3 = y_2 + h\, f(x_2, y_2) = 1.0392 + (0.02)\frac{1.0392 - 0.04}{1.0392 + 0.04} = 1.0577$$

$$y_4 = y_3 + h\, f(x_3, y_3) = 1.0577 + (0.02)\frac{1.0577 - 0.06}{1.0577 + 0.06} = 1.0756$$

$$y_5 = y_4 + h\, f(x_4, y_4) = 1.0756 + (0.02)\frac{1.0756 - 0.08}{1.0756 + 0.08} = 1.0928.$$

Hence, $y = 1.0928$ when $x = 0.1$.

Example 11. Determine the numerical solution at $x = 0.8$ for $\frac{dy}{dx} = \sqrt{x + y}$ given is $y(0.4) = 0.41$. Assume step length $h = 0.2$.

Solution: This problem is solved by the Euler's method.

$$\begin{aligned}
x_0 &= 0.4,\ y_0 = 0.41 \\
x_1 &= x_0 + h = 0.4 + 0.2 = 0.6,\ y_1 = 2 \\
x_2 &= x_0 + 2h = x_1 + h = 0.6 + 0.2 = 0.8,\ y_2 = ? \\
y_1 &= y_0 + h\, f(x_0, y_0) \\
&= 0.41 + 0.2\sqrt{0.4 + 0.4} \\
&= 0.59 \\
y_2 &= y_1 + h\, f(x_1, y_1) \\
&= 0.59 + 0.2\sqrt{0.6 + 0.59} \\
&= 0.808
\end{aligned}$$

Example 12. Solve $\frac{dy}{dx} + y^2 = 0$, given boundary conditions $y = 1$ at $t = 0$. Find approximately the value of y for $t = 1$ by Euler's method in five steps.

Solution:

$$t = 0,\ y_0 = 1$$
$$t = 0\cdot 2,\ y_1 = 1$$
$$t = 0\cdot 4,\ y_2 = 1$$
$$t = 0\cdot 6,\ y_3 = 1$$
$$t = 0\cdot 8,\ y_4 = 1$$
$$t = 1\cdot 0,\ y_5 = 1$$

$$\frac{dy}{dt} = -y^2$$

$$\begin{aligned} y_1 &= y_0 + h\,f(t_0, y_0) \\ &= 1 + 0.2\,[-(1)^2] \\ &= 0.8 \\ y_2 &= y_1 + h\,f(t_1, y_1) \\ &= 0.8 + 0.2\,[-(0.8)^2] \\ &= 0.672 \\ y_3 &= y_2 + h\,f(t_2, y_2) \\ &= 0.672 + 0.2\,(-(0.672)^2) \\ &= 0.5811 \\ y_4 &= y_3 + h\,f(t_3, y_3) \\ &= 0.5811 + 0.2\,(-(0.5811)^2) \\ &= 0.514 \\ y_5 &= y_4 + h\,f(t_4, y_4) \\ &= 0.514 + 0.2\,(-(0.514)^2) \\ &= 0.4610 \end{aligned}$$

Example 13. Solve the equation $\dfrac{dy}{dx} = 1 - y$

with $x = 0$, $y = 0$, using Euler's algorithm and tabulate the solutions of $x = 0.1$, 0.2, 0.3, 0.4.

Solution: Given: $\dfrac{dy}{dx} = f(x, y)$

Euler's algorithm is: $y_{m+1} = y_m + hf(x_m, y_m)$, (1)

Here, $\dfrac{dy}{dx} = 1 - y$, hence, $f(x, y) = 1 - y$

and when $x = 0$, $y = 0$, i.e. $x_0 = 0$, $y_0 = 0$, $h = 0.1$

From (1), taking $m = 0$

$$\begin{aligned} y_1 &= y_0 + h\,f(x_0, y_0) \\ &= 0 + (0.1)\,(1 - y_0) \\ &= 0.1 \times 1 = 0.1 \end{aligned}$$

Next, we have

$$x_1 = x_0 + h = 0 + 0.1 = 0.1$$

$$\begin{aligned} y_2 &= y_1 + h\,f(x_1, y_1) \\ &= 0.1 + 0.1 \times (1 - y_1) \\ &= 0.1 + 0.1 \times (1 - 0.1) \\ &= 0.1 + 0.1 \times .9 = 0.19 \end{aligned}$$

$$x_2 = x_1 + h = 0.2$$

$$\begin{aligned} y_3 &= y_2 + h\,f(x_2, y_2) = 0.19 + 0.1 \times (1 - y_2) \\ &= 0.19 + 0.1 \times (1 - 0.19) = 0.19 + 0.1 \times 0.81 \\ &= 0.271. \end{aligned}$$

$$x_3 = x_2 + h = 0.3$$

$$\begin{aligned} y_4 &= y_3 + h\,f(x_3, y_3) \\ &= 0.271 + 0.1 \times (1 - y_3) \\ &= 0.271 + 0.1 \times (1 - 0.271) \\ &= 0.271 + 0.1 \times 0.729 \\ &= 0.3439 \end{aligned}$$

MODIFIED EULER'S METHOD

To solve $\dfrac{dy}{dx} = f(x, y),\ y(x_0) = y_0$ (1)

We use the Euler's method for first approximation

$$y_1 = y_0 + h\,f(x_0, y_0) \tag{2}$$

We compute $y(x_i) = y_i$, where $x_i = x_0 + ih$, $i = 1, 2, 3, 4$.

$$y(x_1) = y(x_0 + h) = y_1$$

$$y_1 = y_0 + hy'_0 + \frac{h^2}{2!}\,y''_0 + \ldots \tag{3}$$

Now, from the given equation, $y' = f(x, y)$

$$y'' = f_x(x, y) + f(x, y)\,f_y(x, y)$$

From (3), we have

$$y_1 = y_0 + hy'_0 + \frac{h^2}{2!}[f_x(x_0, y_0) + f(x_0, y_0)\,f_y(x_0, y_0)] + 0(h^3)$$

$$\begin{aligned} y_1 = y_0 + \frac{h}{2}[&f(x_0, y_0) + f(x_0, y_0) + hf_x(x_0, y_0) \\ &+ hf(x_0, y_0)\,f_y(x_0, y_0)] + 0(h^3) \end{aligned}$$

$$y_{n+1} = y_0 + \frac{h}{2}[f(x_n, y_n) + f(x_{n+1}, y_{n+1})] + 0(h^3)$$

$$\text{Error} = 0(h^3)$$

Example 14. Use modified Euler's method to compute $y(0.2)$ with $h = 0.1$, from $y' = y - \frac{2x}{y}$, $y(0) = 1$.

Solution:

$$\tilde{y}_1 = y_0 + h\, f(x_0, y_0)$$

$$= 1 + (0.1)\ [1] = 1.1$$

$$\tilde{y}_2 = y_0 + \frac{h}{2}[f(x_0, y_0) + f(x_1, \tilde{y}_1)]$$

$$= 1 + \frac{(0.1)}{2}\left[1 + 1.1 - \frac{2(0.1)}{1.1}\right]$$

$$= 1.096 \text{ (modified value)}$$

$$\tilde{y}_2 = y_1 + h\, f(x_1, y_1)$$

$$= 1.096 + (0.1)\left[1.096 - \frac{0.2}{1.096}\right]$$

$$= 1.1874$$

$$\tilde{y}_2 = y_1 + \frac{h}{2}[f(x_1, y_1) + f(x_2, \tilde{y}_2)]$$

$$= 1.096 + \frac{(0.1)}{2}\left[\left(1.096 - \frac{0.2}{1.096}\right) + \left(1.1874 - \frac{0.4}{1.1874}\right)\right]$$

$$= 1.1842$$

Example 15. Solve by modified Euler's method:

$$\frac{dy}{dx} = x + y^2,\ y(0) = 1$$

Solution: Take $h = 0.1$, $x_0 = 0$, $y_0 = 1$

$$x_1 = 0 + 0.1 = 0.1$$

$$\tilde{y}_1 = 1 + (0.1)\ (0 + 1^2) = 1.1$$

$$\tilde{y}_2 = 1 + \frac{1}{2}(0.1)\ [(0 + 1^2) + (0.1 + 1.1^2)]$$

$$= 1.1155 \text{ (modified value).}$$

Example 16. Using modified Euler's method, solve

$$\frac{dy}{dx} = x^2 + y, \text{ where } y = 0.94 \text{ when } x = 0, \text{ for } x = 0.1.$$

Solution: Here $f(x, y) = x^2 + y$

$x_0 = 0, y = 0.94$

Let $x_1 = 0.1$

$\therefore$ $h = x_1 - x_0 = 0.1$

Now, from Euler's method, corresponding to $x = x_1$, we have the first approximation value of y, as

$$\begin{aligned} y_1 &= y_0 + h\, f(x_0, y_0) \\ &= y_0 + h\, [x_0^2 + y_0] \\ &= (0.94) + (0.1)\,(0 + 0.94) \\ &= 1.034 \end{aligned}$$

Then by Euler's modified method, we have

$$y_1^{(1)} = y_0 + \frac{h}{2}[f(x_0, y_0) + f(x_1, y_1)]$$

where

$$\begin{aligned} f(x_0, y_0) &= x_0^2 + y_0 = 0.94 \\ f(x_1, y_1) &= x_1^2 + y_1 \\ &= (0.1)^2 + 1.034 \\ &= 1.044 \end{aligned}$$

$\therefore$

$$\begin{aligned} y_1^{(1)} &= (0.94) + \frac{0.1}{2}[0.9 + 1.044] \\ &= 1.0392 \end{aligned}$$

Now

$$y_1^{(2)} = y_0 + \frac{h}{2}[f(x_0, y_0) + f(x_1, y_1)]$$

where

$$\begin{aligned} f(x_1, y_1^{(2)}) &= x_1^2 + y_1^{(1)} \\ &= (0.1)^2 + 1.0392 \\ &= 1.0492 \end{aligned}$$

$\therefore$

$$y_1^{(2)} = (0.94) + 0.1/2\;(0.94 + 1.0492) = 1.0394$$

Example 17. Given that $\frac{dy}{dx} = \log(x + y)$ with the initial condition that $y = 1$ when $x = 0$, use Euler's modified method to find y for $x = 0.2$

Solution: Given $\frac{dy}{dx} = \log(x + y) = f(x, y)$

with $x_0 = 0, y_0 = 1$

$$f(x_0, y_0) = \log 1 = 0$$

Suppose $x_1 = 0.2$, where y_1 is computed $x_1 = x_0 + h$

Taking $h = x_1 - x_0 = 0.2 - 0 = 0.2$

We have
$$y_1 = y_0 + h\, f(x_0, y_0)$$
$$= 1 + 0.2 \times 0 = 1$$

$$\tilde{y}_2 = y_0 + \frac{h}{2}[f(x_0, y_0) + f(x_1, \tilde{y}_1)]$$

$$= 1 + \frac{(0.2)}{2}[0 + \log 1.2] = 1 + 0.0079 = 1.0079$$

$$\tilde{y}_3 = y_0 + \frac{h}{2}[f(x_0, y_0) + f(x, \tilde{y}_2)]$$

$$= 1 + \frac{0.2}{2}[0 + \log 1.2029]$$

$$= 1 + 0.0082 = 1.0082$$

$$\tilde{y}_4 = y_0 + \frac{h}{2}[f(x_0, y_0) + f(x, \tilde{y}_3)]$$

$$= 1 + \frac{0.2}{2}[0 + \log 1.2082] = 1 + 0.0082$$

$\therefore$ $\tilde{y}_3 = \tilde{y}_4$

This implies that $y_1 = 1.0082$

at $x = 0.2$

Example 18. Use modified Euler's method to find an approximate solution at the point 0.2 of the problem

$$y' = -y;\ y(0) = 1, \text{ with step length } h = 0.2$$

Solution: Here $f(x, y) = -y$

$$x_0 = 0$$
$$y_0 = 1$$

The formula

$$\tilde{y}_1 = y_0 + h\, f(x_0, y_0)$$
$$= 1 + (0.2)(-1)$$

$$= 1 - 0.2 = 0.8$$

$$y_1 = y_0 + \frac{h}{2}[f(x_0, y_0) + f(x_1, \tilde{y}_1)]$$

$$= 1 + \frac{0.2}{2}[-1 + (-0.8)]$$

$$= 1 + 0.1\ [-1.8]$$

$$= 1 - 0.18 = 0.82$$

This is the approximation at $x = 0.2$.

Example 19. Solve the simultaneous equations by the modified Euler's method with $h = 0.1$:

$$y' = z,\ y(0) = 1,$$

$$z' = y + xz,\ z(0) = 0.$$

Solution:

$$\tilde{y}_1 = y_0 + h\, f(x_0, y_0, z_0) = 1 + (0.1)\,(0) = 1$$

$$\tilde{z}_1 = z_0 + h\, g(x_0, y_0, z_0) = 0 + (0.1)\,(1) = 0.1$$

$$y_1 = y_0 + \frac{h}{2}[f(x_0, y_0, z_0) + f(x_1, \tilde{y}_1, \tilde{z}_1)]$$

$$= 1 + \frac{0.1}{2}[0 + 0.1] = 1.005.$$

$$z_1 = 0 + \frac{0.1}{2}[g(x_0, y_0, z_0) + g(x_1, \tilde{y}_1, \tilde{z}_1)]$$

$$= 0.05\ [1 + 1 + (0.1)\,(0.1)]$$

$$= 0.1005.$$

Here, we have actually solved the second order equation

$$y'' = y + xy',\ y(0) = 1,\ y'(0) = 0.$$

EXERCISES

1. Solve $y' = x + y$, $y(0) = 1$ by the modified Euler's method. Compute y for $x = 0.05$ and 0.1.
2. Solve $y' = y - \dfrac{2x}{y}$, $y(0) = 1$ by the Euler's method with $h = 0.2$. Compute y for $x = 0.2\ (0.2)\ 1.0$, i.e. for $x = 0.2, 0.4, 0.6, \ldots, 1.0$.
3. Solve $y' = 2xy$, $y(0) = 1$ by the modified Euler's method. Compute y for $x = 0.125\ (0.125)\ 1.000$.

4. Solve $y'' + \frac{y'}{x} + y = 0, y(1) = 0.77, y'(1) = -0.44$. Compute y for $x = 0.1$ and 0.2 by the Euler's method.
5. Solve $y' = 1 + x - y^2, y(0) = 1$. Find $y(0.1)$ by the modified Euler's method with $h = 0.02$.
6. Solve $y' = x \ln y - y \ln x, y(1) = 1$. Find $y(1.6)$ with $h = 0.1$ by the modified Euler's method.
7. Find $y(h)$ from $y' = y, y(0) = 0$ by the modified Euler's method.
8. Using the Euler's method, solve $y' = x + y + xy, y(0) = 1$. Compute $y(0.1)$ with $h = 0.01$.
9. Repeat Ex. 8 by the modified Euler's method with $h = 0.025$.
10. Solve $y' = y + \sin x, y(0) = 2$, by the modified Euler's method. Compute y for $x = 0.1\ (0.1)\ 0.5$.

TAYLOR'S SERIES METHOD

To solve $y' = f(x, y)\ y(x_0) = y_0$ (1)

We get by Taylor's series

$$y(x) = y(x_0) + \frac{y'(x_0)}{\lfloor 1}(x - x_0) + \frac{y''(x_0)}{\lfloor 2}(x - x_0)^2 + \ldots \tag{2}$$

From the initial conditions $y(x_0) = y_0$ and from the given differential equation, $y'(x_0) = f(x_0, y_0)$. Differentiating the given d.e., we get

$$y''(x) = \frac{\partial f}{\partial x} + y'\frac{\partial f}{\partial y} \tag{3}$$

which yields $y''(x_0)$ by substituting $x = x_0, y = y_0$ in the r.h.s. of (3). Similarly, we get the higher derivative $y^{iii}(x_0), y^{iv}(x_0)$ ...etc., and hence (2) gives the desired solution. Obviously, one has to truncate the series in (3) to get an approximate solution as a polynomial. If the series in (2) is truncated after n times, then

$$\text{Error} = \frac{y^{(n)}(\xi)}{\lfloor n}(x - x_n)^n \approx 0(h^y) - (h)$$

where $\quad h = x - x_0$

Example 20. Solve $y^i = x^2 + y^2, y(0) = 1$.

Differentiating the given equation

$$y^{ii} = 2x + 2yy^i$$
$$y^{iii} = 2 + 2yy^{ii} + 2y'^2$$
$$y^{iv} = 6y'y^{ii} + 2yy^{iii}$$

$$y^{v} = 8y'y^{iii} + 6y''^2 + 2yy^{iv}$$
$$y^{vi} = 20y^{ii}y^{iii} + 10y'y^{iv} + 2yy^{v}.$$

Solution: Substituting $x = 0$, $y = 1$, we get $y^{i} = 1$, $y^{ii} = 2$, $y^{iii} = 8$, $y^{iv} = 28$, $y^{v} = 144$, $y^{vi} = 888$

Hence, (2) gives the solution

$$y(x) = 1 + x + x^2 + \frac{4}{3}x^3 + \frac{7}{6}x^4 + \frac{6}{5}x^5 + \frac{37}{30}x^6.$$

Example 21. Using Taylor's series method $\frac{dy}{dx} = 1 + xy$ with $y_0 = 2$

From (i) $y(0.1)$

(ii) $y(0.2)$

and (iii) $y(0.3)$

Solution: The Taylor's series is

$$y_1 = y_0 + \frac{h}{1!}\,y_0^{i} + \frac{h^2}{2!}\,y_0^{ii} + \frac{h^3}{3!}\,y_0^{iii} + \ldots \tag{1}$$

Here $x_0 = 0,\ y_0 = 2,\ h = 0.1;\ \frac{dy}{dx} = 1 + xy$; (Given) (2)

Successively differentiating (2), we get

$$y^{ii} = \frac{d^2y}{dx^2} = xy + xy^{i}$$
$$y^{iii} = 2y^{i} + xy^{ii}$$

Now, $y_0^{i} = y^{i}(x_0, y_0) = 1 + x_0\,y_0 = 1.$

$$y_0^{ii} = (y^{ii})_{(x_0,y_0)} = y_0 + x_0\,y_0^{i} = 2$$

$$y_0^{iii} = (y^{iii})_{(x_0,y_0)} = 2y_0^{i} + x_0 y_0^{ii} = 2$$

Using these in (1), we get

(i) $y_1 = 2 + \frac{(0.1)}{1!} + \frac{(0.1)^2}{2!} \times 2 + \frac{(0.1)^3}{3!} \times 2 = 2.1103$

$\therefore\ y(0.1) = 2.1103$

(ii) The Taylor's series for the next approximation is

$$y_2 = y_i + \frac{h}{1!}\,y_1^{i} + \frac{h^2}{2!}y_1^{ii} + \frac{h^3}{3!}y_1^{iii} \tag{3}$$

Now,

$$y_1^{i} = y'_{(x_1,y_1)} = 1 + x_1 y_1 = 1 + (0.1)\,(2.1103) = 1.21103$$

$$y_1^{ii} = (y'')_{(x_1,y_1)} = y_1 + x_1 y_1^{i} = 2.1103 + (0.1)\,(1.21103) = 2.2314$$

$$y_1^{iii} = (y''')_{(x_1,y_1)} = 2y_1^{i} + x_1\, y_1^{ii} = 2(1.21103) + (0.1)\,(2.2314) = 2.6452$$

∴ (3) becomes

$$y_2 = 2.1103 + \frac{(0.1)}{1!}(1.21103) + \frac{(0.1)^2}{2!}(2.2314) + \frac{(0.1)^3}{3!}(2.6452)$$

$$= 2.2430$$

$$\therefore \quad y(0.2) = 2.2430$$

(iii) The Taylor's series for 3rd approximation is

$$y_3 = y_2 + \frac{h}{1!}y_2^{i} + \frac{h^2}{2!}y_2^{ii} + \frac{h^3}{3!}y_2^{iii} \tag{4}$$

Now, $y_2^{i} = y'_{(x_2,y_2)} = (1 + x_2 y_2) = 1.4486$

$$y_2^{ii} = y''_{(x_2,y_2)} = (y_2 + x_2 y_2^{i}) = 2.53272$$

$$y_2^{iii} = y'''_{(x_2,y_2)} = 2y_2^{i} + x_2 \cdot y_2^{ii} = 3.4037$$

∴ (4) becomes,

$$y_3 = 2.2430 + (0.1)\,(1.4486) + \frac{(0.1)^2}{2!}(2.53272) + \frac{(0.1)^3}{3!}(3.4037)$$

$$= 2.4011$$

$$\therefore y(0.3) = 2.4011$$

EXERCISES

Using Taylor's method find the following.

1. $\frac{dy}{dx} - 2xy = 1$, $y_0 = 0$, $y(0.1) = ?$ (upto 3 decimal places)

   (Ans: 0.0993)

2. $\frac{dy}{dx} = x^2 + y^2$; $y(1) = 2.3$ find y at $x = 1.1$ and 1.2

   (Ans: $y(1) = 3.1209$ $y(1.2) = 4.6823$)

3. $y^1 = xy^{\frac{1}{3}}$, $y(1) = 1$ (using first 3 terms of Taylor's series) (upto three decimal places)

(Ans: $y(1.1) = 1.107$, $y(1.2) = 1.228$, $y(1.3) = 1.364$)

4. Solve $y^1 = x - y^2$, $y(0) = 1$ at the point 0.1 upto 4 decimal places

(Ans: $y(0.1) = 0.9138$)

RUNGE-KUTTA METHODS

Runge's formula (second order)

Euler's modified formula is $y_{n+1} = y_n + \frac{h}{2}[f(x_n, y_n) + f(x_{n+1}, y_{n+1})]$

or,

$$y_{n+1} = y_n + \frac{h}{2}[f(x_n, y_n) + f(x_n + h, y_n + hf_n)] \quad (1)$$

Let
$$k_1 = h\, f(x_n, y_n)$$
$$k_2 = h\, f[x_n + h, y_n + h\, f(x_n, y_n)]$$
or
$$k_2 = h\, f(x_{n+h}, y_{n+k_1})$$

Putting the values of k_1 and k_2, we get

$$y_{n+1} = y_n + \frac{1}{2}(k_1 + k_2) \quad (2)$$

This is known as Runge's formula of order 2.

Example 22. Apply Runge's formula of order 2 approximate the value of y when $x = 1.1$, given $\frac{dy}{dx} = 3x + y^2$ and $y = 1.2$ when $x = 1$.

Solution: Here we have $x_0 = 1$, $y_0 = 1.2$, $h = 0.1$

$$f(x, y) = 3x + y^2$$
$$f(x_0, y_0) = 3(1)^1 + (1.2)^2 = 4.44$$
$$k_1 = h\, f(x_0, y_0)$$
$$= 0.1 \times 4.44$$
$$= 0.444$$
$$k_2 = h\, f(x_0 + h, y_0 + k_1)$$
$$= 0.1\, f(1.1,\ 1.2 + 0.444)$$
$$= 0.1\, f(1.1,\ 1.644)$$
$$= 0.1\, [3 \times 1.1 + (1.644)^2]$$
$$= 0.600$$

$$y_{n+1} = y_n + \frac{1}{2}(k_1 + k_2)$$

$$y_1 = 1.2 + \frac{1}{2}(0.444 + 0.600)$$

$$= 1.722$$

Runge's formula (Third order)

$$y_1 = y_0 + \frac{1}{6}(k_1 + 4k_2 + k_3)$$

where
$$k_1 = h\,f(x_0, y_0)$$
$$k_2 = h\,f(x_0 + h, y_0 + k_1)$$
$$k_3 = h\,f(x_0 + h, y_0 + k_2)$$
$$k_4 = h\,f\left(x_0 + \frac{h}{2}, y_0 + \frac{k_1}{2}\right)$$

This is Runge's formula (third order) with an error of the order h^4.

The fourth order Runge-Kutta formula is most commonly used and is often referred to simply as the Runge-Kutta method.

A fourth order R-K formula for solving IVP, $y^1 = f(x, y)\; y(x_0) = y_0$ is given by

$$k_1 = h\,f(x_n, y_n)$$

$$k_2 = hf\left(x_n + \frac{1}{2}h, y_n + \frac{1}{2}k_1\right)$$

$$k_3 = hf\left(x_n + \frac{1}{2}h, y_n + \frac{1}{2}k_2\right)$$

$$k_4 = h\,f(x_n + h, y_n + k_3)$$

$$y_n + 1 = y_n + \frac{1}{6}(k_1 + 2k_2 + 2k_3 + k_4)\;(x_1 = x_0 + h, y_1 = y_0 + \Delta y)$$

For each y, we compute four quantities k_1, k_2, k_3, k_4.

$$\text{Error} = 0\,(h^5)$$

Note: It is to be noted that the calculations for the first interval are exactly the same as for any other interval. The chance in the formulae for the different intervals is only in the values of x and y to be substituted.

Example 23. Given $y^1 = y$
$$y(0) = 1$$

find $y(h)$.

Solution: Here $f(x, y) = y_0$.
we take $k_1 = h$.

Hence, $$k_2 = h\left(1+\frac{h}{2}\right)$$

$$= h + \frac{h^2}{2}$$

$$k_3 = h\left[1+\frac{h}{2}+\frac{h^2}{4}\right] = h + \frac{h^2}{2} + \frac{h^3}{4}$$

$$= h\left[1+\frac{1}{2}k_2\right]$$

$$k_4 = h\left[1+h+\frac{h^2}{2}+\frac{h^3}{4}\right]$$

$$= h + \frac{h^2}{2} + \frac{h^3}{2} + \frac{h^4}{4}$$

$$y(h) = 1 + \frac{1}{6}[k_1 + 2k_2 + 2k_3 + k_4)$$

$$= 1 + h + \frac{h^2}{2} + \frac{h^3}{6} + \frac{h^4}{24}$$

Note that the solution results from the series: $e^h = 1 + h + \frac{h^2}{2}; + \ldots$ after the fifth term, showing that the error is of the order h^5.

Example 24. Use Runge-Kutta method to approximate y, when $x = 0.1$ and $x = 0.2$, given that $x = 0$ when $y = 1$ and $\left(\frac{dy}{dx}\right) = x + y$.

Solution: Let us take $h = 0.1$,

Here $f(x, y) = x + y$

$$k_1 = hf(x_0, y_0) = h(x_0 + y_0)$$
$$= 0.1 \times (0 + 1.0) = 0.1$$

$$k_2 = h\, f\left(x_0 + \frac{h}{2}, y_0 + \frac{k_1}{2}\right)$$

$$= h\, f\left(0+\frac{0.1}{2}, 1+\frac{0.1}{2}\right)$$

$$= h\,f(0.05,\ 1.05)$$
$$= 0.1[(0.05 + 1.05)] = 0.11$$

$$k_3 = h\,f\left(x_0 + \frac{h}{2}, y_0 + \frac{k_2}{2}\right) = h\,f\left(0.05, 1 + \frac{0.11}{2}\right)$$
$$= 0.1\{0.05 + 1.055\}$$
$$= 0.1105$$
$$k_4 = h\,f(x_0 + h,\ y_0 + k_3)$$
$$= h\,f(0 + 0.1,\ 1 + 0.1105) = h\,f(0.1,\ 1.1105)$$
$$= 0.1\ [0.1 + 1.1105]$$
$$= 0.12105$$

$\therefore$
$$\Delta y = \frac{1}{6}[k_1 + 2k_2 + 2k_3 + k_4]$$
$$= \frac{1}{6}[0.1 + 0.22 + 0.221 + 0.12105]$$
$$= 0.11034$$

Thus, $x_1 = x_0 + h = 0.1$

and $y_1 = y_0 + \Delta y$
$$= 1 + 0.1103$$
$$= 1.1103$$

Now for the 2nd interval, we have

$$k_1 = hf(x_1,\ y_1)$$
$$= 0.1(0.1 + 1.1103) = 0.12103$$

$$k_2 = h\,f\left(x_1 + \frac{h}{2}, y_1 + \frac{k_1}{2}\right)$$
$$= 0.1(0.1 + 0.05 + 1.1103 + 0.06051) = 0.13208,$$

$$k_3 = h\,f\left(x_1 + \frac{h}{2}, y_1 + \frac{k_2}{2}\right)$$
$$= 0.1\ (0.1 + 0.05 + 1.1103 + 0.06604)$$
$$= 0.13263$$
$$k_4 = hf(x_1 + h,\ y_1 + k_3)$$
$$= 0.1\ (0.1 + 0.1 + 1.1103 + 0.13262)$$
$$= 0.14429.$$

$\therefore$
$$\Delta y = \frac{1}{6}\{0.12103 + 0.2641 + 0.26226 + 0.14429\}$$
$$= 0.13246.$$

Hence, $x_2 = 0.2$

$y_2 = y_1 + \Delta y = 1.1103 + 0.13246 = 1.2428$

Example 25. Given $y^1 = f(x)$,

$y(x_0) = y_0$, find y_1

Solution:

Here

$$k_1 = h\, f(x_0)$$

$$k_2 = h\, f\left(x_0 + \frac{h}{2}\right)$$

$$k_3 = h\, f\left(x_0 + \frac{h}{2}\right)$$

$$k_4 = h\, f(x_0 + h)$$

Hence,

$$y_1 = y_0 + \frac{1}{6}[k_1 + 2k_2 + 2k_3 + k_4]$$

$$= y_0 + \frac{h}{6}\left[f(x_0) + 4f\left((x_0) + \frac{1}{2}h\right) + f(x_0 + h)\right]$$

This shows that the solution is obtained by integrating the d.e. using Simpson's $\frac{1}{3}$ rule.

Example 26. Find $y(0.1)$ given

$$y^1 = x^2 + y^2,\; y(0) = 1$$

Solution:

$$k_1 = (0.1)\;[0 + 1] = 0.1,\; h = 0.1$$

$$k_2 = (0.1)\;[(0.05)^2 + (1.05)^2] = 0.1105$$

$$k_3 = (0.1)\;[(0.05)^2 + (1.05525)^2]$$

$$= 0.11161$$

$$k_4 = (0.1)\;[(0.1)^2 + (1.11161)^2]$$

$$= 0.12457$$

$$y(0.1) = y_1 = 1 + \frac{1}{6}[0.1 + 2(0.1105 + 0.111 + 0.12457)]$$

$$= 1.11146$$

$$k_3 = (0.2)\left[1.2626 - \frac{0.6}{1.2626}\right] = 0.1575$$

$$k_4 = (0.2)\left[1.3407 - \frac{0.8}{1.3407}\right]$$

$$= 0.1488$$

$$y(0.4) = 1.1832 + \frac{1}{6}(k_1 + 2(k_2 + k_3) + k_4) = 1.3416$$

Example 27. Compute y for $x = 0.2$, and 0.4 given $y^1 = y - \frac{2x}{y}$, $y(0) = 1$.

Solution: With $h = 0.2$

$$k_1 = 0.2$$

$$k_2 = (0.2)\left[1.1 - \frac{0.2}{1.1}\right]$$

$$= 0.1836$$

$$k_3 = (0.2)\left[1.0918 - \frac{0.2}{1.0918}\right]$$

$$= 0.1817.$$

$$k_4 = (0.2)\left[1.1817 - \frac{0.4}{1.1817}\right]$$

$$= 0.1681.$$

$$y(0.2) = y_1 = 1 + \frac{1}{6}[0.2 + 2(0.1836 + 0.1817) + 0.1686]$$

$$= 1.1832$$

To find $y(0.4) = y_2$, we continue the process,

$$k_1 = (0.2)\left[1.1832 - \frac{0.4}{1.1832}\right]$$

$$= 0.1690.$$

$$k_2 = (0.2)\left[1.2677 - \frac{0.6}{1.2677}\right] = 0.1589$$

EXERCISES

Use the fourth-order Runge-Kutta method to solve the following problems:

1. Find $y\,(h)$ if $y' = x + y$, $y(0) = 1$. $\left(\text{Ans:} 1 + h + h^2 + \frac{1}{3}h^3 + \frac{1}{12}h^4\right)$

2. Find $y\,(h)$ if $y' = -\,2xy$, $y(0) = 1$. $\left(\text{Ans:} 1 - h^2 + \frac{1}{2}h^4 - \frac{1}{6}h^6\right)$

3. Compute y for $x = 1.1\ (0.1)\ 1.5$ if $y' = \dfrac{x^2 + y^2}{2xy}$, $y(1) = 2$.

(Ans: 2.1237, 2.2450, 2.3643, 2.4819, 2.5981

Exact Solution: $\sqrt{x^2 + 3x} = y$)

4. Compute y for $x = 0.1\ (0.1)\ 0.5$ if $y' = \dfrac{xy}{1 + x^2}$, $y(0) = 1$.

(Ans: 0.9950, 0.9806, 0.9578, 0.9285, 0.8944,

Exact Solution: $y = \sqrt{1 + x^2}$)

5. Compute y for $x = 0.1\ (0.1)\ 0.5$ given $y' = x^2 + 0.25y^2$, $y(0) = -1$
(Ans: – 0.9753, – 0.9498, – 0.9215, – 0.8887, – 1.8494)

6. Compute $y\,(2)$ with step size $h = 0.2$ given $y' = \dfrac{y}{x} - y^2$, $y\,(1) = 1$.

(Ans: 0.80).

7. Compute y for $x = 0.1\ (0.1)\ 0.3$ from $y' + y + xy^2 = 0$, $y(0) = 1$.
(Ans: 0.9006, 0.8046, 0.7144)

8. Compute y for $x = 0.1\ (0.1)\ 0.3$, if $y' = (2 \tan x)\ (1 + y)$, $y\,(0) = 1$.
(Ans: 1.0201, 1.0819, 1.1894)

9. Compute y for $x = 0.1\ (0.1)\ 0.3$, given $y' + 2xy = 2x^3$, $y(0) = 0$.
(Ans: 0.00005, 0.00079, 0.00393)

10. Compute y for $x = 0.1\ (0.1)\ 0.3$ given $y' = \cos(x + y) + x - y$, $y\,(0) = 0$.
(Ans: 0.09935, 0.19504, 0.28411)

11. Compute $y(1)$ and $z(1)$ with $h = 0.2$ from
$y' = z + 1$, $y(0) = 1$,
$z' = y - x$, $z\,(0) = 1$. (Ans: 3.72, 2.72)

12. Compute y for $x = 0.1\ (0.1)\ 0.5$ from $y'' = 2yy'$, $y\,(0) = 1$, $y'\,(0) = 1$.
(Ans: 1.1111, 1.2500, 1.4286, 1.6667, 2.0000).

13. Determine y at $x = 0.2\ (0.2)\ 1.2\ (0.4)\ 2.0$ from $y' = \dfrac{1}{x + y}$, $y(0) = 2$.

(Ans: 2.0993, 2.1755, 2.2493, 2.3163, 2.3780,
2.4350, 2.5380, 2.6294)

14. Find y for $x = 0.1\ (0.1)\ 1.0$ if $y' = x^2 + y^2$, $y(0) = 0$.
(Ans: 0.0003, 0.0027, 0.0090, 0.0214, 0.0418,
0.0724, 0.1157, 0.1741, 0.2509, 0.3502)

15. Compute y and z for $x = 0.1$ given $y' = 1 + z$, $y\,(0) = 0$,
$z' = x - y$, $z(0) = 1$. (Ans: 0.1998, 0.9950)

16. Compute y for $x = 0.1\ (0.1)\ 1.0$ if $y' = 1 - x + 4y$, $y(0) = 1$.
(Ans: 1.6089, 2.5050, 3.8294, 5.7928, 8.7093, 13.0477, 19.5071, 29.1306, 43.4740, 64.8581)

17. Compute y (0.2) and y (0.4) from $y' = \frac{1}{2} - x + 2y$, $y(0) = 1$.
(Ans: 1.590, 2.423)

18. Compute y (0.2) and y (0.4) from $y' = x^2 + y^2$, $y(0) = 1$.
(Ans: 1.251, 1.698)

19. Show that the truncation error in second-order Runge-Kutta method is minimum if $\alpha_2 = \frac{2}{3}$.

20. Obtain the third-order Runge-Kutta method for solving $y' = f(x, y)$, $y(x_0) = y_0$.

$$k_1 = h\, f(x_0, y_0).$$

$$k_2 = h\, f\left(x_0 + \frac{1}{2}h,\ y_0 + \frac{1}{2}k_1\right).$$

PREDICTOR-CORRECTOR METHODS

Consider $y^1 = f(x, y)$, $y(x_0) = y_0$ (1)

In order to solve the initial value problem, (1), the modified Euler's method can be stated as

$$y_1^p = y_0 + h\, f(x_0, y_0) \tag{2}$$

$$y_1^c = y_0 + \frac{h}{2}[f(x_0, y_0) + f(x_1, y_1^p)] \tag{3}$$

Here, we predict the value of y_1 by Euler's method and use it in (3) to get a corrected (improved) value. This is a typical case of Predictor-Corrector method. We obtain two important methods, namely, Adam's method and Milne's method. These methods are of 4th order and the error is of order h^5. These methods make use of four starting values of y, namely, y_0, y_1, y_2, and y_3. Hence, these methods are called *multistep methods*. Euler's method, Modified Euler's method and Runge-Kutta method are *single step* methods.

Adam's Method

Consider $y^1 = f(x, y)$ with $y(x_0) = y_0$. Here, we have to compute y_4, provided (y_1, y_2, y_3) are assumed to have been already computed, say, either by Taylor's series method or Runge-Kutta method of order four.

Thus, integrating $y^1 = f(x, y)$ from x_3 to x_4,

we get $$y_4 = y_3 + \int_{x_3}^{x_4} f\,dx$$

$$= y_3 + h\int_3^4 f_k\,dx, \text{ using } x = x_0 + kh$$

$$= y_3 + h\int_3^4 E^{k-3} f_3\,dk$$

$$= y_3 + h\int_3^4 (1-\nabla)^{3-k} f_k\,dk$$

$$= y_3 + h\int_{-1}^0 (1-\nabla)^s f_s\,ds, \text{ (setting } S = 3-k)$$

$$= y_3 + h\left[1 + \frac{1}{2}\nabla + \frac{5}{12}\nabla^2 + \frac{3}{8}\nabla^3\right] f_3, \text{ retaining terms upto } \nabla^3$$

Since $\nabla f_3 = f_3 - f_2$; $\nabla^2 f_3 = f_3 - 2f_2 + f_1$

and $\nabla^3 f_3 = f_3 - 3f_2 + 3f_1 - f_0$,

We find $$y_4 = \frac{h}{24}[-9f_0 + 37f_1 - 59f_2 + 55f_3] \quad (1)$$

The error in the Predictor formula is $h\int_{-1}^0 \frac{s(s-1)(s-2)(s-3)}{4!}\nabla^4 f_0\,ds = \frac{251}{720}$ $h\nabla^4 f_3$ on integration.

Since $\nabla = 1 - E^{-1} = 1 - e^{-hD} = hD$ for small h,

Predictor-error $= \frac{251}{720} h^5 f^{iv}(\xi)$

Again,

$$y_4 = y_3 + \int_{x_2}^{x_4} f\,dx$$

$$= y_3 + h\int_3^4 f_k\,dk$$

$$= y_3 + h\int_3^4 E^{k-4} f_4 \, dk$$

$$= y_3 + h\int_3^4 (1 - \nabla)^{4-k} f_4 \, dk$$

$$= y_3 + h\int_0^1 (1 - \nabla)^s f_4 \, ds \text{ (setting } 4 - k = S)$$

$$= y_3 + h\left[1 - \frac{1}{2}\nabla - \frac{1}{12}\nabla^2 + \frac{1}{24}\nabla^3\right] f_4$$

retaining terms upto ∇^3.

Hence, $$y_4 = y_3 + \frac{h}{24}[f_1 - 5f_2 + 19f_3 + 9f_4] \quad (2)$$

The Runge-Kutta methods are variants of the Taylor's series method. Given the initial value problem,

$$y' = f(x, y) \; y(x_0) = y_0 \quad (1)$$

the Taylor's series method makes use of successive derivatives of y to yield

$$y_1 - y_0 = \sum_{i=1}^{\infty} y_0^{(i)} \frac{h^i}{i!} \quad (2)$$

In the Runge-Kutta methods, the integral in

$$y_1 - y_0 = \int_{x_0}^{x_1} f(x, y) \, dx \quad (3)$$

is replaced by a weighted mean of $f(x, y)$ at various points as

$$y_1 - y_0 = \sum_{i=1}^{n} \lambda_i k_i \quad (4)$$

where
$$\left.\begin{aligned}
k_1 &= h\, f(x_0, y_0) \\
k_2 &= h\, f(x_0 + \alpha_2 h, y_0 + \beta_{21} k_1) \\
k_3 &= h\, f(x_0 + \alpha_3 h, y_0 + \beta_{31} k_1 + \beta_{32} k_2) \\
\ldots & \quad \ldots \quad \ldots \quad \ldots \quad \ldots \\
k_n &= h\, f(x_0 + \alpha_n h, y_0 + \sum_{j=1}^{n-1} \beta_{nj} k_j)
\end{aligned}\right\} \quad (5)$$

The constants α's, β's and λ's are chosen such that the right-hand sides of (2) and (4) agree up to the terms of order h^P for some positive integer p.

Since h is a common factor of all k's, we separate it out so that

$$\sum_{i=1}^{n} \lambda_i k_i = h\, g\,(x_0, y_0, h), \text{ say} \tag{6}$$

giving

$$y_1 - y_0 = h\, g\,(x_0, y_0, h) \tag{7}$$

Expanding the right-hand side of (7) and comparing with the series in (2), we get

$$y_0^{(1)} = i \left.\frac{\partial^{i-1} g_0}{\partial h^{i-1}}\right|_{h=0} \qquad i = 1, 2, \ldots p \tag{8}$$

for some positive integer p. In (8) the subscript zero denotes that the functions are evaluated at $x = x_0$ and $y = y_0$.

Since the solution agrees with the Taylor's series solution up to the term, h^p, the method is said to be of order p and

$$\text{Error} = 0\,(h^{p+1}) \tag{9}$$

In fact, it turns out that $p = n$. It can be shown that the constants α's, β's and λ's satisfy the following two relations:

$$\sum_{i=1}^{n} \lambda_i = 1 \tag{10}$$

$$\sum_{j=1}^{i-1} p_{ij} = \alpha_i,\ i = 2, 3, \ldots n \tag{11}$$

We shall demonstrate (10) and (11) in the case of Runge-Kutta methods of first and second orders.

First-order method: For $n = 1$, (4) and (5) take the form.

$$\begin{aligned} y_1 &= y_0 + \lambda_1 k_1 \\ k_1 &= h\, f(x_0, y_0) \end{aligned} \tag{12}$$

From (6), $g = \lambda_1 f(x_0, y_0)$.

From (8) with $i = 1$, we get $y_0 = g_0 = \lambda_1 f(x_0, y_0)$.

Again integrating (1) from x_2 to x_4, we get

$$y_4 = y_2 + \int_{x_2}^{x_4} f\,dx = y_2 + h\int_{2}^{4} f_k\,dx$$

$$= y_2 + h\int_2^4 E^{k-2} f_2\, dk$$

$$= y_2 + h\int_2^4 (1 + \Delta)^{k-2} f_2\, dk$$

$$= y_2 + h\int_0^2 (1 + \Delta)^s f_2\, ds, \text{ setting } k - 2 = s$$

$$= y_2 + \frac{h}{3}[6f_2 + 6\Delta f_2 + \Delta^2 f_2]$$

Hence, $$y_4 = y_2 + \frac{h}{3}[f_2 + 4f_3 + f_4] \tag{13}$$

is the corrector formula of Milne. Note that this formula is obtained by Simpson's $\frac{1}{3}$ rule of integration.

The error is

$$h\int_0^2 \frac{s(s-1)(s-2)(s-3)}{4!} \Delta^4 f_2\, ds = -\frac{h}{90}\Delta^4 f_2$$

Hence, Corrector-Error $= -\dfrac{h^5}{90} f^{\text{iv}}(\xi)$ (14)

The Milne's predictor-corrector formulae (11) and (13) may be stated as

$$y_4^p = y_0 + \frac{4h}{3}[2f_1 - f_2 + 2f_3] \tag{15}$$

$$y_4^c = y_2 + \frac{h}{3}[f_2 + 4f_3 + f_4^p] \tag{16}$$

where $$f_4^p = f(x_4, y_4^p),\ f_i = f(x_i, y_i) \tag{17}$$

The Adam's method and Milne's method can be extended to a system of first-order equations,

$$y' = f(x, y, z),\ y(x_0) = y_0$$
$$z' = g(x, y, z),\ z(x_0) = z_0 \tag{18}$$

Example 28. Solve $y^1 = y - \frac{2x}{y}$, $y(0) = 1$ given $y(0.1) = 1.0954$; $y(0.2) = 1.1832$; $y(0.3) = 1.2649$, calculate $y(0.4)$ and $y(0.5)$ by Adam's method.

Solution: Since $h = 0.1$, $f(x, y) = y - \frac{2x}{y}$

$$f_0 = 1, f_1 = 0.9129, f_2 = 0.8452, f_3 = 0.7906$$

$$y_4^p = y_3 + \frac{h}{24}[-9f_0 + 37f_1 - 59f_2 + 55f_3]$$

$$= 1.2649 + \frac{0.1}{24}[-9 + 37(0.9129) - 59(0.8452) + 55(0.7906)]$$

$$= 1.3415$$

Hence, $$f_4^c = 1.3415 - \frac{2(0.4)}{1.3415} = 0.7452$$

$$f_4^c = y_3 + \frac{h}{24}[f_1 - 5f_2 + 19f_3 + 9f_4^p]$$

$$= 1.2649 + \frac{0.1}{24}[0.9129 - 5(0.8452) + 19(0.7906) + 9(0.7452)]$$

$$= 1.3416.$$

Hence, we take $y_4 = y(0.4) = 1.3416$,
This gives $f_4 = 0.7454$
Similarly, we calculate

$$y_5^p = 1.3416 + \frac{0.1}{24}[-9(0.9129) + 37(0.8452) - 59(0.7906) + 55(0.7454)]$$

$$= 1.4141 \text{ and } f_5^p = 0.7069$$

$$y_5^c = 1.3416 + \frac{0.1}{24}[0.8452 - 5(0.7906) + 19(0.7454) + 9(0.7069)]$$

$$= 1.4142$$

Hence, $y(0.5) = 1.4142$

Example 29. Solve $y' = x^2 + y^2$, $y(0) = 1$. Compute $y(0.3)$ by Milne's method with $h = 0.1$.

Solution: By Taylor's series method we find $y(-0.1) = 0.9088$, $y(0.1) = 1.1115$, $y(0.2) = 1.2530$.

Now, $y_3^p = y_{-1} + \frac{4h}{3}[2f_0 - f_1 + 2f_2]$

$$= 0.9088 + \frac{0.4}{3}[2 - \{(0.1)^2 + (1.1115)^2\} + 2\ \{(0.2)^2 + (1.2530)^2\}]$$

$= 1.4387.$

$$y_3^c = y_1 + \frac{h}{3}[f_1 + 4f_2 + f_3^p]$$

$$= 1.1115 + \frac{0.1}{3}[(0.1)^2 + (1.1115)^2 + 4\{(0.2)^2 + (1.2530)^2\} + (0.3)^2 + (1.4387)^2]$$

$= 1.4397$. We recompute y_3^c as

$$y_3^c = 1.1115 + \frac{0.1}{3}[(0.1)^2 + (1.1115)^2 + 4\ \{(0.2)^2 + (1.2530)^2\} + (0.3)^2 + (1.4397)^2]$$

$= 1.4398$, which is the desired value of $y(0.3)$.

Example 30. Using Adam's method, find $y(0.4)$ given $y^1 = 1 + xy$, $y(0) = 2$.

Solution: Adam's predictor formula is $y_4^p = y_3 + \frac{h}{24}[55f_3 - 59f_2 + 37f_1' - 9f_0']$

Since $\quad h = 0.1, f(x, y) = 1 + xy$

$x_0 = 0, y_0 = 2 \quad (1)$

By Taylor's series method we have $x_1 = 0.1$, $y(0.1) = f_1 = 2.1103$, $x_2 = 0.2$, $f_2 = 2.243$, $x_3 = 0.3$, $f_3 = 2.4011$.

Now, $\quad f_0' = (f')_{(x_0, y_0)} = 1.$

Similarly, $\quad f_1' = 1.21103, f_2' = 1.4486$

and $\quad f_3' = 1.72033$

Using these in (1), we get

$$y_4^p = 2.4011 + \frac{0.1}{24}[55(1.72033) - 59(1.4486) + 37(1.21103) - 9]$$

$= 2.5884$ (using (1))

Adam's corrector formula is

$$y_4^c = y_3 + \frac{h}{24}[9f_4 + 19f_3 - 5f_2 + f_1] \text{ (using (2))}$$

Now, $\quad y_4' = (y')_{x_4, y_4}$

$$= f_4 = 1 + (0.4)\ (2.5885)$$
$$= 2.0354$$

Therefore, (2) becomes

$$y_4^c = 2.4011 + \frac{0.1}{24}[9(2.0354) + 19(1.72033) - 5(1.4486) + 1.21103]$$
$$= 2.5885$$

Hence, $y(0.4) = 2.5885.$

Example 31. Using the Adam's method, solve

$$y' = xy + z,\ y(0) = 0.$$
$$z' = y - z,\ z(0) = 1.$$

Solution: With $h = 0.1$, find y and z at $x = 0.4$.

Using Taylor's series method, we find

$$y = x - \frac{x^2}{2} + \frac{2}{3}x^3 - \frac{1}{4}x^4 + \frac{23}{120}x^5.$$

$$z = 1 - x + x^2 - \frac{1}{2}x^3 + \frac{7}{24}x^4 - \frac{13}{120}x^5.$$

Hence,

$$y(0.1) = 0.0956,\ y(0.2) = 0.1850,\ y(0.3) = 0.2714.$$
$$z(0.1) = 0.9095,\ z(0.2) = 0.8364,\ z(0.3) = 0.7786.$$
$$f_0 = 1,\ f_1 = 0.9191,\ f_2 = 0.8734,\ f_3 = 0.8600.$$
$$g_0 = -1,\ g_1 = -0.8139,\ g_2 = -0.6514,\ g_3 = -0.5072.$$

$$y_4^p = 0.2714 + \frac{0.1}{24}[-9 + 37(0.9191) - 59(0.8734) + 55(0.8600)]$$
$$= 0.3580.$$

$$z_4^p = 0.7786 + \frac{0.1}{24}[9 - 37\ (0.8139) + 59\ (0.6514) - 55\ (0.5072)]$$
$$= 0.7345.$$

$$f_4^p = 0.4\ (0.3580) + 0.7345 = 0.8777.$$
$$g_4^p = -0.3765.$$

$$y_4^c = 0.2714 + \frac{0.1}{24}[0.9191 - 5\ (0.8734) + 19\ (0.8600) + 9\ (0.8777)]$$

$$= 0.3580.$$

$$z_4^c = 0.7786 + \frac{0.1}{24}[-0.8139 + 5\ (0.6514) - 19\ (0.5072) - 9\ (0.3765)]$$

$$= 0.7345.$$

Hence, $y(0.4) = 0.3580$ and $z\ (0.4) = 0.7345$.

Example 32. Use Milne's method to solve

$$y' = x + z,\ y(0) = 1,$$
$$z' = -xy,\ z\ (0) = 0.$$

Solution: With $h = 0.2$, compute y and z at $x = 0.8$. Obtain the starting values by Taylor's series method.

By Taylor's series method, we get

$$y = 1 + \frac{1}{2}x^2 - \frac{1}{6}x^3 - \frac{1}{40}x^5,$$

$$z = -\frac{1}{2}x^2 - \frac{1}{8}x^4 + \frac{1}{30}x^5.$$

Computing, we find

$$y(0.2) = 1.0187,\quad y\ (0.4) = 1.0691,\quad y\ (0.6) = 1.1421,$$
$$z(0.2) = -0.0202,\quad z(0.4) = -0.0829,\quad z(0.6) = -0.1936.$$
$$f_0 = 0,\quad f_1 = 0.1798,\quad f_2 = 0.3171,\quad f_3 = 0.4064.$$
$$g_0 = 0,\quad g_1 = -0.2037,\quad g_2 = -0.4276,\quad g_3 = -0.6853.$$

$$y_4^p = 1 + \frac{0.8}{3}[2\ (0.1798) - 0.3171 + 2(0.4064)]$$

$$= 1.2281.$$

EXERCISES

1. Given $y' = x^2 + y^2$, $y\ (0) = 1$, compute y for $x = 0.1\ (0.1)\ 0.3$ by the fourth-order Runge-Kutta method and $y\ (0.4)$ by Adam's method.
2. Given $y' = \frac{1}{2} - x + 2y$, $y(0) = 1$, compute $y\ (0.1)$, $y\ (0.2)$, $y(0.3)$ by the fourth-order Runge-Kutta method and $y(0.4)$ by Adam's method.
3. Given $y' = 2y - 1$, $y(0) = 1$, compute y for $x = 0.1, 0.2, 0.3$ by the fourth-order Runge-Kutta method and $y\ (0.4)$ by Adam's method.
4. Compute y for $x = 0.1\ (0.1)\ 0.3$ by the fourth-order Runge-Kutta method and $y\ (0.4)$ by Adam's method, if $y' = 1 - x + 4y$, $y(0) = 1$.

5. Compute $y\,(0.5)$ with $h = 0.1$ by Adam's method if $y' = 2x\,(y - 1)$, $y(0) = 0$. Compute the initial values by Taylor's series method.
6. Solve $y' = x + y$, $y(0) = 1$. Compute $y(0.4)$ by Milne's method. Obtain the starting values by Taylor's series method.
7. Solve $y' = -\dfrac{y}{x} - y^2 \ln x$, $y(1) = 1$. Compute y for $x = 1.5$ with $h = 0.1$ by Milne's method.
8. Solve $y' = 5 \sec x - y \tan x$, $y(0) = 1$. Compute $y(0.5)$ with $h = 0.1$ by Milne's method.
9. Compute $y(0.5)$ with $h = 0.125$ by Milne's method given

$$y' = x^3 - y^2,\ y(0) = 0,$$
$$y_1 = 0.000610351,$$
$$y_2 = 0.000976535,$$
$$y_3 = 0.00494281.$$

10. Compute $y\,(1)$ by Adam's method given $y' = x^2 - y^3$, $y(0) = 1$, $y(0.25) = 0.821028$, $y(0.50) = 0.741168$, $y(0.75) = 0.741043$.

Chapter **10**

Numerical Solution of Partial Differential Equations

INTRODUCTION

Most of the problems of engineering reduce to the solution of partial differential equations. The solution has to be obtained satisfying the given boundary conditions. Theoretical solution of these equations is possible only in some very particular simple cases. So help is taken of solving these equations by numerical methods.

In this chapter, we discuss the finite-difference method of solving the following important partial differential equations.

Two-dimensional Laplace's equation,

$$\frac{\partial^2 u}{\partial x^2} + \frac{\partial^2 u}{\partial y^2} = 0 \tag{1}$$

Two-dimensional Poisson's equation,

$$\frac{\partial^2 u}{\partial x^2} + \frac{\partial^2 u}{\partial y^2} = f(x, y) \tag{2}$$

One-dimensional heat equation,

$$\frac{\partial u}{\partial t} = C \cdot \frac{\partial^2 u}{\partial x^2} \tag{3}$$

One-dimensional wave equation,

$$\frac{\partial^2 u}{\partial t^2} = C^2 \frac{\partial^2 u}{\partial x^2} \tag{4}$$

Equations (1) and (2) are solved in a rectangular region with boundary conditions. Equations (3) and (4) are solved with given initial and boundary conditions.

GRAPHICAL REPRESENTATION OF PARTIAL QUOTIENTS

XY plane is divided into a series of rectangles whose sides are parallel to *x* and *y*-axes such that $\Delta x = h$ and $\Delta y = k$. The grid points or mesh points or lattice points are

$$(x, y), (x + h, y), (x + 2h, y) \dots [(x - h), y], (x - 2h, y) \dots$$

If (x_i, y_i) is any grid point,

$$x_i = x_o + ih, \ y_j = y_o + jk.$$

$$x_i = ih, \ y_j, = jk, \ t, j = 0, 1, 2, \dots \text{ (From the origin)}$$

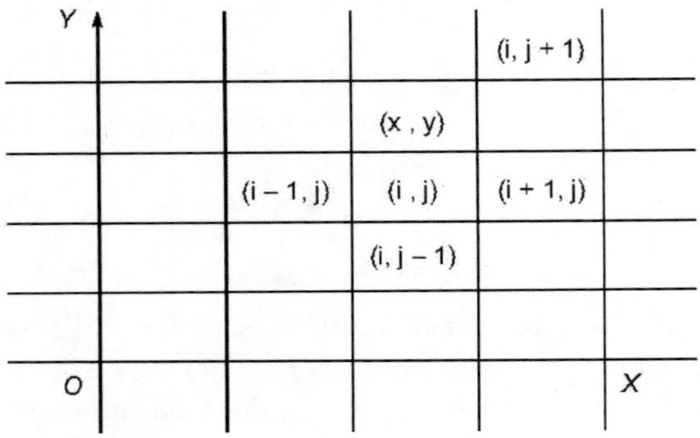

Writing $u(x, y) = u_{ij}$, the above approximations become

$$u_x = \frac{u_{i+1,j} - u_{i,j}}{h} \quad \text{(forward difference)} \quad (1)$$

$$u_x = \frac{u_{i,j} - u_{i-1,j}}{h} \quad \text{(backward difference)} \quad (2)$$

$$u_y = \frac{u_{i,j+1} - u_{i,j}}{k} \quad \text{(forward difference)} \quad (3)$$

$$u_y = \frac{u_{i,j} - u_{i,j-1}}{k} \quad \text{(backward difference)} \quad (4)$$

$$u_{xx} = \frac{u_{i+1,j} - 2u_{i,j} + u_{i-1,j}}{h^2}$$

$$u_{yy} = \frac{u_{i,j+1} - 2u_{i,j} + u_{i,j-1}}{k^2}$$

We can also write

$$u_x = \frac{u_{i+1,j} - u_{i-1,j}}{2h}$$

$$u_y = \frac{u_{i,j+1} - u_{i,j-1}}{2k}$$

GENERAL SECOND ORDER LINEAR PARTIAL DIFFERENTIAL EQUATIONS

A general linear partial differential equation of second order in two independent variables is of the form

$$A(x, y)\frac{\partial^2 u}{\partial x^2} + B(x, y)\frac{\partial u}{\partial y \partial x} + C(x, y)\frac{\partial^2 u}{\partial y^2} + D(x, y)\frac{\partial u}{\partial x} + E(x, y)\frac{\partial u}{\partial y}$$

$$+ F(x, y)\, u = 0$$

i.e.

$$Au_{xx} + Bu_{xy} + Cu_{yy} + Du_x + Eu_y + Fu = 0 \qquad (1)$$

Equation (1) is said to be

(i) elliptic at a point (x, y) in the plane if $B^2 - 4AC < 0$
(ii) parabolic if $B^2 - 4AC = 0$
(iii) hyperbolic if $B^2 - 4AC > 0$

Examples

	Elliptic	*Parabolic*	*Hyperbolic*
1.	$\frac{\partial^2 u}{\partial x^2} + \frac{\partial^2 u}{\partial y^2} = 0$ (Laplace equation in two dimensions)	$\frac{\partial^2 u}{\partial x^2} = \frac{1}{\alpha^2}\frac{\partial u}{\partial t}$ (One-dimensional heat equation)	$\frac{\partial^2 u}{\partial x^2} = \frac{1}{\alpha^2}\frac{\partial^2 u}{\partial t^2}$ (One-dimensional wave equation)

2. $\frac{\partial^2 u}{\partial x^2} + \frac{\partial^2 u}{\partial y^2} = f(x, y)$

(Poisson's equation)

(i) Classify the following equations:

(i) $\frac{\partial^2 u}{\partial x^2} + 2\frac{\partial^2 u}{\partial x \partial y} + \frac{\partial^2 u}{\partial y^2} = 0$

(ii) $x^2 f_{xx} + (1 - y^2) f_{yy} = 0$

Ans: (i) Here, $A = 1$, $B = 2$, $C = 1$, $B^2 - 4AC = 4 - 4 = 0$, $\forall\ x, y$. Hence, the equation is parabolic at all the points.

Ans: (ii) $A = x^2$, $B = 0$, $C = 1 - y^2$

$B^2 - 4AC = -4x^2(1 - y^2)$

$= 4x^2(y^2 - 1)$

$\forall x$, except $x = 0$, x^2 is positive,

If $-1 < y < 1$, $y^2 - 1$ is negative

$\therefore$ $B^2 - 4AC$ is $-$ ve if $-1 < y < 1$, $x \neq 0$

$\therefore$ For $-\infty < x < \infty$ $(x \neq 0)$, $-1 < y < 1$, the equation is elliptic

For $-\infty < x < \infty$, $x \neq 0$, $y < -1$ or $y > 1$, the equation is hyperbolic

For $x = 0$ for all y or for all x, $y = \pm 1$, the equation is parabolic

Elliptic Equations

An important and frequently occurring elliptic equation is Laplace's equation

$$\frac{\partial^2 u}{\partial x^2} + \frac{\partial^2 u}{\partial y^2} = 0 \text{ i.e. } \nabla_u^2 = 0 \text{ or } u_{xx} + u_{yy} = 0$$

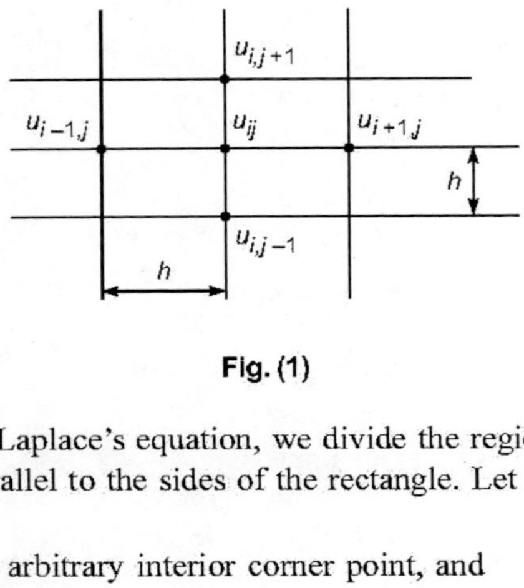

Fig. (1)

In order to solve Laplace's equation, we divide the region into small squares by drawing lines parallel to the sides of the rectangle. Let h be the side of each square.

Let (x_0, y_0) be an arbitrary interior corner point, and

$$x_i = x_0 + ih, \; y_j = y_0 + jk \tag{1}$$

For brevity, we use the notation

$$u_{ij} = u(x_i, y_j) \tag{2}$$

Since $u_{i+1,j} = u(x_i + h, y_j)$, we have by Taylor's series

$$u_{i+1,j} = u_{ij+h} \frac{\partial u}{\partial x} ij + \frac{h^2}{2!}\frac{\partial^2 u_{ij}}{\partial x^2} + \frac{h^3}{3!}\frac{\partial^3 u_{ij}}{\partial x^3} + \ldots$$

and $$u_{i-1,\,j} = u_{ij} - h\,\frac{\partial u}{\partial x}\,ij + \frac{h^2}{2!}\frac{\partial^2 u_{ij}}{\partial x^2} - \frac{h^3}{3!}\frac{\partial^3 u_{ij}}{\partial x^3} + \ldots$$

Adding these two equations, we get

$$u_{i+j,j} + u_{i-i,j} = 2u_{ij} + h^2\frac{\partial^2 u_{ij}}{\partial x^2} + \frac{h^4}{12}\frac{\partial^4 u_{ij}}{\partial x^4} + \ldots$$

Dropping h^4 term, we get the approximation

$$\frac{\partial^2 u_{ij}}{\partial x^2} = \frac{1}{h^2}(u_{i+1,j} - 2u_{ij} + u_{i-i,j}) \tag{3}$$

Similarly,

$$\frac{\partial^2 u_{ij}}{\partial y^2} = \frac{1}{h^2}(u_{i,j+1} - 2u_{ij} + u_{i,j-1}) \tag{4}$$

Error in (3) and (4) are of order h^2.

To solve Laplace's equation

$$\frac{\partial^2 u}{\partial x^2} + \frac{\partial^2 u}{\partial y^2} = 0 \tag{5}$$

We set $x = x_i$ and $y = y_j$ in (5) and use (3) and (4).

Thus, we obtain

$$\frac{u_{i+1,j} - 2u_{ij} + u_{i-1,j}}{h^2} + \frac{u_{i,j+1,} - 2u_{ij} + u_{i,j-1}}{h^2} = 0$$

or $$u_{ij} = \frac{1}{4}(u_{i+1,j} + u_{i-1,j} + u_{i,j+1} + u_{i,j-1}) \tag{6}$$

This shows that the value of u at the point (i, j), ... is the average of its value at four neighbouring points in north, east, west and south directions respectively This can be remembered by the mnemonic

$$u_{ij} = \frac{1}{4}(\mathrm{N} + \mathrm{E} + \mathrm{W} + \mathrm{S}) \tag{7}$$

Equation (6) is called "standard five point formula".

Diagonal Five-point Formula

To solve the Poisson's equation

$$\frac{\partial^2 u}{\partial x^2} + \frac{\partial^2 u}{\partial y^2} = f(x, y) \tag{8}$$

We use (3) and (4) in (8) to obtain

$$u_{ij} = \frac{1}{4}(u_{i+1,j} + u_{i-1,j} + u_{i,j-1} + u_{i,j-1} - h^2 f_{ij}) \tag{9}$$

where $$f_{ij} = f(x_i, y_j)$$

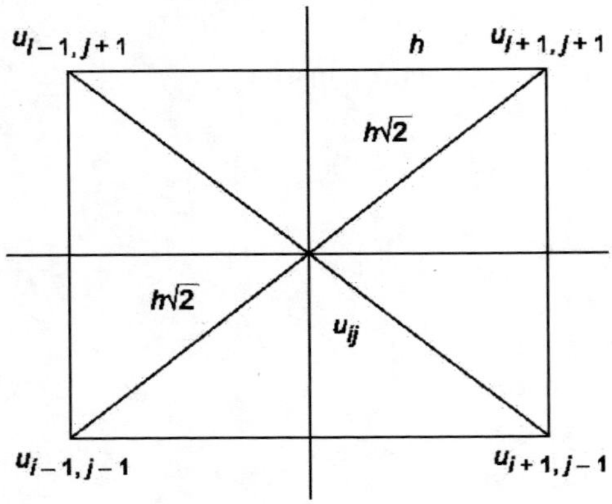

This can be remembered by the mnemonic.

$$u_{ij} = \frac{1}{4}(\text{N} + \text{E} + \text{W} + \text{S} - h^2 f_{ij}) \tag{10}$$

The use of (7) and (10) at each interior corner point of the square mesh yields a system of simultaneous equations which when solved gives the values of u at the grid points.

The use of Gauss-Seidel iteration methods to solve the system of equations obtained by finite-difference method is called Liebmann's method.

$$u_{ij} = \frac{1}{4}(u_{i-1,\,j-1} + u_{i-1,\,j+1} + u_{i+1,\,j-1} + u_{i+1,\,j+1})$$

This is called diagonal five-point formula.

Example 1. Solve $\dfrac{\partial^2 u}{\partial x^2} + \dfrac{\partial^2 u}{\partial y^2} = 0$ in the region shown in the figure which is indicated boundary conditions.

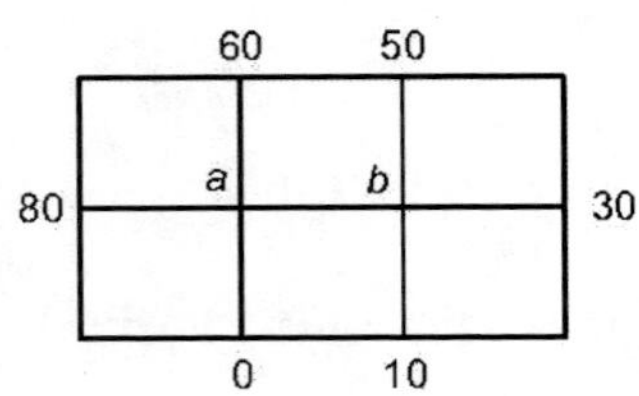

Solution: The unknown values of u at the two interior points are a and b.

Using $$u_{ij} = \frac{1}{4}(\text{N} + \text{E} + \text{W} + \text{S})$$

Hence, $$a = \frac{1}{4}(60 + b + 80 + 0)$$

$$= \frac{1}{4}(b + 140)$$

$$b = \frac{1}{4}(50 + 30 + a + 10)$$

$$= \frac{1}{4}(a + 90)$$

On solving, we get, $\boxed{\begin{array}{l} a = 43.3 \\ b = 33.3 \end{array}}$

Example 2. Solve $\dfrac{\partial^2 u}{\partial x^2} + \dfrac{\partial^2 u}{\partial y^2} = 0$ in $0 < x < 1$, $0 < y < 1$

$u(x, 1) = u(0, y) = 0$, $u(1, y) = 9(y - y^2)$ $u(x, 0) = 9(x - x^2)$

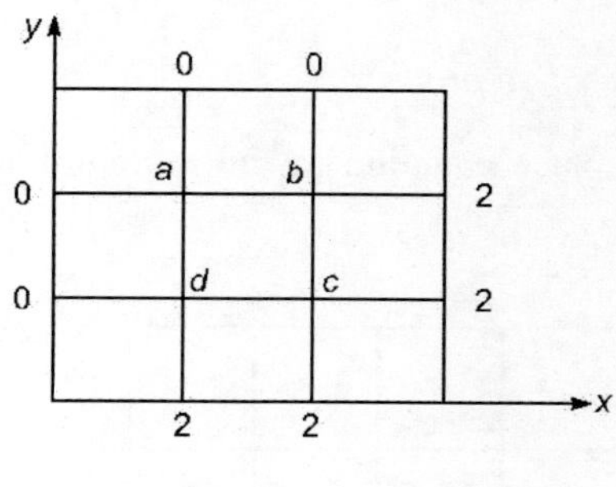

Solution: With $h = \dfrac{1}{3}$, the unknown values of u at the grid points are a, b, c, d.

By symmetry, $d = b$

Using (7), we have

$$a = \frac{2b}{4}, \quad b = \frac{a + c + 2}{4}, \quad c = \frac{2b + 4}{4}$$

Solving, $a = \dfrac{1}{2}$, $b = 1$, $c = \dfrac{3}{2}$

(A) Gauss-Jacobi Method

Let $u_{i,j}^{(n)}$ denote the nth iterative value of $u_{i,j}$ then the iterative procedure to solve (the standard five-point formula as exhibited in the figure shown)

$$u_{i,j} = \frac{1}{4}[u_{i-1,j} + u_{i+1,j} + u_{i,j-1} + u_{i,j+1}]$$

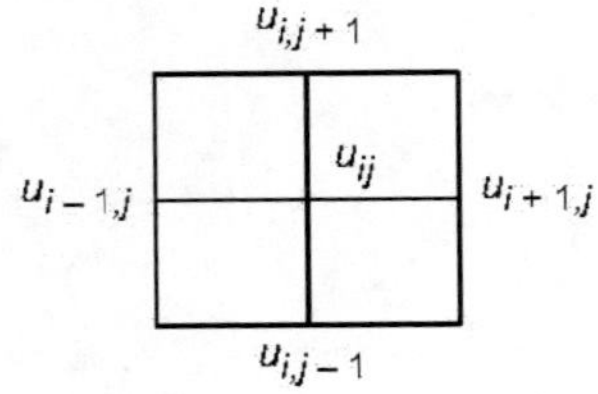

is given by

$$u_{i,j}^{(n+1)} = \frac{1}{4}\left[u_{i-1,j}^{(n+1)} + u_{i+1,j}^{(n)} + u_{i,j-1}^{(n+1)} + u_{i,j+1}^{(n)}\right] \tag{1}$$

for the interior mesh points.

Here, we use the latest available iterative values and hence the rate of convergence will be twice as fast as the Jacobi method.

(B) Gauss-Seidel Method

The iterative procedure to solve 'standard five-point formula' by using Gauss-Seidel iteration method is given by equation (1)

i.e.,
$$u_{i,j}^{(x+1)} = \frac{1}{4}\left[u_{i-1,j}^{(n+1)} + u_{i+1,j}^{(n)} + u_{i,j-1}^{(n+1)} + u_{i,j+1}^{(n)}\right]$$

is known as '*Leibmann's iterative process*'.

Example 3. Solve the Laplace equation in the region shown (in figure) by Liebmann's method.

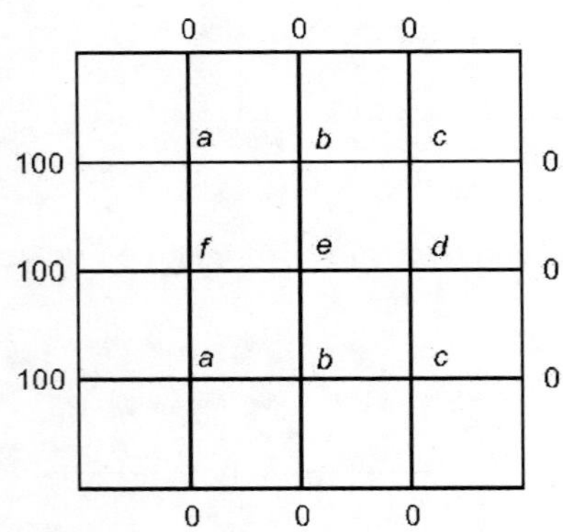

Solution: By symmetry, there are only six unknowns a, b, c, d, e, f.

Using
$$u_{ij} = \frac{1}{4}(u_{i+1,j} + u_{i-1,j} + u_{i,j+1} + u_{i,j-1})$$

(standard five-point formula)

or $u_{ij} = \frac{1}{4}(\text{N} + \text{E} + \text{W} + \text{S})$, we have the six equations

$$a = \frac{b+f+100}{4}, \qquad b = \frac{a+c+e}{4}, \qquad c = \frac{b+d}{4},$$

$$d = \frac{e+2c}{4}, \qquad e = \frac{2b+d+f}{4}, \qquad f = \frac{2a+e+100}{4}$$

Solving by Gauss-Seidel method, we obtain

$a = 42.86,$ $\quad b = 18.75,$ $\quad c = 7.14,$
$d = 9.82,$ $\quad e = 25.00,$ $\quad f = 52.68$

Example 4. Solve $\dfrac{\partial^2 u}{\partial x^2} + \dfrac{\partial^2 u}{\partial y^2} = 0$ for figure given below by Jacobi's method.

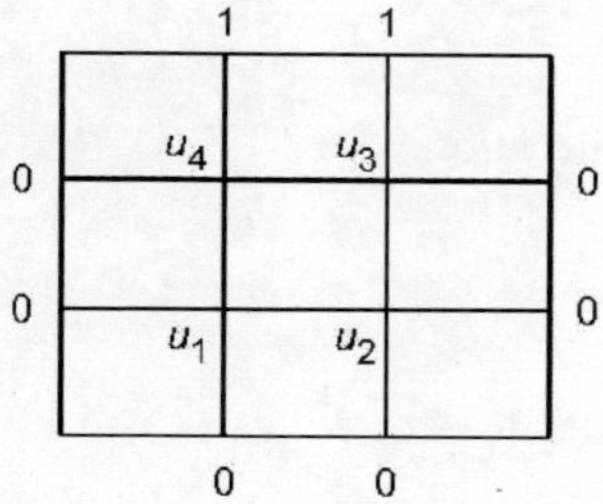

Solution: With the help of given value starting assumption

Let $\quad u_1^{(0)} = 0,$
$u_2^{(0)} = 0$
$u_3^{(0)} = 1,$
$u_4^{(0)} = 1$

Using standard five-point formula

$$u_1^{(1)} = \frac{1}{4}[0 + 0 + u_2^{(0)} + u_4^{(0)}] = \frac{0+0+0+1}{4} = 0.25$$

1st iteration

$$u_2^{(1)} = \frac{1}{4}[0 + 0 + u_1^{(0)} + u_3^{(0)}] = \frac{0+0+0+1}{4} = 0.25$$

$$u_3^{(1)} = \frac{1}{4}[1 + u_2^{(0)} + u_4^{(0)} + 0] = \frac{1+0+1+0}{4} = 0.50$$

$$u_4^{(1)} = \frac{1}{4}[0 + 1 + u_1^{(0)} + u_3^{(0)}] = 0.50$$

Using Jacobi's Method

$$u_{ij}^{(n+1)} = \frac{1}{4}[u_{n-1,j}^{(n)} + u_{i+1,j}^{(n)} + u_{i,j-1}^{(n)} + u_{i,j+1}^{(n)}]$$

2nd iteration

$$u_1^{(2)} = \frac{1}{4}(0 + 0 + 0.25 + 0.50) = 0.1875$$

$$u_2^{(2)} = \frac{1}{4}(0 + 0 + 0.25 + 0.50) = 0.1875$$

$$u_3^{(2)} = \frac{1}{4}(0 + 1 + 0.25 + 0.50) = 0.4375$$

$$u_4^{(2)} = \frac{1}{4}(0 + 1 + 0.25 + 0.50) = 0.4375$$

3rd iteration,

$$u_1^{(3)} = u_2^{(3)} = 0.15625;$$
$$u_3^{(3)} = u_4^{(3)} = 0.40625$$

4th iteration

$$u_1^{(4)} = u_2^{(4)} = 0.14062;$$
$$u_3^{(4)} = u_4^{(4)} = 0.39062$$

5th iteration

$$u_1^{(5)} = u_2^{(5)} = 0.13281;$$
$$u_3^{(5)} = u_4^{(5)} = 0.38281$$

6th iteration

$$u_1^{(6)} = u_2^{(6)} = 0.12891;$$
$$u_3^{(6)} = u_4^{(6)} = 0.37891$$

Example 5. With the step size $h = \frac{1}{2}$, solve

$$\frac{\partial^2 u}{\partial x^2} + \frac{\partial^2 u}{\partial y^2} = -1, \; |x|<1, \; |y|<1$$

$$u(\pm 1, y) = u(x, \pm 1) = 0$$

Solution:

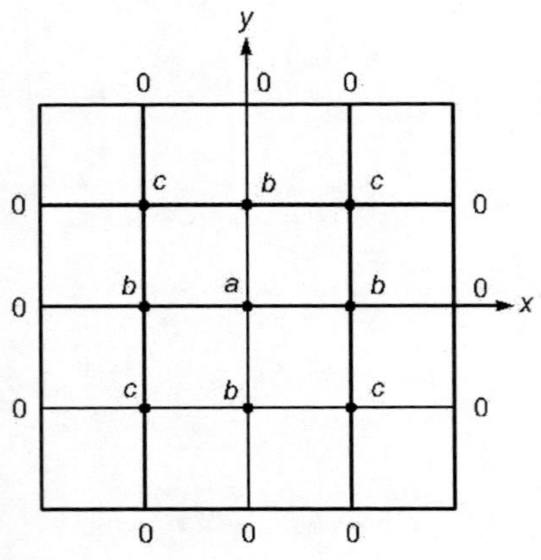

By symmetry, there are only three unknowns: a, b, c. Applying $u_{ij} = \frac{1}{4}(\text{N} + \text{E} + \text{W} + \text{S} - h^2 f_{ij})$, we get

$$a = \frac{1}{4}\left(4b + \frac{1}{4}\right)$$

$$b = \frac{1}{4}\left(a + 2c + \frac{1}{4}\right)$$

$$c = \frac{1}{4}\left(2b + \frac{1}{4}\right)$$

On solving, we get $a = \frac{9}{32}$, $b = \frac{7}{32}$, $c = \frac{11}{64}$.

Example 6. Solve $\nabla_u^2 = -81xy$, $0 < x < 1, 0 < y < 1$ with $h = \frac{1}{3}$,

$$u(0, y) = u(x, 0) = 0.$$

$$u(1, y) = u(x, 1) = 100$$

Solution:

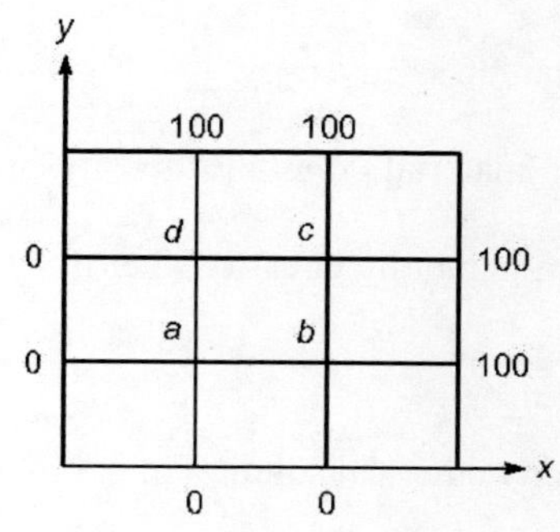

By symmentry, there are only three unknowns a, b, c using

$$u_{ij} = \frac{1}{4}(\text{N} + \text{E} + \text{W} + \text{S}),$$

we get
$$a = \frac{1}{4}(2b + 1)$$

$$b = \frac{1}{4}(a + c + 102)$$

$$c = \frac{1}{4}(2b + 204)$$

Solving, we get $a = 26$, $b = 51$, $c = 76$

EXERCISES

1. Solve $\nabla_u^2 = -10(x^2 + y^2 + 10)$ over the square with sides $x = 0 = y$; $x = 3 = y$; with $u = 0$ on the boundaries and mesh length 1.

   (Ans: $u_1 = 75 = u_4$, $u_2 = 82.5$, $u_3 = 67.5$)

2. Solve $\nabla_u^2 = 8x^2y^2$ in the square mesh given $u = 0$ on the four boundaries dividing the square into 16 subsquare of length 1 unit

   (Ans: $u_1 = u_3 = u_7 = u_9 = -3$
   $u_2 = u_4 = u_6 = u_8 = -2$
   and $u_5 = -2$)

3. Solve the following Poisson equation over the square region bounded by the lines $x = 0$, $y = 0$, $x = 3$, $y = 3$ given that $u = 0$ throughout the boundaries taking $h = 1$.

   (i) $\nabla_u^2 = -(x + y)^2$
   (ii) $\nabla_u^2 = x^3 + y^3$

Heat Equation (Parabolic Equation)

The one-dimensional heat equation is given by $\dfrac{\partial u}{\partial t} = c^2 \dfrac{\partial^2 u}{\partial x^2}$ where $c^2 = \dfrac{k}{s\rho}$.

(S is the specific heat of the material, ρ is the density and k is the thermal conductivity). Clearly, this is a parabolic equation. Let $c^2 = \dfrac{1}{a}$. Then the above equation takes the form

$$u_{xx} - au_t = 0 \tag{1}$$

We develop two methods to find numerical solution of (1) subject to the following boundary and initial conditions

$$u(0,\ t) = T_0 \tag{2}$$

$$u(l,\ t) = T_1 \tag{3}$$

$$u(x,\ 0) = f(x) \tag{4}$$

Bender Schmidt Method

Consider a rectangular mesh in the $x - t$ plane with side lengths h and k in the x and t-directions respectively. Let us denote the mesh point $(x, y) = (ih, jk)$ by i,j. Then, we have

$$u_{xx} = \frac{u_{i+1,j} - 2u_{i,j} + u_{i-1,j}}{h^2}; \text{ and } u_t = \frac{u_{i,j+1} - u_{i,j}}{k}$$

Substituting in (1), we get

$$\frac{u_{i+1,j} - 2u_{ij} + u_{i-1,j}}{h^2} = \frac{a}{k}(u_{i,j+1} - u_{i,j})$$

$$\therefore \qquad u_{i,j+1} - u_{ij} = \frac{k}{ah^2}(u_{i+1,j} - 2_{ui,j} + u_{i-1,j})$$

$$= \lambda(u_{i+1,j} - 2u_{ij} + u_{i-1,j})$$

where
$$\lambda = \frac{k}{ah^2}$$

$$\therefore \qquad u_{i,j+1} = \lambda u_{i+1,j} + (1 - 2\lambda)\, u_{ij} + \lambda u_{i-1,j} \tag{5}$$

Now, the boundary conditions (2), (3) can be written in the difference notation as

$$u_{0,j} = T_0$$

and

$$u_{n,j} = T_1$$

where
$$nh = 1$$

and the initial condition (4) as

$$u_{i,0} = f(ih);\ i = 1, 2, \ldots,$$

We choose k in such a way that the coefficient of $u_{i,j}$ in (5) is zero.
Thus, (5) becomes $u_{i,j+1} = \lambda\,[u_{i+1;} + u_{i-1,j}]$
Coefficient of $u_{i,j} = 0 \Rightarrow 1 - 2\lambda = 0$

$$\Rightarrow \lambda = \frac{1}{2}$$

Further $\lambda = \dfrac{k}{ah^2} \Rightarrow \dfrac{1}{2} = \dfrac{k}{ah^2}$

$$\Rightarrow k = \frac{ah^2}{2}$$

$$\therefore \qquad u_{i,j+1} = \frac{1}{2}[u_{i+1,j} + u_{i-1,j}] \tag{6}$$

(6) is called *Bender Schmidt Recurrence Equation.*

Example 7. Solve $\dfrac{\partial^2 u}{\partial x^2} - 2\dfrac{\partial u}{\partial t} = 0$ given

$$u(0, t) = 0,\ u(4, t) = 0,\ u(x, 0) = x(4 - x).$$

Assume $h = 1$. Find the values of u upto $t = 5$.

Solution:

Given: $u_{xx} = au_t$

$\therefore \quad a = 2$

To use Bender-Schmidt equation, $K = \dfrac{a}{2}h^2 = 10$, step size in time $= K = 1$. The values of $u_{i,j}$ are tabulated below.

j \ i	0	1	2	3	4	
0	0	3	4	3	0	← $u\ (x, 0) = x\ (4 - x)$
1	0	2	3	2	0	
2	0	1.5	2	1.5	0	
3	0	1	1.5	1	0	
4	0	0.75	1	0.75	0	
5	0	0.5	0.75	0.5	0	

Analysis: Range for x: (0, 4); for t : (0, 5)

$u(x, 0) = x(4 - x)$. This gives $u(0, 0) = 0$, $u(1, 0) = 3$
$u(2, 0) = 4$, $u(3, 0) = 3$, $u(4, 0) = 0$.

For all t, at $x = 0$, $u = 0$ and $\forall\ t$ at $x = 4$, $u = 0$.
Using these values we fill up column under $x = 0$, $x = 4$ and row against

$t = 0$ 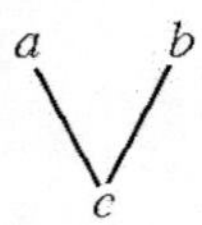

$\Rightarrow$ $\boxed{c = \dfrac{a+b}{2}}$

The values of u at $t = 1$ are written by seeing the values of u at $t = 0$ and using the average formula.

Example 8. Solve $32\dfrac{\partial u}{\partial t} = \dfrac{\partial^2 u}{\partial x^2}$, $0 < x < 1$, $t > 0$

$$u(x, 0), \quad u(0, t) = 0, \quad u(1, t) = t$$

Assume $h = \dfrac{1}{4}$.

Solution: Here, $c = \dfrac{1}{32}$, $h = \dfrac{1}{4}$.

Hence $k = \dfrac{h^2}{2c} = 1$. The calculation is as shown in figure.

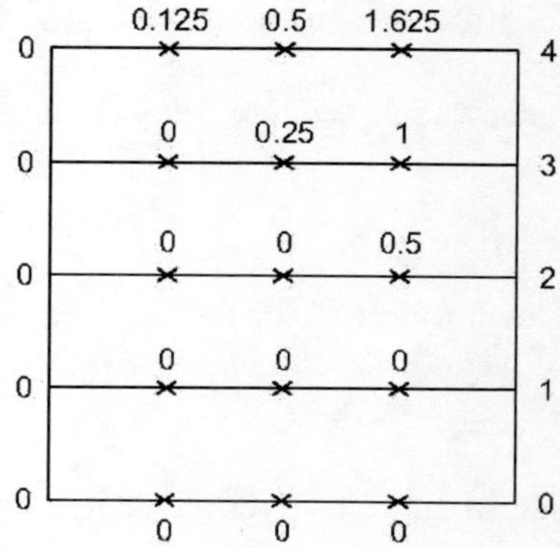

For example, $u(0.5, 4) = 0.5$

Example 9. Compute u for four steps with $h = 0.25$

$$\frac{\partial u}{\partial t} = \frac{\partial^2 u}{\partial t^2},\ 0 < x < 1,\ t > 0$$

$$u(x, 0) = 100(x - x^2)$$

$$u(0, t) = u(1, t) = 0.$$

Solution:

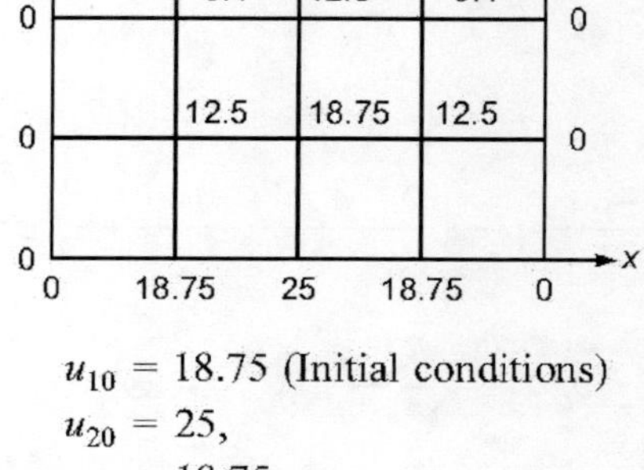

$$u_{10} = 18.75 \text{ (Initial conditions)}$$

$$u_{20} = 25,$$

$$u_{30} = 18.75$$

Apply $u_{i,j+1} = \dfrac{1}{2}(u_{i+1,j} + u_{i-1,j})$

which is used to find $t = t_{j+1}$

$$u_{11} = 12.5$$
$$u_{21} = 18.75$$
$$u_{31} = 12.5$$

Rest of the calculations are shown in the figure. Here the time step $k = \frac{h^2}{2} = \frac{1}{32}$

Hyperbolic Equations

(2) Derive "Explicit three-level formula" for the wave equation

$$a^2 \frac{\partial^2 u}{\partial x^2} = \frac{\partial^2 u}{\partial t^2}$$

Let us solve by the method of finite differences. The equation

$$a^2 u_{xx} - u_{tt} = 0 \tag{1}$$

with the boundary conditions

$$u(0, t) = 0 \qquad \text{(i)}$$
$$u(l, t) = 0 \qquad \text{(ii)}$$

and the initial conditions

$$u(x, 0) = f(x) \qquad \text{(iii)}$$
$$u_t(x, 0) = 0 \qquad \text{(iv)}$$

We select a spacing h for the variable x and a spacing k in the time direction. We know that

$$u_{xx} = \frac{u_{i+1,j} - 2u_{i,j} + u_{i-1,j}}{h^2}$$

$$u_{tt} = \frac{u_{i,j+1} - 2u_{ij} + u_{i,j-1}}{k^2}$$

and

$$u_t = \frac{u_{i,j+1} - u_{ij}}{k}$$

Equation (i) becomes

$$a^2\left(\frac{u_{i+1,j} - 2u_{ij} + u_{i-1,j}}{h^2}\right) - \left(\frac{u_{i,j+1} - 2u_{ji} + u_{i,j-1}}{k^2}\right) = 0$$

i.e.,

$$u_{i+1,j} - 2u_{i,j} + u_{i-1,j} - \frac{h^2}{k^2 a^2}(u_{i,j+1} - 2u_{ij} + u_{i,j-1}) = 0$$

Let $\lambda = \frac{k}{h}$. Then the above gives

$$\lambda^2 a^2(u_{i+1,j} - 2u_{i,j} + u_{i-1,j})$$
$$- u_{i,j+1} + 2u_{i,j} - u_{i,j-1} = 0$$

$\therefore$
$$u_{i,j+1} = 2(1 - \lambda^2 a^2)u_{i,j}$$
$$+ \lambda^2 a^2(u_{i+1,j} + u_{i-1,j}) - u_{i,j-1} = 0 \quad (2)$$

This equation is called the explicit three-level formula for the wave equation (1) and is the solution of the wave equation (1).

Note (1) If $k < h$, this formula is convergent.

(2) The coefficient of u_{ij} in (2) will be zero if $(1 - \alpha) = 0$

i.e.,
$$(1 - \lambda^2 a^2) = 0$$

i.e.,
$$\lambda^2 = \frac{1}{a^2} \quad \text{or} \quad \frac{k^2}{h^2} = \frac{1}{a^2}$$

or
$$\frac{k}{h} = \frac{1}{a}$$

In that case (2) becomes

$$u_{i,j+1} = u_{i+1,j} + u_{i-1,j} - u_{i,j-1} \quad (3)$$

(Parabolic Equation)

(3) Derive the Crank-Nicholson's difference scheme for the parabolic equation

$$u_{xx} = au_t \text{ with } B, C\text{'s}$$
$$u(0, t) = T_0,\ u(l, t) = T_e \text{ and the}$$
$$\text{I. } C\text{'s } u(x, 0) = f(x)$$
$$u_{xx} = au_t \quad \text{(Given)}$$

At the point u_{ij}, we have the Finite Difference approximation for u_{xx}

$$u_{xx} \approx \frac{u_{i+1,j} - 2u_{ij} + u_{i-1,j}}{h^2}$$

Similarly, at the point $u_{i,j+1}$

$$u_{xx} \approx \frac{u_{i+1,j+1} - 2u_{i,j+1} + u_{i-1,j+1}}{h^2}$$

Averaging these two approximations,

$$u_{xx} \approx \frac{u_{i+1,j+1} - 2u_{i,j+1} + u_{i-1,j+1} + u_{i+1,j} - 2u_{ij} + u_{i-1,j}}{2h^2}$$

For u_t, we use the forward difference approximations,

$$u_t \approx \frac{u_{i,j+1} - u_{i,j}}{k}$$

Hence, substituting in (1), we have

$$u_{i+1,j+1} - 2u_{i,j+1} + u_{i-1,j+1} + u_{i+1,j} - 2u_{ij} + = a\left(\frac{u_{i,j+1} - u_{ij}}{k}\right)$$

$$\frac{k}{2h^2 a}[u_{i+1,j+1} - 2u_{i,j+1} + u_{i-1,j+1} + u_{i+1,j} - 2u_{ij,} + u_{i-1,j}]$$

$$\equiv u_{i,j+1} - u_{ij}$$

Putting $\frac{k}{h^2 a} \equiv \lambda$, we have

$$\frac{1}{2}\lambda u_{i+1,j+1} - \lambda u_{i,j+1} + \frac{1}{2}\lambda u_{i-1,j+1} - u_{i,j+1}$$

$$= \lambda u_{ij} - \frac{1}{2}\lambda u_{i+1,j} - \frac{1}{2}\lambda u_{i-1,j} - u_{ij}$$

i.e. $\frac{1}{2}\lambda u_{i-1,j+1} - (\lambda + 1)\, u_{i,j+1} + \frac{1}{2}\lambda u_{i+1,j+1}$

$$= -\frac{1}{2}\lambda u_{i-1,j} + (\lambda - 1)\, u_{ij} - \frac{1}{2}\lambda u_{i+1,j}$$

It is called the Crank-Nicholson's Difference scheme.

Example 10. Solve (wave equation)

$\frac{\partial^2 u}{\partial x^2} = \frac{\partial^2 u}{\partial t^2}$ subject to the initial conditions $u(x, 0) = \sin \pi x$

$\frac{\partial u}{\partial t}(x, 0) = 0$, $(0 < x < 1)$ and the b.c's $u(0, t) = 0$,

$u(1, t) = 0$, $t \geq 0$,

given $h = \frac{1}{4}$, $k = \frac{1}{5}$, as the step length in the x- and t-directions respectively.

Find the 3rd level solution in the solution.

Solution: Since $h = \frac{1}{4}$ and $(0 < x < 1)$

$\therefore$ $nh = 1$

$\therefore$ $n = 4$

and $k = \frac{1}{5}$, $\therefore$ (x_i, t_j) are

$$x_i = ih = \frac{i}{4} \quad i = 0, 1, 2, 3, h.$$

$$t_j = jk = \frac{j}{5}, j = 0, 1, 2, 3, \ldots$$

Now $u(0, t) = u_{0,j} = 0$ (Finite Difference Form)

The *b.c*'s $u(1, t) = u_{n,j} = 0$

$$u_{4,j} = u(1, jk) = 0, j = 0, 1, 2, \ldots$$

Thus, $$u_{0,0} = u_{0,1} = u_{0,2} = u_{0,3} = \ldots = 0$$

$$u_{4,0} = u_{4,1} = u_{4,2} = u_{4,3} = \ldots = 0$$

Next, the 3rd level solutions are:

$$u_{i,0} = u(x_i, 0) = \sin(\pi x_i) = \sin(i\pi/4) \; i = 1, 2, 3,$$

and $u_{i,j} - u_{i,0} = 0$, so that

$$u_{i,1} = u_{i,0} = \sin\left(\frac{i\pi}{4}\right), i = 1, 2, 3$$

Thus, $$u_{i,0} = u_{1,1} = \sin\left(\frac{\pi}{4}\right) = \frac{1}{\sqrt{2}} = 0.7071$$

$$u_{2,0} = u_{2,1} = \sin\left(\frac{\pi}{2}\right) = 1$$

$$u_{3,0} = u_{3,1} = \sin\left(\frac{3\pi}{4}\right) = \frac{1}{\sqrt{2}} = 0.7071.$$

Example 11. Solve numerically the hyperbolic equation $4u_{xx} = u_{tt}$ with the boundary conditions $u(0, t) = 0$, $u(4, t) = 0$ and the initial conditions $u_t(x, 0) = 0$ and $u(x, 0) = x(4 - x)$ taking $h = 1$ (for 4 time steps).

Solution:

Table for calculation.

t \ x	0	1	2	3	4
0	0	3	4	3	0
0.5	0	2	3	2	0
1	0	0	0	0	0
		(3 + 0 − 3)	(2 + 2 − 4)	(3 + 0 − 3)	
1.5	0	−2	−3	−2	0
2	0	−3	−4	−3	0
2.5	0	−2	−3	−2	0
3	0	0	0	0	0
3.5	0	2	3	2	0
4	0	3	4	3	0

$$a^2 = 4,\ h = 1,\ k = \frac{h}{a} = \frac{1}{2} \tag{1}$$

$\therefore$ Taking $k = 1/2$, we use the formula

$$u_{i,j+1} = u_{i-1,j} + u_{i+1,j} - u_{i,j+1} \tag{2}$$

From $u(0, t) = 0 \Rightarrow$ along $x = 0$ are all zero

From $u(4, t) = 0 \Rightarrow u$ along $x = 4$ are all zero

$u(x, 0) = x(4 - x)$

$\Rightarrow u(0, 0) = 0,\ u(1, 0) = 3,\ u(2, 0) = 4,\ u(3, 0) = 3.$

Now, along the row $t = 0$, using the above values,

$$u_t(x, 0) = 0 \Rightarrow u_{i,1} = \frac{u_{i+1,0} + u_{i-1,0}}{2} \tag{3}$$

We draw the above for calculation, for the same we require $u_{i,1} = \frac{u_{2,0} + u_{0,0}}{2}$

$$= \frac{4+0}{2} = 2.$$

$$u_{2,1} = \frac{u_{3,0} + u_{1,0}}{2} = \frac{3+3}{2} = 3$$

$$u_{3,1} = \frac{u_{4,0} + u_{2,0}}{2} = 2$$

$$u_{4,1} = 0$$

period = 4 seconds or $8(k)$

$$= 8\left(\frac{1}{2}\right) = 4\,\text{sec}$$

Example 12. Solve by Crank-Nicholson's scheme:

$$\frac{\partial u}{\partial t} = \frac{1}{16}\frac{\partial^2 u}{\partial x^2},\ 0 < x < 1,\ t > 0$$

$$u(x, 0) = 0$$

$$u(0, t) = 0$$

$$u(1, t) = 100t$$

Compute u for one step with $h = \frac{1}{4}$.

Solution: Here $c = \frac{1}{16},\ h = \frac{1}{4},\ k = \frac{h^2}{c} = 1$

$t = 1$	0	a	b	c	100
$t = 0$	0	0	0	0	0
	$x = 0,$	$\frac{1}{4}$	$\frac{1}{2}$	$\frac{3}{4}$	1

Here, (a, b, c) are three unknowns

$$u\left(\frac{1}{4},1\right) = a; \quad u\left(\frac{1}{2},1\right) = b; \quad u\left(\frac{3}{4},1\right) = c$$

Applying, that is,

$$-u_{i-1,j+1} + 4u_{i,j+1} - u_{i+1,j+1} = u_{i-1,j}\ u_{i+1,j}$$

We get

$$-0 + 4a - b = 0 + 0$$
$$-a + 4b - c = 0 + 0$$
$$-b + 4c - 100 = 0 + 0$$

or,

$$b = 4a,\ 4b = a + c,\ 4c = b + 100$$

on solving, $a = 1.79$. $b = 7.14$, $c = 26.79$.

EXERCISES

1. Using Schmidt's process solve $25\ u_{xx} = u_t$ when $0 < x < 1$ and $t > 0$ with $b.c$'s $u(0, t) = 0 = u(10, t)$;

$$u(x, 0) = \frac{x(10-x)}{25} \text{ and}$$

choosing $h = 1$ and k suitably.
Find $u_{i,j}$ for $i = 1, 2, 3, \ldots 9$ and $j = 1, 2, 3, 4$.

2. Solve the equation $u_{xx} = u_t$ with the conditions

$$u(0, t) = 0\ u(x, 0) = x(1 - x);\ u(x, t) = 0$$

Assume that the region between $x = 0$ and $x = 1$ is divided into 10 equal increments of $h = 0.1$ Tabulate u for $t = k, 2k, 3k$ choosing an appropriate value for k.

3. Using Crank-Nicholson's method

$$u_t = u_{xx} \text{ subject to}$$
$$u(x, 0) = 0 = u(0, t) \text{ and } u(1, t) = t$$

taking $h = 0.5$ and $k = \frac{1}{8}$.

4. Solve

$$\frac{\partial^2 u}{\partial t^2} = \frac{\partial^2 u}{\partial x^2},\ 0 < x < 1,\ t > 0.$$

$$u(x, 0) = 1 - x^2, \quad \frac{\partial u}{\partial t}(x, 0) = 0$$

$$u(0, t) = 1 - t^2, \; u(1, t) = 0.$$

Choosing $h = 0.25$, compute u for 3 time steps.

5. Solve

$$\frac{\partial^2 u}{\partial t^2} = \frac{\partial^2 u}{\partial x^2}, \; 0 < x < 2, \; t > 0.$$

$$u(x, 0) = 0, \quad \frac{\partial u}{\partial t}(x, 0) = 100(2x - x^2).$$

$$u(0, t) = u(2, t) = 0.$$

Choosing $h = \frac{1}{2}$, compute u for 4 time steps.

6. Solve

$$\frac{\partial^2 u}{\partial t^2} = \frac{\partial^2 u}{\partial x^2}, \; 0 < x < 1, \; t > 0.$$

$$u(x, 0) = 100 \sin \pi x, \quad \frac{\partial u}{\partial t}(x, 0) = 0$$

$$u(0, t) = u(1, t) = 0.$$

Taking $h = \frac{1}{4}$, compute u for 4 time steps.

Index